Praise for

THE PANIC VIRUS

"Should be required reading at every medical school in the world. Seth Mnookin's *The Panic Virus* is a lesson on how fear hijacks reason and emotion trumps logic. . . . A brilliant piece of reportage and science writing."

—Michael Shermer, *The Wall Street Journal*

"Mr. Mnookin's passionate defense of vaccination may be just what the public needs, in equal parts because of what it says and because of who is saying it. . . . Parents who want to play it safe, but are not altogether sure how, should turn with relief to this reasoned, logical and comprehensive analysis of the facts."

—Abigail Zuger, MD, *The New York Times*

"This important book should be read by anyone who has a child, cares about public health, or is interested in the state of discourse in twenty-first-century America. It is a terrific and terrifying call to action."

—Jonah Lehrer, author of *How We Decide*

"A disturbing and well-told chronicle of the childhood vaccine wars in the United States and England."

—Sandra G. Boodman, *The Washington Post*

"There have been hundreds of recent outbreaks of ailments like whooping cough and measles that we thought would be eradicated by now—and might have been, if not for the anti-vaccine obfuscation. Bravo Seth Mnookin for digging for the truth and telling eloquent stories of what happens when lies, half-truths, and self-interest collide with fear."

—Laurie Garrett, Pulitzer Prize–winning writer and author of *Betrayal of Trust: The Collapse of Global Public Health*

"Seth Mnookin has given us a nonfiction story worthy of Michael Crichton—an absorbing, disturbing, and scrupulously researched account of a contagion of human unreason run wild. This time the hysteria was over autism; the next panic virus could be even more dangerous."

—Jonathan Mahler, author of *The Challenge* and *Ladies and Gentlemen, the Bronx Is Burning*

"The braiding of well-researched scientific information with riveting takes on the people fueling recent vaccine wars is one of the book's chief strengths. . . . I have seen a number of young adults seeking vaccines that their parents once blocked. Someday, perhaps one of them will pen an account of life inside the maelstrom of anti-vaccine anger and fear. For now, however, *The Panic Virus* remains the definitive text."

—Claire Panosian Dunavan, *Los Angeles Times*

"With rationality and science under siege these days, Seth Mnookin has produced a riveting and important chronicle of one life-and-death realm in which passionate, panicky belief has dangerously trumped reason—and put millions of children at risk."

—Kurt Andersen, host of *Studio 360* and author of *Heyday* and *Reset*

"In his disturbing chronicle, U.S. writer Seth Mnookin looks into the anti-vaccine movement. His analysis is serious and gripping. . . . Mnookin's careful science and compassion for both sides are examples for all journalists, and *The Panic Virus* should be read and pondered."

—Melvin Konner, *Nature*

"Insightful and frightening. . . . One of the many things Mnookin does well is show how desperate parents are for answers to this bewildering affliction, and how eager they are for someone to listen to them. He never forgets who the victims are."

—John Wilkens, *San Diego Union-Tribune*

"Mnookin takes on the anti-vaccine movement with the skill of a journalist and the intellectual concerns of a sociologist. In this riveting book, filled with fascinating human interest stories, he also manages to explain why, even though science has so clearly shown that the evidence argues against a relationship between vaccines and autism, so many people still believe there is one. . . . *The Panic Virus* is a superb case study in the crisis of science in a democratic society."

—Roy Richard Grinker, author of *Unstrange Minds*,
in *The American Journal of Psychiatry*

"*The Panic Virus* is sure to attract attention. . . . Mnookin's book is an unsparing brief against the vaccine skeptics. But in a larger sense, this volume is less about the insurrection against inoculations than it is about the democratization of information. . . . Less about the contagion of ideas than about the contagion of misinformation and mistrust that metastasizes in the new technology."

—David M. Shribman, *The Boston Globe*

"An accomplished journalist, Seth Mnookin takes an objective look at both sides of the vaccine/autism controversy and lands squarely on the side of science. With humor and wit, *The Panic Virus* examines the often bizarre events that led some families to become distrustful of science and erroneously conclude that vaccines might cause autism. This book will leave you scratching your head in pure amazement that this issue could get so out of hand when the science is so clear."

—Alison Singer, president, Autism Science Foundation

"Fueled by the web, the deliberate dissemination of disinformation is a feature of our time. Never has there been a greater need for good, honest investigative journalism. Seth Mnookin looks closely at the spurious link between vaccination and autism and shows us just how dangerous it can be when emotion drives good people to abandon critical thought."

—Peter C. Doherty, Nobel Laureate in Medicine

"In plain language, Seth Mnookin provides an excellent narrative and evaluation that helps clarify for readers how and why vaccine controversies have arisen over the years as well as sensible ways for readers to understand the science that supports vaccine usage. Vaccines are the most effective public health measure since clean water."

—Judith Palfrey, MD, FAAP, professor of pediatrics, Harvard Medical School, and author of *Child Health in America*

"Seth Mnookin understood there was something more to the cruelly misled and dangerously misleading vaccines-cause-autism movement than just an unhappy group of parents with a need to blame someone. He saw the connection between this deathless conspiracy theory and the proliferating irrationality of a society that has supersized its information diet while starving its capacity to think straight. For that reason alone—not to mention the deft, often charming characterizations woven into its skillful and fascinating narrative—this is an important, powerful, and bracing book."

—Arthur Allen, author of *Vaccine* and *Ripe*

"Mnookin, employing reason, logic and an investigative reporter's shoe leather, has written an old-fashioned book. And by old-fashioned, I mean that he adheres to the principles of the Enlightenment."

—Sarah Vowell, *New York Post*

"Engaging, provocative. . . . Hard to put down. . . . Mnookin provides his readers plenty of prods with which to jab those who don't vaccinate their kids, and some of these prods are awfully sharp."

—Anna B. Reisman, *Slate*

"A must-read for parents and parents-to-be."

—Trine Tsouderos, *Chicago Tribune*

ALSO BY SETH MNOOKIN

Feeding the Monster: How Money, Smarts, and Nerve Took a Team to the Top

Hard News: The Scandals at The New York Times *and Their Meaning for American Media*

THE PANIC VIRUS

The True Story Behind the Vaccine-Autism Controversy

Seth Mnookin

With a New Afterword by the Author

Simon & Schuster Paperbacks

NEW YORK • LONDON • TORONTO • SYDNEY • NEW DELHI

Simon & Schuster Paperbacks
A Division of Simon & Schuster, Inc.
1230 Avenue of the Americas
New York, NY 10020

First Simon & Schuster trade paperback edition January 2012

SIMON & SCHUSTER PAPERBACKS and colophon are registered trademarks of Simon & Schuster, Inc.

For information about special discounts for bulk purchases, please contact Simon & Schuster Special Sales at 1-866-506-1949 or business@simonandschuster.com.

The Simon & Schuster Speakers Bureau can bring authors to your live event. For more information or to book an event, contact the Simon & Schuster Speakers Bureau at 1-866-248-3049 or visit our website at www.simonspeakers.com.

Designed by Paul Dippolito

Manufactured in the United States of America

7 9 10 8 6

The Library of Congress has cataloged the hardcover edition as follows:

Mnookin, Seth.
The panic virus : a true story of medicine, science, and fear / Seth Mnookin.
p. cm.
Includes bibliographical references and index.
1. Vaccination—History. 2. Vaccination—Psychological aspects.
3. Health behavior. 4. Mass media and culture. I. Title.
[DNLM: 1. Vaccination—history. 2. Vaccination—psychology.
3. Health Knowledge, Attitudes, Practice. 4. Mass Media. 5. Panic. WA 115]
RA638.M675 2011
614.4'7—dc22
2010036579

ISBN 978-1-4391-5865-4
ISBN 978-1-4391-6567-6 (ebook)

For Sara and Max

"A lie will go round the world while truth is pulling its boots on."

—PROVERB POPULARIZED BY BAPTIST PREACHER
CHARLES HADDON SPURGEON IN AN 1855 SERMON
AND OFTEN ATTRIBUTED TO MARK TWAIN

CONTENTS

CAST OF CHARACTERS

Advocacy Organizations

AutismOne: Parent-led group that believes most cases of autism are caused by vaccines

Autism Research Institute: Early advocacy group; founded by Bernard Rimland in 1967

Autism Science Foundation: Founded in 2009 by Alison Singer and Karen London to fund and promote scientific research into autism

Autism Speaks: Largest autism charity in the United States; founded by former NBC Universal chariman Bob Wright and his wife, Suzanne, in 2005

Cure Autism Now: Founded in 1995 to fund research into autism; merged with Autism Speaks in 2006

Defeat Autism Now!: Offshoot of ARI founded in 1995 to promote the use of nonstandard treatments for autism

Generation Rescue: Also known as "Jenny McCarthy's Autism Organization"; promotes the view that vaccines cause autism and other neurological disorders

National Alliance for Autism Research: Founded in 1994 to fund scientific research into autism; merged with Autism Speaks in 2005

National Vaccine Information Center: Founded in 1982 by Barbara Loe Fisher and other parents who believe their children suffered brain injuries caused by the DPT vaccine

SafeMinds: Founded in 2000 to fight against "mercury-induced neurological disorders"

Talk About Curing Autism: Founded in California in 2000; went national in 2007

Parents and Family Members

Lisa Ackerman: Executive director of Talk About Curing Autism; introduced Jenny McCarthy to the autism advocacy movement

Sallie Bernard: Mercury Mom; one of the leaders of SafeMinds; lead author of "Autism: A Novel Form of Mercury Poisoning"

Vicky Debold: Board member of the National Vaccine Information Center; director of SafeMinds

Barbara Loe Fisher: President of the National Vaccine Information Center

Jane Johnson: Co-managing director of the Thoughtful House Center for Children; director of the Autism Research Institute

Eric and Karen London: Co-founders of the National Alliance for Autism Research

Jenny McCarthy: Actress; believes vaccines caused her son's autism; promotes the use of gluten- and dairy-free diets

Lyn Redwood: Mercury Mom; one of the leaders of SafeMinds; collaborated with David Kirby on *Evidence of Harm*

Bernard Rimland: Father of the modern-day autism advocacy movement

Alison Singer: Former executive vice president of Autism Speaks; co-founder of the Autism Science Foundation

Bob and Suzanne Wright: Founders of Autism Speaks

Katie Wright: Daughter of Bob and Suzanne Wright

Doctors and Researchers

Mark Geier: Frequent expert witness in Vaccine Court lawsuits

Jay Gordon: Former pediatrician to Jenny McCarthy's son, Evan

Neal Halsey: Director of the Institute for Vaccine Safety at Johns Hopkins University

Leo Kanner: Austrian psychiatrist; coined the term "autism" in a 1943 research paper

Arthur Krigsman: Pediatrician; former colleague of Andrew Wakefield's; treated Michelle Cedillo

Paul Offit: Co-inventor of a rotavirus vaccine; chief of the division of Infectious Diseases and director of the Vaccine Education Center at Children's Hospital of Philadelphia

Bob Sears: California-based pediatrician; author of *The Vaccine Book*

Andrew Wakefield: Lead author of a 1998 paper hypothesizing a connection between the MMR vaccine and autism

Journalists and Writers

Brian Deer: British investigative journalist; wrote a series of articles on Andrew Wakefield and his research into the MMR vaccine

Richard Horton: Editor of *The Lancet*

David Kirby: Author of *Evidence of Harm*; frequent contributor to *The Huffington Post*

Lea Thompson: Reporter on the 1982 television special "Vaccine Roulette"

The Omnibus Autism Proceeding

Michelle Cedillo: Autistic girl whose Vaccine Court claim was an Omnibus test case

Theresa and Michael Cedillo: Parents of Michelle Cedillo

Sylvia Chin-Caplan: Lawyer for the Cedillo family

George Hastings: Special Master who presided over the Cedillo trial

Other

Richard Barr: British personal injury lawyer; worked with Andrew Wakefield on lawsuits related to the MMR vaccine

Mary Leitao: Founder of the Morgellons Research Foundation

Lora Little: Early twentieth-century anti-vaccine activist

Lorraine Pace: Founder of the West Islip Breast Cancer Coalition

ABBREVIATIONS

Governmental Agencies

ACIP—Advisory Committee on Immunization Practices (U.S.)
CBER—Center for Biologics Evaluation and Research (U.S.)
CDC—Centers for Disease Control and Prevention (U.S.)
EIS—Epidemic Intelligence Service (U.S.)
EPA—Environmental Protection Agency (U.S.)
GMC—General Medical Council (U.K.)
FDA—Food and Drug Administration (U.S.)
HRSA—Health Resources and Services Administration (U.S.)
MRC—Medical Research Council (U.K.)
NIH—National Institutes of Health (U.S.)
NIMH—National Institute of Mental Health (U.S.)
IOM—Institute of Medicine (U.S.)
SSI—Statens Serum Institut (Denmark)
WHO—World Health Organization

Medical Terms

ASD—autism spectrum disorder
IBD—inflammatory bowel disease
PDD-NOS—pervasive developmental disorder, not otherwise specified

Organizations

ARI—Autism Research Institute
CAN—Cure Autism Now
DAN!—Defeat Autism Now!
JABS—Justice, Awareness and Basic Support (U.K.)
NAAR—National Alliance for Autism Research
NVIC—National Vaccine Information Center
TACA—Talk About Curing Autism

Professional Associations

AAP—American Academy of Pediatrics
AAPS—Association of American Physicians and Surgeons
AMA—American Medical Association
APA—American Psychiatric Association

Publications

DSM—*Diagnostic and Statistical Manual of Mental Disorders*
JAMA—Journal of the American Medical Association
NEJM—The New England Journal of Medicine

Vaccines

DPT—diphtheria-pertussis-tetanus
Hib—*Haemophilus influenzae* type b
MMR—measles-mumps-rubella

Other

NCVIA—National Childhood Vaccine Injury Act
PSC—Petitioners' Steering Committee (Omnibus Autism Proceeding)
VAERS—Vaccine Adverse Event Reporting System
VSD—Vaccine Safety Datalink

THE PANIC VIRUS

Introduction

On April 22, 2006, Kelly Lacek looked around her dinner table and smiled: Dan, her husband of thirteen years, was there, along with the couple's three children, Ashley, Stephen, and Matthew. Kelly's parents had also come over: There was a father-daughter dance at the local church that evening, and Kelly and her dad were double-dating with Dan and Ashley. As the four of them were getting ready to leave, Kelly couldn't resist needling her mother. "You're stuck with the boys," she said. "But don't worry—we won't be out too late." She kissed Stephen goodbye, and then bent down to say good night to Matthew. He was three years old, and Kelly marveled at how quickly he was growing up: It seemed as if it was only moments ago that he'd been an infant, and now he was already being toilet-trained. (Dan and Kelly both agreed that it was adorable how proudly he announced that he had to go to the bathroom.)

For a brief moment, Kelly says, she wondered if Matthew was okay—he seemed a little out of sorts, and earlier that afternoon, he'd complained of a sore throat—but then she figured he'd probably just tired himself out wrestling with his older brother.

Kelly and Dan returned home that night around eight o'clock. They'd barely walked in the door when Kelly's mother rushed over: "It's Matthew," she said. "He's running a fever—and his breathing seems a little shallow." The Laceks realized right away that something was seriously wrong. "He was just sort of hunched over," Kelly says. "We didn't know what to do." Since there was no way to get in touch with Matthew's doctor, they decided to make the ten-minute drive from their home in Monroeville, about fifteen miles east of Pittsburgh, to the Forbes Regional Campus of the Western Pennsylvania Hospital.

When the Laceks arrived at the emergency room, the attending physician told them there was nothing to worry about. In all likelihood, he said, Matthew had a case of strep throat. Worst-case scenario, it was asthma; regardless, they'd be home in no time. Two hours later, they were feeling much less assured: Matthew's fever was still rising, and when a doctor tried to swab his throat, he began to choke. By eleven P.M. Matthew's temperature had risen to 104 degrees and his breathing seemed to be growing shallower by the minute.

It was around that time that a doctor the Laceks hadn't met before walked over. He was older—probably in his sixties, Kelly thought—and as soon as he saw Matthew, he began to suck nervously on his teeth. He turned to the Laceks: Had Matthew received all his shots? Actually, Kelly said, he hadn't. Matthew had been born in March 2003, several years after rumors of a connection between autism and vaccines had begun to gain traction in suburban enclaves around the country. That May, Kelly's chiropractor warned her about the dangers of vaccines. "He asked if we were going to get [Matthew] vaccinated and I said yes," Kelly says. "And then he told me about mercury. He said, 'There's mercury in there.' " Kelly had already heard rumors that the combined measles-mumps-rubella (MMR) vaccine was dangerous, but this was something new. "He was really vocal about it causing autism. He said there was this big report over in Europe and blah blah blah. And I thought, Well, I'm surrounded by people who have autistic children. What if this happened to Matthew?" If Kelly was unconvinced, the chiropractor said, she should make Matthew's pediatrician prove to her that the vaccines Matthew was scheduled to receive were one hundred percent safe.

"So that's what I did," Kelly says. "I asked my doctor if she could give me a label that says there's no mercury and she said, 'No.' She said she wouldn't give it to me." It was as if, Kelly says, her pediatrician was hiding something. The doctor tried to tell Kelly that she would be putting Matthew at serious risk by not immunizing him, but, Kelly says, "I don't think I heard anything else she might have said, quite honestly. At that point I had lost faith."

From that day forward, Matthew didn't receive any of his scheduled vaccinations, including one for a bacterial disease called *Haemophilus influenzae* type b, or Hib. Oftentimes, a Hib infection is not particularly threatening—if the germs stay in the nose and throat, it's likely the child won't get sick at all—but if the infection travels into the lungs or the bloodstream, it can result in hearing loss or permanent brain damage. Hib can also cause severe swelling in the throat due to a condition called epiglottitis, which, if not treated immediately, results in infected tissue slowly sealing off the victim's windpipe until he suffocates to death. As recently as the 1970s, tens of thousands of children in America had severe Hib infections each year. Many of those suffered from bacterial meningitis, and between five hundred and one thousand died. After the Hib vaccine was put into widespread use, the disease all but disappeared in the United States: In 1980, approximately 1 in 1,000 children caught Hib; today, fewer than 1 in 100,000 do. In fact, the immunization had been so effective that out of everyone working in the Monroeville ER, the doctor who'd asked Kelly Lacek about her son's vaccine history was the only one who had been practicing long enough to have seen an actual Hib infection in a child.

Until that night, Kelly had never given much thought to the potential repercussions of her decision not to have Matthew vaccinated. "I must have read somewhere that after he turned three, he would have been okay for many of those diseases," she says. "I thought he was in the clear." She was wrong. "I have never seen a doctor panic so quickly," she says. If, as the doctor was all but certain was the case, Matthew had been infected, then everything that had been done to him in the hospital that night—the examinations, the swabs, the breathing treatments—had served only to further inflame his throat.

It wasn't until Kelly saw her son's X-rays that she realized just how dire the situation was: It looked as if Matthew had a thumb lodged in his throat. "I started to shake," Kelly says. "There was just a tiny bit of airway left for him to breathe."

Within minutes, the entire emergency room was thrown into a frenzy. Kelly heard someone shout out, "Page Children's!" Then she heard a second command: "Get Life Flight here right away." Finally, a doctor pulled the Laceks aside and explained the situation to them. "If we don't get Matthew on a helicopter [to the Children's Hospital in Pittsburgh] right now, your son is probably going to die," he said. "It could be within minutes." While they were waiting, the doctor said, Kelly had to make sure Matthew remained calm. "I do not want you crying," the doctor said. "I do not want you reacting to anything. If you are upset, Matthew will be upset, and that will make his throat close up more. If that happens he will suffocate." As if in a daze, Kelly went and picked up her son. It wasn't until she heard her teeth chattering that she realized she was shaking. She focused all her energy on trying to remain still.

While Kelly was holding Matthew, Dan Lacek was conferring with the hospital staff. It had rained earlier in the evening, and now the entire area was covered in fog, which made it too dangerous to land a helicopter. Matthew was going to have to make the trip to Pittsburgh in an ambulance—but before he could be moved, he'd have to be intubated. If that didn't work—if there was not enough room in Matthew's throat for a breathing tube—the doctors would try to perform a tracheotomy, which involves cutting into the windpipe in an effort to form an alternate pathway for air to get into the lungs. (The procedure is not without risk: The physicist Stephen Hawking lost his speech when the nerves that control the vocal cords were damaged during an emergency tracheotomy.) Once again, it fell to Kelly to keep her son calm. Fortunately, the tube slid down Matthew's throat. Unless it closed up so much that the tube was forced out, they'd bought themselves a few more hours.

It was almost four in the morning when the Laceks arrived in

Pittsburgh. Matthew was immediately placed in a medically induced coma. All the doctors could promise was that he'd live through the night. "They said something about not catching it quickly enough with the antibiotics," Kelly says. "Even if he did recover, there was a good chance he would have permanent brain damage, or, best-case scenario, he would have hearing loss."

For forty-eight hours, Dan and Kelly Lacek's son remained in stable condition. "You're in shock," Kelly says. "You never let your guard down. You're just so focused on him getting better." Then, on Tuesday, just as they were growing more hopeful, Matthew's blood pressure plummeted. The only thing the Laceks could think to do at that point was to ask their friends to pray for them.

When Kelly Lacek's chiropractor told her that vaccines had been linked to autism, he was repeating the most recent of hundreds of years' worth of fears about vaccinations. The roots of this latest alarm dated back to 1998, when a British gastroenterologist named Andrew Wakefield claimed to have discovered a new gut disorder associated with the MMR vaccine—and with autism. Wakefield based his conclusions on a case study of a dozen children who'd been brought to his clinic at the Royal Free Hospital in London. Almost immediately, Wakefield's research methods and his interpretations, which had been published in the medical journal *The Lancet*, came under fire. Wakefield's response was to appeal to the public rather than to his colleagues: The medical establishment was so determined to discredit him, he said, because he threatened their hegemony by taking parents' concerns seriously. The media took the bait, and despite Wakefield's lack of proof and his track record of dubious assertions and unverified lab results, they began churning out stories about how a maverick doctor was trying to protect innocent children from corrupt politicians and a rapacious pharmaceutical industry. Within months, vaccination rates across Western Europe began to fall.

Then, a year later, the Centers for Disease Control and Prevention (CDC) and the American Academy of Pediatrics (AAP) publicly recommended the removal of a widely used mercury-based preservative called thimerosal from childhood shots. The move had been hotly debated; in the end, one of the factors that had tipped the balance was a concern that following the Wakefield brouhaha, any connection, real or rumored, between vaccines and neurodevelopmental disorders had a chance of unraveling public confidence in vaccines.

That fear proved to be well founded, in no small part because of the growing hold autism had on the public's consciousness. In the half-century since "infantile autism" had been defined as a discrete medical condition, it had gone from being a source of shame for parents, who were blamed for their children's conditions, to becoming a seemingly omnipresent concern, especially among those well-educated, upper-middle-class families for whom child rearing had become an all-encompassing obsession.

In spite of this increased attention, researchers in the 1990s were barely any closer to understanding autism's origins or devising effective therapies for its treatment than their predecessors had been fifty years earlier. For parents of autistic children, this lack of reliable information resulted in feelings of hopelessness and frustration; for parents in general trying to determine the best course of action for the future, it fueled a sense that medical experts and health authorities couldn't be counted on to look out for their families' well-being.

Together, these reactions prepared the ground for new hypotheses to take root, regardless of how speculative or scientifically dubious they were. In the year following the CDC/AAP recommendations regarding thimerosal, a small group of parents decided that some of the symptoms of mercury poisoning seemed to match the behavior they saw in their autistic children—and they suddenly realized that their children had appeared to be fine before they'd received their vaccines. These parents began posting their observations online, sparking hundreds more parents to confirm that they'd noticed the exact same thing. With a network of nontraditional doctors and alternative health practitioners urging them on, they became more and more convinced that the common threads that ran through their stories were too odd and too widespread to be mere happenstance.

The more these newly politicized parents learned, the more outraged they became. Why were children with weak immune systems injected with vaccines just as potent as those used on children in perfect health? Why was everyone instructed to receive the same number of inoculations, regardless of their medical histories or family backgrounds? Why, for that matter, were more and more shots being added all the time? Was a chicken pox vaccine really necessary? Or one for the flu?

Just as had been the case with the MMR vaccine, there was no concrete evidence linking thimerosal to autism, and the anecdotal corroboration often seemed more impressive than it actually was. (To take but one example: Despite superficial similarities, the motor difficulties exhibited by people with mercury poisoning bear little resemblance to the repetitive movements typical of autistics.) That didn't stop the American media from reacting much the same way their colleagues across the Atlantic had when Andrew Wakefield had published his assertions, as the emotional pull of stories featuring sick children and devoted parents outstripped anything as boring as hard data or the precautionary principle. In a matter of months, an ad hoc coalition of "Mercury Moms" transformed itself into a potent political force: Senators spoke at their rallies, public health officials tried to assuage their concerns, and federal agencies included them in discussions on how to spend tens of millions of dollars. Soon, vaccination rates began to fall in the United States as well.

By the beginning of the new millennium, Wakefield's supporters and the proponents of the thimerosal link had joined forces to create an international cadre of vaccine skeptics whose message had an undeniable appeal: Parents trying to do nothing so much as raise their children had been taken advantage of by a society they had trusted—and now they were determined to make it right.

• • •

Over the past two decades, the instant accessibility of information has dramatically reshaped our relationship to the world of knowledge. Five hundred years after Gutenberg's introduction of the printing press and Martin Luther's translation of the Bible let common people bypass the priestly class, the vernacular of twenty-four-hour news channels and Internet search engines is freeing us to take on tasks that we'd long assumed were limited to those with specialized training. Why, after all, should we pay commissions to real estate brokers or stock analysts when we can find online everything we need to sell our houses or manage our investments? And why should we blindly follow doctors when we can diagnose our own ailments?

One of the first effects of this hyper-democratization of data was to unmoor information from the context required to understand it. On the Internet, facts float about freely and are recombined more according to the preferences of intuition than the rules of cognition: Mercury is toxic, toxins can cause development disorders, mercury is in vaccines; ergo, vaccines cause autism. Combined with the self-reinforcing nature of online communities and a content-starved, cash-poor journalistic culture that gravitates toward neat narratives at the expense of messy truths, this disdain for actualities has led to a world with increasingly porous boundaries between facts and beliefs, a world in which individualized notions of reality, no matter how bizarre or irrational, are repeatedly validated.

Take the "birther" movement, which contends that Barack Obama was born in Kenya and therefore is not eligible to be president. In the summer of 2009, Orly Taitz, a Russian-born dentist/lawyer/real estate agent, almost single-handedly turned her one-woman media blitz into a national preoccupation. Taitz, who believes that the Federal Emergency Management Agency is building internment camps to house anti-Obama activists and that Venezuelan president Hugo Chávez controls the software that runs American voting machines, makes for undeniably good television: She looks like a young Carol

Channing, sounds like an overexcited Zsa Zsa Gabor, and has the ability to make absurd accusations with a completely straight face. By midsummer, Taitz was appearing regularly on CNN, Fox News, and MSNBC, a decision the news channels justified with the risible pretext of needing to be fair to those on "both sides" of an issue about which there was nothing up for debate—at least not in the real world. Before long, mainstream on-air personalities like Lou Dobbs were pimping the story as hard as Taitz or any of her allies were, to equally comical effect.

This type of cognitive relativism—or "truthiness," as fictional talk show host Stephen Colbert termed it—has become the defining intellectual trend of our time. Colbert coined truthiness as a way to define former president George W. Bush's disdain for "those who think with their head" as opposed to "those who know with their heart." Its pervasiveness was most tragically illustrated in Iraq: By inventing a set of facts to support the overthrow of Saddam Hussein, the Bush administration changed a discussion of whether Iraq had weapons of mass destruction to whether the theoretical presence of WMDs was sufficient justification for war. In the fall of 2004, after both WMDs and easy victory were revealed as mirages, a presidential aide made an astounding admission to *The New York Times Magazine*. The White House, he said, didn't waste time worrying about those "in what we call the reality-based community" who "believe that solutions emerge from your judicious study of discernible reality." That, the aide said, "is not the way the world really works anymore. . . . When we act, we create our own reality." Orly Taitz couldn't have put it any better herself.

My interest in the controversies surrounding childhood inoculations began in 2008. My wife and I were newly married, and though we didn't yet have children, we found ourselves initiates in a culture in which people obsessed over issues about which we'd previously been unaware, such as the political implications of disposable diapers and the merits of home births. Another common preoccupation, we discovered, was the fear that widespread fraud was being perpetrated by the medical establishment. These people were our peers: They gravitated toward fields like journalism or law or computer programming or public policy; they lived in college towns like Ann Arbor and Austin or sophisticated urban centers like Boston and Brooklyn; they drove Priuses and shopped at Whole Foods. They tended to be self-satisfied, found it difficult to conceive of a world in which their voices were not heard, and took pride in being intellectually curious, thoughtful, and rational.

And, we soon learned, a good number of them didn't trust the American Medical Association (AMA) or the American Academy of Pediatrics—or at least didn't trust them enough to adhere to their recommended immunization schedules, which included vaccinations for diphtheria, hepatitis B, Hib, influenza, measles, mumps, pertussis, pneumococcal, poliovirus, rotavirus, rubella, and tetanus, all in the first fifteen months of a child's life. This caught us by surprise: The AAP wasn't high on the list of organizations we thought likely to be part of a widespread conspiracy directed against the nation's children.

That fall, we were at a dinner party when the subject of vaccines came up for what felt like the millionth time. I asked the parents at the table how they went about making decisions concerning their children's health. Did they talk to their pediatricians? Other parents? Were they reading books? Poking around online? One friend, a forty-one-year-old first-time father, said there was so much conflicting information out there he hadn't known what to do.* In the end, he said, he and his wife decided to delay some shots, including the ones for the MMR vaccine, which he'd heard was particularly dangerous. "I don't know what to say," he told me. "It just feels like a lot for a developing immune system to deal with."

At the time, I had no idea what the evidence supported. Still, I

* This character, and the subsequent conversation, is drawn from an amalgam of discussions I had over a period of several months. These quotes represent my best recollection; as a rule, I try to avoid taking notes during dinner parties.

cringed when my friend said he'd made his decision based on what he *felt* rather than by trying to assess the balance of the available evidence. Anecdotes and suppositions, no matter how right they feel, don't lead to universal truths; experiments that can be independently confirmed by impartial observers do. Intuition leads to the flat earth society and bloodletting; experiments lead to men on the moon and microsurgery.

The more I pushed my friend, the more defensive he grew. Surely, I said, there had to be something tangible, some experiment or some epidemiological survey, that informed his decision. There wasn't; I was even more taken aback when he said he likely would have done the same thing even if he'd been presented with conclusive evidence that the MMR vaccine was safe. "Let's say that there haven't been any studies that have uncovered a problem," he said. "That doesn't mean they won't find one someday." He was, of course, technically correct: It is always impossible to prove a negative. That's why gravity is still a "theory"—and why you can't prove with absolute certainty that I won't wake up tomorrow with the ability to fly. (As Einstein said, "No amount of experimentation can ever prove me right; a single experiment can prove me wrong.") Finally, he offered up this rationalization: "If everyone agreed that vaccines are so safe, we wouldn't even be having this discussion." By that point, my wife was kicking me under the table. I let the subject drop.

But when I got home that night I couldn't stop thinking about that conversation. The issue didn't affect me directly: No one close to me had a personal connection with autism and I didn't know any vaccinologists or government health officials. What nagged at me, I realized, was the pervasiveness of a manner of thinking that ran counter to the principles of deductive reasoning that have been the foundation of rational society since the Enlightenment.

I began work on this book the next day. After reading hundreds of academic papers and thousands of pages of court transcripts, I couldn't help but agree with the ruling in an omnibus proceeding in which thousands of families with autistic children requested compensation for what they claimed were vaccine injuries: This was "not a close case."

Once I'd arrived at the conclusion that there was no evidence supporting a causal link between childhood inoculations and autism spectrum disorders, I had to confront a set of issues that get to the heart of social dynamics and human cognition: Why, despite all the evidence to the contrary, do so many people remain adamant in their belief that vaccines are responsible for harming hundreds of thousands of otherwise healthy children? Why is the media so inclined to air their views? Why are so many others so readily convinced? Why, in other words, are we willing to believe things that are, according to all available evidence, false?

In an effort to answer those questions, I interviewed scientists and doctors, healers and mystics, government appointees and elected officials. I also spoke with dozens of parents who watched helplessly as their autistic children were enveloped by worlds outsiders could not penetrate. Some of these children were in obvious physical pain, some were sullen and unresponsive, some were violent and uncontrollable, and some moved from one extreme to another. The suffering of parents who feel unable to protect their children is almost impossible to describe—and helplessness only begins to cover the range of emotions they endured. There was guilt: Despite everything they were told, it was impossible for some parents to fully rid themselves of a feeling that somehow their child's condition was their fault. There was resentment: Many were tired of having their lives taken over by a disease about which so little is known and so little can be done. There was bitterness: How could a society that propped up foreign governments and bailed out failing banks not pay for adequate services for disabled children? And there was anger: Surely someone or something was to blame for the ways in which their lives had been upended.

But more than anything else, parents spoke of their isolation. Those split seconds of synchronicity that freckle people's days—the half-smile a new mom gives a pregnant woman on the street, the glance shared by two strangers reading the same book on the subway—those are missing from a lot of these parents' lives. Those with children on the more extreme end of the autism spectrum tend to feel the most alone: There are no knowing winks when a child won't stop screaming, no "I've been there" grins when he defecates in public. No one thinks it's cute when a child scratches his mother until she bleeds and strangers don't chuckle when a ten-year-old wants to know why the woman who just got on the bus is so fat.

This sense of being cut off from the world helps to explain why tens of thousands of parents have gravitated to a close-knit community that stretches around the globe. The fact that the community's most vocal and active members believe that vaccines cause autism and that autism can be cured by "biomedical" treatments like gluten-free diets and hormone injections is of secondary importance—what's paramount is the sense of fellowship and support its members receive. Every spring, between fifteen hundred and two thousand of these parents travel to Chicago's Westin O'Hare hotel for the annual conference of a grassroots organization called AutismOne, which claims to be the single largest producer of information about the disorder in the world. For those three or four days, the dynamic that shapes many of these parents' lives is turned on its head: Here, it's people whose lives haven't been affected by autism who feel out of place.

In order to protect this space, AutismOne discourages outsiders from attending. In incidents over the past several years, the organization has barred journalists identified as unsympathetic, kicked out parents who were perceived as being impertinent, and asked security to remove a public health official. This gatekeeping is severe but the worry behind it—that only people with a vested interest in the organization's survival can be trusted to take a generous view of its beliefs—is not misplaced. Even after I was granted permission to attend one of the group's conferences, I always had the feeling that my temporary visa did not come with the right to ask about the apparent contradictions highlighted by the weekend's proceedings.

The most obvious of these is the insistence by AutismOne's

founders that they promote a "pro-science" and not an "anti-vaccine" agenda, a claim that is hard to reconcile with the group's mission statement: "The great majority of children suffering from autism regressed into autism after routine vaccination. . . . Autism is caused by too many vaccines given too soon."* If anything, the conference's speakers have become more extreme as an ever-growing body of evidence disproves their claims: Included among the 150 presentations at the conference I attended was a four-hour-long "vaccine education" seminar, a lecture on "autism and vaccines in the US [legal system]," an environmental symposium on "the toxic assault on our children," and a presentation on "Down syndrome, vaccinations, and genetic susceptibility to injury." During her talk, Barbara Loe Fisher, the grande dame of the American anti-vaccine movement, explained how vaccines are a "de facto selection of the genetically vulnerable for sacrifice" and said that doctors who administer vaccines are the moral equivalent of "the doctors tried at Nuremberg." (That parallel, she said, had been pointed out to her by Andrew Wakefield, in whose honor the 2009 conference was held.) One night, there was the premiere of a documentary called *Shots in the Dark,* which examined "current large-scale vaccination policies" in light of the "onset of side effects such as autism or multiple sclerosis." This list could go on for pages.

If you assume, as I had, that human beings are fundamentally logical creatures, this obsessive preoccupation with a theory that has for

---

* The term "anti-vaccine" has been the subject of extremely contentious disputes. Throughout the book, I have used it to describe groups or individuals whose efforts to discredit vaccines depend on claims that are not supported by, and in many cases are directly contradicted by, the available scientific evidence. It's worth noting that activist groups have been largely successful in pressuring the media to adopt a standard that relies on these groups' own beliefs about what qualifies as being "anti-vaccine." A correction that appeared in *The New York Times* in April 2010 is an example of this: "A picture caption on Tuesday . . . referred incorrectly to the rally in Washington in 2008 at which the actors Jenny McCarthy and Jim Carrey were shown. Participants were calling for the elimination of what *they said were* toxins in children's vaccines and for a reassessment of mandatory vaccination schedules for children; it was not 'an anti-vaccine' rally." (Emphasis added.)

all intents and purposes been disproved is hard to fathom. But when it comes to decisions around emotionally charged topics, logic often takes a back seat to what are called cognitive biases—essentially a set of unconscious mechanisms that convince us that it is our feelings about a situation and not the facts that represent the truth. One of the better known of these biases is the theory of cognitive dissonance, which was developed by the social psychologist Leon Festinger in the 1950s. In his classic book *When Prophecy Fails*, Festinger used the example of millennial cults in the days after the prophesied moment of reckoning as an illustration of "disconfirmed" expectations producing counterintuitive results:

> Suppose an individual believes something with his whole heart; suppose further that he has a commitment to this belief, that he has taken irrevocable actions because of it; finally, suppose that he is presented with evidence, unequivocal and undeniable evidence, that his belief is wrong; what will happen? The individual will frequently emerge, not only unshaken, but even more convinced of the truth of his beliefs than ever before. Indeed, he may show a new fervor about convincing and converting other people to his view.*

In this light, another seeming paradox of the anti-vaccine movement—its extreme paranoia about ulterior motives on the part of anyone promoting vaccination combined with an almost willful

---

* One of the case studies on which Festinger based his theory focused on Dorothy Martin, a housewife and former follower of L. Ron Hubbard's Dianetics movement. Martin claimed that inhabitants of the planet Clarion had told her that Chicago would be destroyed in a flood just after midnight, early on December 21, 1954. Festinger and his colleagues observed Martin's followers as they quit their jobs, left their spouses, and gave away their money in preparation for their rescue by a flying saucer. By 4:45 A.M. on the morning of the 21st, Martin and her disciples had to acknowledge that Chicago was not underwater and that they had not been rescued by aliens. At that point, Martin received a new message: The apocalypse had been canceled. As Festinger wrote, "The little group, sitting all night long, had spread so much light that God had saved the world from destruction."

blindness to the conflicts of interest of the profiteers in their midst—also makes more sense. In a speech delivered at eight in the morning of the first full day of the conference I attended, Lisa Ackerman, the head of a group called Talk About Curing Autism (TACA), ran through a long list of things parents "need" to do for their children, including testing for mineral deficiencies, installing water filtration systems, eating organic chickens, and throwing out all flame-retardant clothing, mattresses, and carpeting. "If you buy clothes and pajamas from Target and Wal-Mart, almost all of those have a flame retardant applied," she said. "If you're building or doing home improvement, that's like the biggest toxic exposure you can give a child. New carpets are one of the worst things you can do. Sorry if you just did it." Ackerman also talked about "supplementation [and] nutritional therapies," including "vitamin B_{12} shots in the buttocks" and antioxidant IVs, and described a range of other alternative treatments, many of which were available from the eighty-plus vendors who'd set up shop in one of the hotel's exhibition halls. (According to Ackerman, twenty-second treatments in one of the types of hyperbaric oxygen chamber for sale—"They're not just for people like Michael Jackson; they're really cool!"—had transformed her son from a "caveman" into a verbal child capable of having normal conversations.)

The following afternoon, the father-son team of Mark and David Geier stood on stage in the same lecture hall for the first of their two presentations on "New Insights into the Underlying Biochemistry of Autism." The most recent insight of the Geiers, who've been stalwarts of the anti-vaccine movement for decades, involved a treatment called the "Lupron protocol," which is based on a theory so odd it sounds like a joke: Autism, the Geiers were claiming, is the result of a pathological reaction between mercury and testosterone, and Lupron, an injectable drug used to chemically castrate sex offenders, is the cure. Before determining whether patients are candidates for their "protocol," the Geiers order up dozens of lab tests at a cost of more than $12,000. The treatment itself, which consists of daily injections and bimonthly deep-tissue shots, can run upward of $70,000 a year. It also is excruciatingly painful. (In an article in the *Chicago Tribune*, an acolyte of the Geiers' described giving a shot to one child: "His dad is a big guy like myself, [and] it took both of us to hold him down to give him the first injection. It reminded me of . . . a really wild dog or a cat.") At the time of the 2009 conference, the Geiers had already opened eight Lupron clinics in six different states. Mark Geier, who calls Lupron a "miracle drug," told a reporter that was just the beginning of their expansion aspirations: "We plan to open everywhere."*

Outside the conference rooms of AutismOne and without a child suffering from the disorder, it can be hard to fathom how something as bizarre as Lupron ever gains momentum. But when you watch the transaction happening in real time it's not hard to understand its appeal. "If someone like Mark Geier comes up to you at a conference, and he's got twenty impressive PowerPoint slides, and he's got a Ph.D. and a long string of letters after his name, you're going to listen to him because you've been taught that someone like that is someone who knows what he's talking about," says Kevin Leitch, a British blogger and the parent of an autistic child. "And if this same parent reads in *The Guardian* or the London *Times* or *The New York Times* that a new study has been published in *Science* about a gene that *might* be associated with 15 percent of cases of autism, hooray. They look at that and think, 'Screw that. One of them is mildly interesting and the other gives me a load of hope.' "

The vast majority of parents, of course, don't bring such strong predilections to the topic of vaccines. What parents do want is to protect their children from infectious diseases while also being conscientious and informed about what is being injected into their bodies. A

* Over the course of more than two decades, judges have ruled that Mark Geier's expert testimony in vaccine-related lawsuits was "below the ethical standards" required of lawyers, "intellectually dishonest," and "not reliable, or grounded in scientific methodology and procedure." The Geiers counter such criticism by insisting that there are mainstream scientists who support their work. One person they've cited is a British clinical psychologist named Simon Baron-Cohen. When the *Chicago Tribune* asked Baron-Cohen about the Geiers' "protocol," he said that administering Lupron to autistic children "fills me with horror."

lot of parenting decisions come down to our gut reactions—science can't tell us what's an appropriate curfew for a sixteen-year-old or whether it's better to indulge or resist a child who says he wants to quit violin lessons—and when it comes to vaccines, most of the "commonsense" arguments appear to line up on one side of the equation: Vaccines contain viruses, viruses are dangerous, infants' immune systems aren't fully developed, drug companies are interested only in profit, and the government can't always be trusted. The problem, as psychologist and Nobel laureate Daniel Kahneman and his longtime research partner Amos Tversky demonstrated in a series of groundbreaking papers in the 1970s, is that in many situations regarding risk perception and data processing, "commonsense" arguments are precisely the ones that lead us astray.* Because the risks associated with foregoing vaccines *feel* so hypothetical, and because the infinitesimally remote possibility that vaccines could hurt our children is so scary, and because there's nothing in our daily experience to indicate that a little fluid administered through a needle would protect us from a threat we can't even see, it's very hard for parents working by intuition alone to know what's best for their children in this situation.

This leaves us with two choices: We can either take it upon ourselves to do a systematic analysis of all the available information—which becomes ever less feasible as the world grows more complex—or we can trust experts and the media to be responsible about the information and advice they provide. When they're not, whether it's because they're naive or underresourced or lazy or because they've become true believers themselves, the consequences can be severe indeed. A recent Hib outbreak in Minnesota resulted in the deaths

---

* Virtually anything having to do with technology provides a good example of how often commonsense assumptions end up being wrong. Think about how recently it would have sounded ludicrous to propose that an invisible worldwide communication network would be capable of beaming movies into a device smaller than a deck of cards, or that shooting lasers into people's eyeballs could improve their sight. Politics is also an area in which the fantastical has a way of becoming reality: Twenty years ago, a scenario in which the Terminator was elected governor of California would have seemed possible only in a science fiction movie.

of several children—including one whose parents said they do not "believe" in vaccination. In 2009, there were more than 13,000 cases of pertussis (more commonly known as whooping cough) in Australia, which is the highest number ever recorded. Among those infected was Dana McCaffery, whose parents *do* believe in vaccination, but who was too young to get the pertussis vaccine. She died when she was thirty-two days old. Six months later, Dana's mother got an e-mail from a woman in Dallas, Texas, named Helen Bailey. Bailey was looking for someone who might understand her grief: Her son, Stetson, died of pertussis when he was just eleven weeks old. If anything, the situation is getting even worse: In 2010, a yearlong pertussis outbreak in California was so severe that in September some foreign governments began warning their citizens of the dangers of traveling to the region.

Then there's measles, which is the most infectious microbe known to man and has killed more children than any other disease in history. A decade after the World Health Organization (WHO) declared the virus effectively eradicated everywhere in the Americas save for the Dominican Republic and Haiti, declining vaccination rates have led to an explosion of outbreaks around the world. In Great Britain, there's been more than a thousandfold increase in measles cases since 2000. In the United States, there have been outbreaks in many of the country's most populous states, including Illinois, New York, and Wisconsin. A recent outbreak in California began when a grade-schooler whose doctor supports "selective vaccination" was infected while on a family vacation in Europe. In an anonymously published article in *Time* magazine, that child's mother said she "felt safe in making the choice to vaccinate selectively" because she lives in "a relatively healthy first-world country" with a well-functioning health care system. "Looking at the diseases mumps, measles and rubella in a country like the US . . . it doesn't tend to be a problem," she said. "Children will do fine with these diseases in a developed country that has good nutrition. And because I live in a country where the norm is vaccine, I can delay my vaccines."

That statement could not be more false. Measles remains deadly everywhere in the world. (Before the MMR vaccine was introduced, its annual death toll in the United States reached into the hundreds, and during the 1964–1965 rubella epidemic, there were more than 11,000 miscarriages or therapeutic abortions, two thousand infant deaths, and 20,000 children born blind, deaf, or developmentally disabled.) This mother's conviction also perfectly encapsulates one of the most vexing paradoxes about vaccines: The more effective they are, the less necessary they seem.

On the fourth morning of Matthew Lacek's coma, a doctor told his parents that he appeared to have stabilized from the drop in his blood pressure the day before. There was a chance, the doctor said, that the antibiotics were winning the fight against the infection that had taken over Matthew's body—but there was no way to know for sure until he woke up. For the rest of the day, Kelly and Dan sat by their son's bedside. Late that afternoon, he began to breathe on his own—slowly at first, but then more regularly. It was getting dark outside when he blinked open his eyes. The first words out of his mouth were, "I want to go potty."

In April 2010, about a year after we first spoke, Kelly Lacek e-mailed me a picture of Matthew and his older brother. "We just celebrated [Matthew's] 7th birthday," she wrote. "It takes everything for me not to cry each day, let alone his birthday. We are so blessed to still have him. He had strep throat 4 times since January and each time his tonsils swell up, [he gets a] high fever and my husband Dan and I are reminded of that day." Kelly told me about her sons' Little League games and how the whole family loves board games and sitting around campfires. "I hope that helps, Seth," she wrote. "Please let me know if you need anything else. . . . [It's important] to make sure families are making the right decisions, based on fact and not by fear or misinformation."

Part One

CHAPTER 1

The Spotted Pimple of Death

To understand the roots of modern-day fears of vaccines, it is necessary to understand vaccines themselves, and to do that requires us to look back briefly at the deadly diseases they protect against. The most consequential of these is smallpox. In the three thousand years since the first recorded smallpox epidemic in 1350 B.C., no virus has affected humanity more profoundly than *Variola vera*—a term that comes from the Latin for "spotted pimple." (The term "smallpox" was coined in the fifteenth century in an effort to distinguish *Variola* from syphilis, which was known as "the great pox.") Smallpox's telltale scars mark the mummified face of Ramses V, an Egyptian pharaoh who died in 1157 B.C. The Plague of Antonine, with a death toll of between three and a half and seven million, hastened the decline of the Roman Empire. The collapse of the Aztec and Incan kingdoms was expedited by the introduction of smallpox by Old World conquistadors. In eighteenth-century Europe, 400,000 people a year died from smallpox, and those who survived accounted for a third of all cases of blindness on the continent. Between 1694 and 1774, eight reigning sovereigns—Queen Mary II of England, King Nagassi of Ethiopia, Emperor Higashiyama of Japan, Emperor Joseph I of Austria, King Louis I of Spain, Czar Peter II of Russia,

Queen Ulrika Eleanora of Sweden, and King Louis XV of France—died of the disease; the Habsburg line of succession changed four times in four generations because of smallpox deaths.

The virulence of smallpox brought about the first attempts at inoculation, which medical historians estimate occurred more than two thousand years ago in the Far East, although the earliest known records are from eighth-century India. In either case, it wasn't until the eighteenth century that the practice spread to the West: In 1717, Lady Mary Wortley Montagu, the wife of the British ambassador to the Ottoman Empire, was amazed to discover that "the small-pox, so fatal, and so general amongst us is here entirely harmless, by the invention of *ingrafting*"—a crude and painful process by which pus from an individual with a relatively mild case of the disease was spread on an open wound of an as yet uninfected person. Contrary to Montagu's enthusiastic claim, however, inoculation—which was also known as variolation, in honor of the disease it sought to combat—did *not* render smallpox entirely harmless: Inoculees still got sick; that, after all, was the whole point of undergoing the procedure in the first place, since a bout of smallpox was the only known way to achieve lifelong immunity. What's more, even though an inoculation-induced case of the disease was usually less severe than one resulting from a natural infection, there were still those who became alarmingly ill and even died. Montagu, who'd lost a brother to smallpox, clearly considered those risks worth taking, and she successfully inoculated her five-year-old son while still in Constantinople and her four-year-old daughter shortly after arriving home. In London, her doctor was given permission to test the procedure on a half-dozen prisoners who'd been condemned to hang. All six survived—and were eventually granted their freedom. (Authorities in Georgian-era England might not have been overly concerned with prisoners' rights, but they apparently did have a sense of fair play.) Within a year, the Prince of Wales, mindful of the fate that had befallen so many of his royal contemporaries, had inoculated his own daughters.

In New England, the Puritans were also learning about inoculation, but there acceptance of the procedure was slower. In 1706, a "Coromantee" slave of Cotton Mather's named Onesimus described for the minister being inoculated as a child in Africa. In the following years, Mather told friends about his slave, who "had undergone an Operation, which had given him something of y[e] *Small-Pox*, & would forever praeserve him from it." Mather, whose wife and three youngest children had died of measles, soon became a passionate advocate of the procedure; still, it wasn't until 1721, in the midst of an epidemic in which eight hundred Bostonians died and half the city became ill, that he was able to persuade a local doctor named Zabdiel Boylston to inoculate a pair of slaves, along with Mather's and Boylston's young sons. After a brief period of illness, all recovered.

Mather, a Puritan minister best known for his involvement in the Salem Witch Trials several decades earlier, began preaching to anyone who would listen that inoculation was a gift from God. This view, he quickly discovered, was not a popular one: Not long after he and Boylston publicly announced their results, Mather's house was firebombed. An accompanying warning read, "COTTON MATHER, You Dog, Dam you. I'l inoculate you with this, with a Pox to you." Some of the procedure's most vocal opponents feared that inoculation would spread smallpox rather than guard against it. Others cited biblical passages—especially apropos was Job 2:7, which read, "So went Satan forth from the presence of the Lord and smote Job with boils, from the sole of his foot, unto his crown"—as proof that smallpox was a form of a divine judgment that should not be second-guessed or interfered with. (It was for this reason that gruesome vaccination scars came to be known as the "mark of the beast.")

The tendency of our forefathers to view smallpox as an otherworldly affliction is easy to understand: It is one of the world's all-time nastiest diseases. After a dormant period in the first several weeks following infection, the virus erupts into action, causing bouts of severe anxiety, lacerating headaches and backaches, and crippling nausea. Within days, small rashes begin to cover the hands, feet, face, neck, and back. For an unlucky minority, those rashes lead to internal hemorrhaging that causes victims to bleed out from their eyes, ears, nose, and gums.

Most of the time, however, the progression of the disease is not so swift. Over a period of about a week, the initial rashes transform first into pimples and then into small, balloonlike sacs, which render some of the afflicted all but unrecognizable. Three weeks after infection, the vesicles begin to fill with oozing pus. Several days later, after these increasingly foul boils are stretched to capacity, they burst. The resulting stench can be overpowering: One eighteenth-century account described "pox that were so rotten and poisonous that the flesh fell off . . . in pieces full of evil-smelling beasties." Throughout the 1700s, between 25 and 30 percent of all smallpox victims died. Even those survivors who were not permanently blinded did not escape unscathed, as the vast majority of them were left with scars across their cheeks and noses.

The fact that for many people the threat of being afflicted with smallpox was not enough to overcome an innate resistance to having infected pus smeared on an open wound can likely be attributed in part to a phenomenon called the "disgust response." In a 2001 paper, sociologists Valerie Curtis and Adam Biran speculated about a possible evolutionary explanation for what the cognitive scientist Steven Pinker has called human beings' "intuitive microbiology": "Bodily secretions such as feces, phlegm, saliva, and sexual fluids, as well as blood, wounds, suppuration, deformity, and dead bodies, are all potential sources of infection that our ancestors are likely to have encountered," Curtis and Biran wrote. "Any tendency towards practices that prevented contact with, or incorporation of, parasites and pathogens would have carried an advantage for our ancestors."* Looked at

* As part of their research, Curtis and Biran tried to identify universal objects of disgust. While there were some notable differences—in India, people were sickened by the thought of food cooked by menstruating women, while in the United Kingdom, subjects were repulsed by cruelty to horses—everyone listed bodily fluids and decaying or spoiled food near the top of their lists. Curtis and Biran also discovered a near-universal physical reaction that accompanies disgust: moving back the head, wrinkling the eyes and nose, and turning down of the mouth. Don't believe them? Imagine being trapped in an overflowing outhouse.

from this perspective, it's a testament to smallpox's sheer hellishness that anyone willingly underwent the crude vaccination efforts of the early eighteenth century.

There were, to be sure, other explanations for opposition to inoculation, including the colonists' hair-trigger resistance to anything that was perceived as infringing upon individual liberties. Even the smallpox epidemic that engulfed Boston in 1752, in which 7,669 of the city's 15,684 residents were infected, did not sway the procedure's most fervent opponents. In the years to come, these reactions would, at least for a brief while, become secondary to the fear of losing the struggle that defined America's very existence: the Revolutionary War.

On September 28, 1751, nineteen-year-old George Washington and his half-brother, Lawrence, left the family plantation at Mount Vernon for Barbados. This was no vacation: Lawrence hoped the West Indian island's warmer climate would help cure his tuberculosis. The day after completing their five-week trip, George and Lawrence were persuaded to go to dinner with a local slave trader. "We went," Washington wrote, "myself with some reluctance, as the smallpox was in his family." As Elizabeth Fenn recounts in *Pox Americana*, her history of smallpox outbreaks during the American Revolution, Washington's concern was justified: Exactly two weeks after that dinner he wrote in his diary that he had been "strongly attacked with the small Pox." Washington, overcome with the anguish of the disease, would not make another journal entry for close to a month.

Twenty-four years later, with Washington newly installed as commander in chief of the Continental Army, the colonies were struck with the deadliest smallpox epidemic in their brief history. In Boston, the death toll reached five per day, then ten, then fifteen; by the time it was at thirty, the city's churches no longer even bothered to ring their funeral bells. It was in the midst of this environment that Washington mounted an ill-conceived wintertime attack on the British forces that were holed up in the walled city of Quebec. Throughout December, ragtag American battalions and straggling troops marched to Canada from as far south as Virginia. Some reached the American encampment already infected with smallpox; others, weakened by their travels, found themselves thrust into a prime breeding ground for the virus. Many of the new conscripts had ignored recommendations that they get inoculated before reporting for duty; once ill, they regularly disregarded procedures for alerting commanding officers about their infections.

As 1775 drew to an end, Benedict Arnold, then the leader of the northern forces, warned Washington that any further spread of smallpox could lead to the "entire ruin of the Army." Since the disease was endemic in Europe, many British soldiers had been infected when they were boys—an age when survival rates were relatively high—and were now, like Washington, immune, which allowed them to occupy smallpox-stricken towns and fight against smallpox-infected regiments without fear of falling ill.

The Americans had no such luxury. By Christmas Day, roughly a third of the troops massed outside Quebec had fallen ill. Still, the Continental forces stuck to their plan, and launched an assault on New Year's Eve. The American and Canadian coalition maintained its largely ineffectual siege until May, when newly arriving British troops forced an embarrassing retreat. It was the first battlefield defeat in America's history.

Determined not to repeat the mistakes of Quebec, Washington spent much of 1776 torn about whether to require variolation for new conscripts. It was a torturous decision: On the one hand, inoculation would protect his soldiers and dampen what was rapidly becoming a full-fledged, smallpox-induced fear of enlistment. On the other, it would mean that significant numbers of American troops would be out of commission for weeks on end. More than once in 1776, Washington issued decrees requiring inoculation, only to change his mind days later.

In the end, repeated rumors of Tory biowarfare tipped the balance. "It seems," Josiah Bartlett told a fellow congressional delegate in early 1777, that the British plan was "this Spring to spread the small pox through the country."* That February, Washington ordered his commanders to "inoculate your men as fast as they are enlisted." "I need not mention the necessity of as much secrecy as the nature of the Subject will admit of," he wrote, "it being beyond doubt that the Enemy will avail themselves of the event as far as they can." Washington's plan did remain secret—and from that point forward, American soldiers could focus their energies on defeating the British instead of on maintaining their health.

* These were not paranoid fears: In 1763, the commander of the British forces in North America recommended giving rebellious Native Americans blankets that had been sprinkled with ground-up smallpox-infected scabs.

CHAPTER 2

Milkmaid Envy and a Fear of Modernity

By the time the United States Constitution was adopted in Philadelphia in 1787, the benefits of inoculation were clear: When naturally occurring, smallpox was lethal up to a third of the time; when the result of variolation, that ratio dropped to under 2 percent.* At the time, there was only a vague understanding of precisely why inoculation was so effective. Today, that process is much better understood. As soon as the immune system realizes the body has been attacked by a foreign body, a type of white blood cell protein called an antibody jumps into action. After identifying the interloper, the antibody carefully traces its contours in order to manufacture an exact mirror image of the invader's perimeter. Once this has been completed, the immune system is able to disarm the antigen by enveloping it in much the same way that a lock envelops a key. (Strategically, antibodies are more generals than front-line troops: They enlist a specialized type of white blood cell for these nitty-gritty, surround-and-destroy missions.) What makes this defense system so effective is that antibodies have an excellent "memory": Once they've

* The actual differences in these ratios, while significant, were likely not quite so stark: People were usually inoculated when they were in good health and could receive adequate medical care, thus increasing their chances of survival.

successfully defeated a pathogen, they retain their operating instructions so that they are ready to spring into action if the same disease returns. It's an elegant system that beautifully demonstrates the sophistication of the human organism, but there is a rub: Because the body must get sick before it can produce antibodies, variolation carried with it the risk of death from the very disease it was meant to protect against.

The discovery that essentially benign viruses could spur the production of antibodies that protected against lethal diseases changed that calculus dramatically. Like so many history-changing scientific breakthroughs, this one came about through the examination of a seemingly prosaic fact of ordinary life. For centuries, people had observed that milkmaids almost never came down with smallpox. (One popular rhyme played off the fact that virtually everyone else had been scarred by the disease: "Where are you going to, my pretty maid? I'm going a-milking sir, she said. What's your fortune, my pretty maid? My face is my fortune, sir, she said.") It wasn't until the eighteenth century that gentlemen farmers across Europe began more actively exploring the reasons why this might be the case. An English scientist and naturalist named Edward Jenner, among others, speculated it could have to do with the milkmaids' frequent contact with open blisters on the udders of cowpox-infected cows.

In 1796, Jenner enlisted a milkmaid named Sarah Nelmes and an eight-year-old boy named James Phipps to test his theory. Jenner transferred pus from Nelmes's cowpox blisters onto incisions he'd made in Phipps's hands. The boy came down with a slight fever, but nothing more. Later, Jenner gave Phipps a standard smallpox inoculation—which should have resulted in a full-blown, albeit mild, case of the disease. Nothing happened. Jenner tried inoculating Phipps with smallpox once more; again, nothing.*

---

* Nelmes was infected by a cow named Blossom, a saintly contrast to Mrs. O'Leary's Cow, the purported protagonist of the Great Chicago Fire of 1871. To this day, Blossom's hide hangs on the library wall in St. George's Medical School, where Jenner studied. Phipps and Nelmes were, thankfully, allowed to rest in peace.

The implication of Jenner's work was momentous. If cowpox-induced antibodies protected against smallpox—and if cowpox could be transferred directly from person to person—an all-powerful weapon against one of the most ruthless killers in history was suddenly at hand. What's more, if vaccination were not limited to those who could afford to convalesce while receiving high-quality medical care, preventive health measures could be practiced on a much more egalitarian basis.

These new realities not only exponentially increased the number of people who could be vaccinated, they also opened up the possibility of "herd immunity," a mechanism whereby individuals for whom a given vaccine does *not* work are protected by the successful immunization of those around them. (There are a number of reasons vaccines might not be beneficial to an individual. The main ones are limitations of the vaccine itself—no vaccine is effective one hundred percent of the time—and specific reasons a patient is unable to receive a vaccine, such as poor health or a preexisting immune deficiency.) Herd immunity occurs when a high-enough percentage of a population has been successfully vaccinated to create a barrier in which the immune members of society protect the unimmunized by making it impossible for a virus to spread in the first place. It is like a herd of buffalo encircling its weakest members to protect them from predators.

The relative safety of the cowpox vaccine also made state-sponsored immunization drives more appealing. Spain instituted mass vaccination programs in its colonies as early as 1803, and the Netherlands, the United Kingdom, and parts of the United States soon followed. Initially, with smallpox's deadly power still fresh in people's minds, these efforts were widely accepted, but over time, as the disease began to be perceived as less of a threat, resistance to vaccination grew. By the middle of the century, compliance had dropped to the extent that laws mandating vaccination were being passed in the U.K. and in a number of the American states. These compulsory vaccination programs, in turn, fueled even more impassioned resistance, creating a vicious cycle that continues to this day.

Looked at in a vacuum, it's remarkable how static the makeup, rhetoric, and tactics of vaccine opponents have remained over the past 150 years. Then, as now, anti-vaccination forces fed on anxiety about the individual's fate in industrialized societies; then, as now, they appealed to knee-jerk populism by conjuring up an imaginary elite with an insatiable hunger for control; then, as now, they preached the superiority of subjective beliefs over objective proofs, of knowledge acquired by personal experience rather than through scientific rigor.

But if we think about vaccine resistance as prompted by what Harvard history of science professor Anne Harrington calls the feeling of being "broken by modern life," the parallels are less startling. Two hundred years ago, in the early stages of the Second Industrial Revolution, there was growing anxiety throughout the Western world about the dehumanizing nature of modernity. In the United States, the American Revolution's promise of a country where every person would be free to pursue the good life had been replaced by the reality of fetid, squalid cities where horse carcasses rotted in open-air sewers and newly arrived immigrants were crammed together in abasing, lawless tenements.

One manifestation of this widespread disillusionment was the flowering of all manner of utopian and spiritualist movements, ranging from Idealism and Occultism to Swedenborgianism and Transcendentalism. By celebrating individualism and holding intuition above empiricism, these philosophies promised a more authentic and meaningful form of existence amid a harsh, chaotic world. Alternative medical practitioners—who proselytized that physical suffering was a symptom of spiritual disorder and implied that anyone opposing their methods did so because he believed in a "one size fits all" approach to medicine—were able to exploit these conditions with particular effectiveness. (The apotheosis of this approach came in 1866, with Mary Baker Eddy's founding of Christian Science, a religion that eschewed medical interventions in favor of prayer-induced psychosomatic cures.) When the American Medical Association was founded in 1847, a hodgepodge of newly marginalized "irregular physicians" gravitated toward the Eclectic Medicine movement, which embraced everyone from homeopaths to hydropaths, mail-order herbalists to Native American healers.

Today, a similarly colorful collection of practitioners has seized on the anxiety created by the bureaucracy of managed care and the generally effacing qualities of modern society in order to peddle a brand of medicine whose primary appeal lies in its focus on the patient as a whole person. Defeat Autism Now! (DAN!), a group that accredits doctors who want to treat patients according to its nontraditional protocols, typifies a medically permissive movement that lists the AMA and the AAP among its primary bogeymen. (A recent DAN! conference included exhibits and presentations by mail-order herbalists, energy healers, and purveyors of home purification systems and hyperbaric oxygen chambers.) As was the case a century and a half ago, a distinct advantage of this avowed break from the mainstream is the built-in defense it offers from accusations of misconduct: Any criticism can be dismissed as a power play by the medical establishment against alternative practitioners who challenge it.

The person who best embodies the philosophical continuity of vaccine resistance movements is the early twentieth-century activist Lora Little, whom journalist and author Arthur Allen described in his book *Vaccine* as a "granola-belt Mother Jones who promoted whole foods and naturopathy and denounced . . . white male medical practitioners before it was fashionable to do so." According to Little, vaccines (along with mechanization, Western medicine, establishment doctors, processed foods, and white sugar) contributed to most modern ills. During her anti-vaccination campaigns, Little passionately decried the alienation brought about by the efficiencies of industrialization. Speaking of New York City, which even then was the prototypically modern city, she said, "Every doctor there has become a cog in the medical machine. And once the machine gets its grip on you, you cannot escape, you are drawn in and ground through the mill."

Little also predicted the current-day conspiracy theories that lash together doctors, government officials, and vaccine manufacturers in their quest for riches. In her 1906 tract "Crimes in the Cowpox Ring," Little wrote:

> The salaries of the public health officials in this country, reach the sum of $14,000,000 annually. One important function of the health boards is vaccination. Without smallpox scares their trade would languish. Thousands of doctors in private practice are also beneficiaries in "scare" times. And lastly the vaccine "farmers" represent a capital of $20,000,000, invested in their foul business.

Substitute "measles" for "smallpox" and that same paragraph could appear in pamphlets put out by groups like the National Vaccine Information Center (NVIC) or the Australian Vaccine Network. Here's Barbara Loe Fisher, the founder and president of the NVIC, talking about pharmaceutical companies in a 2009 speech in which she compared the United States government's vaccine policy to medical experiments conducted by Nazis during World War II:

> They make trillions of dollars in the global market. . . . Literally, every known disease that you can imagine, there's a vaccine being created for it. Their intention is to have laws passed requiring businesses to use it because that's how they make their money. So, what is the situation on the ground, with parents . . . who take their children to expensive pediatricians whose practices are economically heavily dependent on the [vaccine] schedules? . . . Pediatricians are now saying children must get forty-eight doses of fourteen vaccines by age six.

For all the tactical and rhetorical similarities, the most poignant link between early activists such as Little and their modern-day descendants is their tendency to locate the cause of their personal tragedies in some larger evil. For Little, that tragedy was the death of her seventeen-month-old son, who, according to his medical records, suffered from simultaneous measles and diphtheria infections. Little was convinced the cause of his death lay elsewhere. Her son, she decided, could not have been the victim of the vagaries of human existence; that would be too senseless. Instead, as she told thousands of people over the years, he was killed by the smallpox vaccine. For Fisher, a television program about vaccines that she saw a year and a half after her own son received a combined diphtheria-pertussis-tetanus (DPT) shot convinced her that his physical and developmental difficulties began the very day he had been vaccinated.

Despite her popularity, Little's fears were hardly representative of the general mood during the first half of the twentieth century, when science notched one victory after another in what had been previously a one-sided fight between bacteria and viruses and the human beings they attacked. At times, it must have seemed as though scientists needed only identify the cause of a disease in order to cure it: In 1910, Paul Ehrlich discovered that an arsenic-based compound could wipe out syphilis; in the 1930s, sulfur-based medicines were proven to be effective against everything from pneumonia and puerperal sepsis to staph and strep; and in 1941, Howard Florey and Ernst Boris Chain showed that humans could use penicillin without fear of death.

The implications of these medical triumphs were stunning. Even as deadlier weapons were being created, warfare was becoming safer: In the Spanish-American War in 1898, thirteen soldiers fell ill for each combat-related death. By World War I, that ratio was 1:1; by World War II, it had fallen to 1:85. Between 1920 and 1955, the lifespan of an average American increased by more than 25 percent—an increase that was primarily due to fewer young people dying of disease. Over that same span, California went from having 110,000 cases of diphtheria and seven hundred deaths annually to forty-two cases and two deaths a year. By mid-century, the notion of a world free of infectious diseases seemed, for the first time in human history, to be a possibility and not a pipe dream.

These victories led to a transnational excitement and pride. In the 1930s, after witnessing state-sponsored institutes in Europe and Asia make one breakthrough after another, the American government for the first time assumed a central role in funding and conducting medical research. Doctors and scientists were repeatedly cited as among the most trusted members of society. It was perhaps inevitable that the march of accomplishments led to the scientific establishment's growing increasingly mesmerized by its own power. Vaccine proponents, be they doctors, politicians, or self-styled intellectuals, held fast to their own credo and accused their opponents of taking part, as a *New York Times* editorial put it, "in a futile attempt to head off human progress." This smug sense of superiority was mixed with a condescending bewilderment at what physician Benjamin Gruenberg described as the hoi polloi's insistence "upon the right to hold opinions (and to act according to these opinions) upon such highly technical questions as the efficacy of vaccination, the value of serums, or the causation of cancer."

One of the most shocking examples of this hubris occurred during the early years of World War II. Even before America's involvement in the conflict, the public health infrastructure in the United States had set in place a plan to give yellow fever vaccinations to any troops headed to tropical climates. By the time the country joined the Allied cause in late 1941, some high-ranking military officials had become so overwrought about the threat of biological warfare that they decided that *all* troops should be vaccinated, regardless of their assignment. The resulting scramble to develop vast quantities of the vaccine—at one point, the Rockefeller Foundation was producing tens of thousands of doses every week—led, not surprisingly, to shoddy quality control. Within months, large numbers of troops were showing signs of jaundice; eventually, up to 10 percent of soldiers in the most severely affected units were hospitalized. It turned out that batches of the vaccine had been contaminated with hepatitis B. By the time the vaccinations were halted, 300,000 troops had been infected and more than sixty had died.

The entire yellow fever campaign was, as Arthur Allen wrote, a dark page in the history of public health: "None of the 11 million Americans vaccinated against yellow fever during the war got yellow fever. Then again, none was challenged with yellow fever. [The] fear of biologically trained killer mosquitoes was not realized. It was somehow all in vain."

Amid the killing of World War II, the deaths of five dozen soldiers from hepatitis B did not attract a lot of notice. In the 1950s, the fight to conquer polio made vaccines one of the biggest news stories of the decade. Here, the threat was real and the potential victims came from all parts of society. The result was an unprecedented campaign with almost universal support—and a case where national exuberance led health officials to forget everything they should have learned about the risks of poor oversight. This time the ensuing tragedy wouldn't escape the public's attention.

CHAPTER 3

The Polio Vaccine: From Medical Miracle to Public Health Catastrophe

On June 6, 1916, the first two polio cases of the summer were diagnosed in New York City. At the time, polio—or infantile paralysis, as it was often called—was a mysterious and frightening, although not terribly common, childhood illness. It had been a quietly persistent presence in human populations for thousands of years, but continual, widespread epidemics didn't appear until the 1880s. The virus quickly made up for lost ground, and in the first decade of the twentieth century, countries around the world were ravaged by outbreaks, especially during warmer weather. One reason for the disease's sudden virulence was undoubtedly the crowded living conditions of modern cities: Polio, like typhoid, cholera, and hepatitis A, is transmitted through what's clinically referred to as the fecal-oral route, which typically occurs as a result of inadequately treated drinking water, improper food handling, and poor sewage methods.*

Even with the growing number of epidemics, polio victims in the first decade and a half of the twentieth century generally did not suffer from lifelong consequences—in fact, in the vast majority of cases, the infection was so minor as to be barely noticeable. Approximately 5 percent of the time, however, the virus reached the central nervous system. Usually, even those infections resulted in relatively benign symptoms—headaches, diarrhea, muscle pain—that disappeared in a matter of weeks. In fewer than 2 percent of cases, however, the virus attacked motor neurons in the spinal cord and brain. At first, those patients also exhibited flulike symptoms such as achiness or general discomfort. Soon the muscles became weak and spastic, before eventually becoming completely unresponsive. The result was paralysis or, if the muscles of the chest became permanently incapacitated, death.

The epidemic of 1916 completely upended those percentages, and the result was unlike anything New York City, or the country, had experienced before. To start with, the virus was spreading exponentially faster than it previously had. Even more terrifying: The fatality rate among those with clinical infections spiked to between 20 and 25 percent, five times higher than expected. Since 1911, when New York's Department of Health began keeping statistics, the city had averaged 280 polio patients a year, and the annual death toll had fluctuated from a high of seventy to a low of thirteen. By the end of June, more children were dying of polio every week than had died in the previous five years combined.

The epidemic that overwhelmed New York City that summer was

* This method of viral transmission occurs much more frequently than we'd like to admit. Viruses that are transmitted through the ingestion of animal fecal matter are also quite common. Take toxoplasmosis, a parasitic disease that occurs most commonly in cats: In the United States, approximately one-third of the population has been infected; in France, the rate is closer to two-thirds. Food for thought the next time you don't wash your hands before a meal.

so unexpected, so inexplicable, and so lethal that it affected the public's perception of health and medicine for years to come. Every funeral highlighted doctors' helplessness; every paralyzed child was a stark reminder of the lack of effective treatment. One local expert told the media that polio seemed to "pick the strong and well children in preference to the weak"—an anti-Darwinian truism that mocked scientific progress and knowledge. A mishmash of confusing and contradictory emergency measures only heightened the panic and confusion. Sometimes infected children were told to remain at home; others were instructed to go immediately to the nearest hospital. Quarantines were put into effect in some neighborhoods but not others. Some parents of sick children were unable to find help: In Staten Island, after watching his son die in his car on the way to the hospital, a father proceeded to drive around "with the boy's body for hours looking for someone who would receive it." Others fought to resist quarantine efforts: In one instance, a local newspaper reported that it took "the authority and strength of four deputy sheriffs and two physicians to get a child from its father." Before long, surrounding communities closed their doors to New York City residents: In Hoboken, New Jersey, "policemen were stationed at every entrance to the city—tube, train, ferry, road and cow path—with instructions to turn back every van, car, cart, and person."

By early August, with the weekly death count approaching four hundred and parts of New York City poised to descend into anarchy, the police department gave its blessing to the Home Defense League, a "volunteer vigilance force" whose 21,000 members were authorized to patrol the streets and barge into homes looking for "violations of the sanitary code that might mean a spread of infantile paralysis." Conspiracy theories took hold: In the town of Oyster Bay, on Long Island, city counselors accused John D. Rockefeller and Andrew Carnegie of using their millions to corrupt "men and microbes" in order to create "causeless hysteria and . . . needless hardships."

By the end of November, when the epidemic had run its course, 26,212 people had been infected in a total of twenty-eight states.

In New York City alone, there were 9,023 total victims and 2,448 deaths. The vast majority of them were under ten years old.

For the next thirty years, the spread of polio waxed and waned without any discernible logic. In 1923, only 695 cases were reported nationwide. In 1931, that number rose to 14,105. Eleven years later, it was back under 3,000. Then, just as the United States was recovering from its involvement in World War II, the country was battered again. In 1946, the total number of reported polio cases approached those recorded thirty years earlier. Two years later, the number of victims hit an all-time high. The year after that, it rose again—this time by an astounding 35 percent, to 40,076. With parents no clearer than they'd been decades earlier on how to protect their children, each summer's outbreak undercut the country's sense of postwar victory and security. Polio might not have been the biggest public health threat—for much of the 1940s, an average of 190,000 Americans died of the flu each year—but it took up the most space in the public's imagination.

By 1952, when more than 58,000 people were infected, polio ranked second only to the atomic bomb as the thing Americans feared most. This outsized anxiety was, to be sure, partly due to what a leading medical writer of the time called "polio's uncommon nastiness." It was also fueled by the popular press, which then, as now, relied on dramatic stories to draw in readers. Perhaps most responsible of all was the National Foundation for Infantile Paralysis, which had been founded in 1937 by President Franklin D. Roosevelt, polio's most famous victim.* By 1945, when Roosevelt died just months into his fourth term in office, the foundation had become the best-known charity in the country. Year after year, its celebrity-studded March of Dimes campaigns highlighted its prodigious fund-raising prowess. Outside the federal government, it was perhaps the only organization that had the prestige and the national apparatus in place

* In 2003, an academic paper presented convincing evidence that Roosevelt's age when he first got sick (he was thirty-nine) and the disease's progression made it likely that he had actually suffered from Guillain-Barré syndrome, an autoimmune disorder that attacks the nervous system and was all but unknown at the time.

to unite the public around a single cause. When, in November 1953, a foundation-funded virologist at the University of Pittsburgh named Jonas Salk announced that he'd developed a polio vaccine, millions of citizens had already been primed to do whatever they could to fight this national menace. Within days, parents were lining up to volunteer their children for what would be the largest medical field trial in history.

The Salk trials began in the spring of 1954; by May, 1.8 million children had been injected in 211 counties spread over forty-four states—and there were still six months of tests remaining. (At the end of the year a Gallup poll found that more people knew about Salk's field trials than knew the president's full name.) This degree of public involvement and scrutiny created inevitable pressures. While Salk, the foundation, and the government worked together, they all stood to gain something different. For Salk, the success of the vaccine would make him among the most revered people on the planet; for the foundation, it would validate the billions of dollars it had raised and spent over the years; and for the government, success would be a fulfillment of an implicit promise to defend its citizens, thereby solidifying its place as a positive force in people's lives.

That summer and fall, another 25,000 Americans came down with polio, but for the first time there was the expectation of a future free of paralyzed children confined to iron lungs. When the foundation announced that it was committing another $9 million to buy 25 million doses of the vaccine, the press surmised that the trials had already been proven effective; why else, they reasoned, would anyone spend all that money? At one point, the *New York World-Telegram and Sun* reported that Salk's vaccine had worked in one hundred percent of the test subjects. Almost nobody bothered to point out that that was a scientific impossibility; with the next March of Dimes campaign set to start in January, there was nothing to be gained by tempering expectations.

As the year drew to a close, Thomas Francis, a professor of epidemiology and a postdoc advisor of Salk's, was chosen to review the reams of data that had been collected and to write an independent report analyzing the results. With the entire world looking over his shoulder, he set to work. The following spring, mere months before that year's polio season would begin, he announced that he was ready with his conclusions.

On April 12, 1955—the tenth anniversary of FDR's death—five hundred doctors and scientists, and hundreds more reporters, gathered in the University of Michigan's Rackham Auditorium to watch Francis, a short, stocky man with a neat mustache, present his findings. Francis used the same measured tone and somnolent cadence as he did when talking to colleagues or students; even so, one newspaper described the morning as being infused with a sense of "fanfare and drama far more typical of a Hollywood premiere than a medical meeting." At the back of the auditorium, a special riser supported the sixteen television cameras covering the event; three flights up, a media center housed another two hundred members of the press.* In ballrooms and conference halls around the country, 54,000 doctors had assembled to watch Francis's presentation via closed-circuit broadcasts. Six thousand crammed into Manhattan's Waldorf-Astoria alone.

At 10:20 A.M., as spotlights clicked on and cameras whirred, Francis strode purposefully to a lectern. Salk's vaccine, he announced, was "safe, effective, and potent." He spoke for another ninety minutes—detailing the tests, describing his analyses, cautioning against over-optimism—but those three words, which had been so eagerly anticipated for so long, were all anyone needed to hear. The seeming inevitability of Francis's announcement did nothing to temper the

---

* Reporters on site had received an embargoed copy of the report at 9:15 A.M., which was delivered under the protection of four policemen. Proving that the media's frenzy for beating competitors by mere minutes is not a product of the Internet age, NBC immediately broke the embargo, and was just as quickly denounced by its competitors as forever tainting the sanctity of agreements made between reporters and their sources.

jubilation that swept the country: Air raid sirens were set off; traffic lights blinked red; churches' bells rang; grown men and women wept; schoolchildren observed a moment of silence.

The rest of the day brought more of the same. The chairman of the board of trustees of the AMA said the success of Salk's vaccine was "one of the greatest events in the history of medicine." *The New York Times* devoted almost all of its front page and five full interior pages to coverage of the event. "Gone are the old helplessness, the fear of an invisible enemy, the frustration of physicians," its editors wrote in an editorial that called for "world-wide rejoicing and thanksgiving. . . . Science has enriched mankind with one of its finest gifts." Oveta Culp Hobby, the secretary of the Department of Health, Education, and Welfare, said, "It's a wonderful day for the whole world. It's a history making day." In the second half of the twentieth century, only the assassination of John F. Kennedy, the moon landing, the resignation of Richard Nixon, and the fall of communism commanded as much of the nation's attention. Never again would science be held in such unequivocal admiration.

At five that afternoon, Hobby and the surgeon general, Leonard Scheele, licensed six preapproved pharmaceutical companies to manufacture the vaccine, a move they said would enable production of enough vaccine to inoculate every child in the country before the polio season began in earnest. Injections were taking place by the following morning.

Over the next hours and days, no accolade would be too lofty, no honor out of reach, no prediction too optimistic. Salk would surely win a Nobel Prize. The foundation, confident that victory over polio was within its grasp, discussed where to focus its energies in the future. Questions and concerns about national vaccine programs that had been present for more than a century seemed to melt away.

It didn't take long for that initial bout of euphoria to fade. The foundation had poured incredible effort into convincing citizens that fighting polio was their patriotic duty, but it had neglected to educate the public about the inherent limitations of its planned eradication campaign. Depending on the strain of the polio virus, the Salk vaccine failed to generate immunity anywhere from 10 to 40 percent of the time. Another factor was polio's standard incubation period, which meant there would be people who'd been infected in the weeks immediately preceding vaccination but wouldn't show symptoms until later. There was also a period of vulnerability between receiving the vaccine and protection taking hold. Finally, there was the inevitable reality that out of the millions of Americans who received the polio shot, some would have a negative reaction.

One illustration of the public's overhyped expectations was the panic brought about by rumors of widespread shortages. (Partisan political jockeying didn't help matters, as Democrats in Washington blamed any and all problems on President Dwight Eisenhower's incompetence while Republicans spoke sotto voce of a mysterious, Democrat-fueled "black market" that was siphoning off valuable supplies.) Ten days after Francis's celebratory press conference, Surgeon General Scheele told reporters that there would likely not be enough doses for everyone to get vaccinated before the end of the year. The day after that he retracted his forecast and promised that by August 1, there would be enough doses for all children ages one to nine.

In the midst of this frantic activity, the first reports of children who had become sick with polio *after* receiving the vaccine began to trickle into Washington. This was not necessarily cause for alarm, but as soon as the distribution of those cases began to be analyzed, it became clear that something had, in all probability, gone wrong. By April 26, exactly two weeks after Francis's announcement, the Public Health Service had identified six children who'd been paralyzed after receiving doses from Berkeley, California's, Cutter Laboratories—a significantly higher number than would have been expected to occur under normal circumstances. Before dawn the next morning, Scheele told Cutter to halt production and ordered a recall of all Cutter-produced doses that had already been distributed. Publicly, he put on a brave face. "This action does not indicate even that the vaccine was in any way faulty," he said, stressing that the program should continue and parents should not be concerned.

Within twenty-four hours, that attitude of insouciance had evaporated. What everyone had hoped was a statistical aberration quickly became a full-blown disaster, as dozens of children who'd received Cutter-produced vaccines were paralyzed or killed. As if on cue, the parties that had been so quick to claim credit for the vaccine's success began sniping at each other through official statements and thinly veiled, off-the-record comments: The foundation said it had "no control of the manufacturing of the vaccines" and that testing was the "sole responsibility" of the government; the same AMA officials who had so recently celebrated "one of the greatest events in the history of medicine" now announced that they had not been shown an advance copy of Francis's report and therefore had never given their imprimatur to Salk's vaccine; and President Eisenhower, at a loss to explain the situation, speculated at one of his weekly press conferences that the vaccine might have a "provocative effect" that activated a "latent" form of the virus already present in some people. "The actual puncture of the skin with this, to give the shot might—and they have not proven this—but it isn't impossible that that might cause some trouble," he said. That this conjecture went against virtually every piece of scientific evidence did not stop Eisenhower from telling reporters that perhaps it made sense to suspend the program until *after* the peak periods of transmission that summer.

Eisenhower's emphasis on damage control over sound science extended to discussions occurring behind the scenes. Before continuing with the nationwide effort, his aides said, public health officials needed to promise that there would not be any more children who were diagnosed with polio after being vaccinated. That, as anyone with an elementary understanding of immunology knew, was an impossible guarantee to provide, and so, instead of trusting people to understand and accept that there are risks with every medical procedure and that correlation does not equal causation, or trying to explain that the problems appeared to be related to the specific conditions under which the infected batches had been produced and not with the safety of the vaccine generally, the government took the one step guaranteed to undermine public confidence: On May 7, Scheele announced that the polio vaccine program was being shut down so that the government, "with the help of the manufacturers," could undertake "a reappraisal of all of their tests and procedures."* "The Public Health Service believes that every single step in the interest of safety must be taken," he said. "We believe—and I am sure the American people join us in believing—that in dealing with the lives of our children, it is impossible to be too cautious."

That was most definitely not what the "American people" wanted to hear. Did these new safety steps mean that previously the Public Health Service had *not* believed it had to exercise the utmost caution? Why, after insisting that the doses of the vaccine that were already at distribution centers were safe to use, did Scheele reverse course? Would the vaccinations start again in two weeks, as some officials were saying? Would it be a month, as others claimed? Or in the fall, as Eisenhower had hinted? (Both the president and members of his administration continued to be particularly inept messengers. At one point Eisenhower attributed the premature release of the vaccine to unspecified "pressure," and said scientists had likely taken "shortcuts" on some safety tests. His appointees, meanwhile, had a habit of

---

* The difficulty in determining whether correlation equals causation causes an enormous number of misapprehensions. Until a specific mechanism demonstrating how A causes B is identified, it's best to assume that any correlation is incidental, or that both A and B relate independently to some third factor. An example that highlights this is the correlation between drinking milk and cancer rates, which some advocacy groups (including People for the Ethical Treatment of Animals) use to argue that drinking milk *causes* cancer. A more likely explanation is that cancer diagnoses and milk consumption both have a positive correlation with increased age: On average, milk drinkers live longer than non–milk drinkers, and the older you are, the more likely you are to develop cancer. This does not, however, mean that drinking milk actually causes people to live longer: It could be that people who drink milk have better access to high-quality health care or eat more healthily than those who do not.

making baldly inaccurate statements, such as when Secretary Hobby told a Senate committee, "No one could have foreseen the public demand for the vaccine.") As it happened, it took only a week for the government to certify another million doses—but by then, the public's initial elation had been transformed into mistrust and apprehension. "The nation is now badly scared," read an article in *The New York Times*. "Millions of parents fear that if their children don't get the vaccine they may get polio, but if they get the vaccine, it might give them polio."

Before the end of the month, when the first of many investigations into what became known as the Cutter Incident had been completed, it was already obvious that the problem had been the result of a combination of inadequate safety guidelines and a lack of official oversight. The government had not required the pharmaceutical companies making the vaccine to divulge any safety issues that arose during manufacturing, turning what should have been a collaborative effort into one driven by competitive commercial interests. As a result, neither the independent scientists working on the project nor the public health officials in charge of implementing it had known that Cutter had discarded a full third of the batches it had produced because of failed safety tests.

Among those listening to officials' conflicting and confused pronouncements with increasing horror were Robert Gottsdanker and his wife, Josephine. The couple was better acquainted than most with the ravages of the polio virus: The children of several friends of theirs had been infected with the disease, giving them a firsthand view of the suffering that it could cause. The Gottsdankers were academics with jobs at the University of California's Santa Barbara campus—he was a psychologist, she was a counselor—and they'd always believed in the ability of science to make the world a better place. Even before the results of the Salk trials had been announced, they'd done everything possible to ensure that their two children were among the first to receive the vaccine. Within days of Thomas Francis's April 12 press conference, the Gottsdanker family physician arrived at the family's home to personally inject five-year-old Anne and seven-year-old Jerry.

At first, the Gottsdankers were relieved that their lobbying efforts had paid off: With a vacation planned for the week of April 18, they didn't want to risk missing out due to the rumored nationwide shortages. It didn't take long for them to start to question their zeal. By the time the family reached Calexico, a small border town about 120 miles east of San Diego, both children had fallen ill. Jerry soon recovered, and in later years he could barely remember being sick at all. His sister was not so lucky. By the time the family arrived back in Santa Barbara, Anne had lost the use of both of her legs. A week after being vaccinated, the Gottsdankers' younger child had developed full-blown, paralytic polio.

By the end of the month, Anne's parents learned that she had been among the children injected with the contaminated doses produced by Cutter Labs. (Her case was one of the ones that set off those first alarm bells in Washington.) The eagerness with which the Gottsdankers had sought out the vaccine only exacerbated the betrayal they felt after their daughter was infected, and they were one of the first of more than forty families to sue Cutter for damages. On November 22, 1957, just nineteen months after Salk's vaccine was first made available to the public, opening arguments in *Gottsdanker v. Cutter Laboratories* were heard in Alameda County Superior Court. From the outset, the trial was an odd one. Each side readily conceded central aspects of its opponent's argument: Walter Ward, Cutter's medical director, agreed that Anne's paralysis was in all likelihood the result of her having been injected with live virus, while Melvin Belli, the Gottsdankers' attorney, acknowledged that the contamination had resulted from fundamental flaws in the government's safety protocols and not from negligence on Cutter's part.

The case, then, would hinge on whether the Gottsdankers could show that Cutter had violated what in legal terms is referred to as an "implied warranty." If anyone was capable of convincing a jury that this was the case, it was Belli, a man who'd been dubbed the King of

Torts by the press and Melvin Bellicose by his enemies.* Belli had first achieved national prominence thirteen years earlier, when, at age thirty-five, he'd represented a waitress named Gladys Escola, who sued the Coca-Cola Bottling Co. after a glass soda bottle exploded in her hand for no reason and without warning. (The blood vessels, nerves, and muscles in Escola's thumb were severed in the incident.) In order to win that case, Belli had successfully overturned a century's worth of precedents regarding the doctrine of privity, which holds that there needs to be a contract between two parties in order for one of them to file suit. (In the Escola case, for example, both the waitress and Coca-Cola had agreements with the diner, but not with each other.) Belli's victory pushed privity aside in favor of the doctrine of foreseeability, which has been the cornerstone of modern product-liability law ever since.

With *Gottsdanker v. Cutter*, Belli was trying to radically expand the parameters he'd set with his 1944 victory. In that case, it was clearly negligent of Coca-Cola to produce bottles that shattered willy-nilly; the debate was who (or what) had standing to sue the company for its negligence. Cutter, on the other hand, had just as clearly *not* been negligent: It adhered to the government-dictated safety standards every step of the way. (That the lab had faith in its work was evidenced by Cutter's lead pathologist injecting his own children with vaccine produced by the company.) Belli, in an effort to square that circle, argued that the mechanism that caused the vaccine to be unsafe was

* Belli's tort-related work, which resulted in more than $600 million in damages by the time he died in 1996, was not all he was known for: He served as Jack Ruby's pro bono counsel in his trial for the murder of Lee Harvey Oswald; he received a letter from the Zodiac Killer; in 1969, he helped set up the Rolling Stones–headlined concert at Northern California's Altamont Speedway at which a member of the Hell's Angels stabbed a fan to death; over the course of his career, he represented everyone from Mae West to Zsa Zsa Gabor and Chuck Berry to Muhammad Ali; he accused his fifth (of six) wife of throwing their pet dog off the Golden Gate Bridge; and he played an evil overlord named Gorgan in a *Star Trek* episode titled "And the Children Shall Lead." Ironically, one of his last victories led to his financial ruin: In 1995, Belli won a class-action lawsuit against breast implant manufacturer Dow Corning, but his firm was unable to recover the $5 million it had spent on the trial when the company declared bankruptcy.

beside the point—the bottom line was that whether the company was aware of it or not, it had abused the Gottsdankers' trust. "There is no doubt in my mind," Belli told the jury, "and there should be none in yours, that the process could and should be perfect." (This abstract notion of perfection is, of course, just that: an abstraction. In its most extreme application, a strict adherence to that standard would prohibit all medical care, since there is always the possibility that a given remedy will be improved, and every treatment carries with it at least some measure of risk.)

After more than five weeks of testimony, Belli's case remained shaky, at best, and Cutter likely would have prevailed had it not been for instructions from Judge Thomas J. Ledwich. Cutter's compliance with government regulations, Ledwich said, had no bearing on the question of liability: "If you find that the vaccine contained infectious amounts of live virus and that the vaccine caused [Anne Gottsdanker] to become infected," he said, "implied warranty is applicable."

These instructions, which remain controversial to this day, threw the jurors into turmoil. On January 17, 1958, they returned the verdict they felt they'd been all but forced to deliver: Cutter was liable for Anne Gottsdanker's paralysis. The jurors made their frustration known by including an impassioned (and highly unusual) statement with their decision, in which they wrote that "a preponderance of the evidence" had convinced them that "the defendant, Cutter Laboratories, was *not* negligent either directly or by inference."

The $147,300 the Gottsdankers were awarded did not end the family's pain, which ultimately extended far beyond Anne's paralysis. For the remainder of their lives, the Gottsdankers seemed to simultaneously obsess over Anne and deny the realities of her condition, refusing even to install a wheelchair ramp in their home. Jerry Gottsdanker grew increasingly resentful of the attention his parents lavished on his sister; he later admitted that he had been "cruel" to Anne while growing up. Robert Gottsdanker, meanwhile, seemed to lose faith in the world. "He was a scientist," his son said in a 2005 interview. "And he felt science had let him down."

For the Gottsdankers and the families of the other fifty-six children who'd been paralyzed after receiving contaminated doses of the polio vaccine, the Cutter Incident was a life-altering personal tragedy. It also sullied the reputations of everyone from Salk and Scheele to the leaders of the AMA and the National Foundation for Infantile Paralysis. As a result of their obfuscation and equivocation, rumors whipped around the country: Health officials and drug makers had conspired to lie to the public; the nation's schoolchildren were the subject of massive experiments; the vaccine had *never* been safe. The outcome was as predictable as it was unnecessary: Confused and unsure about what to believe, citizens who had just weeks earlier fought to make sure they weren't passed over began to wonder if it was such a good idea to get the vaccine after all.

If anything positive were to come from the entire affair, it would depend on future leaders' capacity to learn from these mistakes. Instead, the precise formula that resulted in such chaos a half-century ago has been repeated time and time again. Perhaps nothing illustrates this better than a contemporaneous account that ran in the August 1955 issue of *Harper's*—an account that could just as easily characterize numerous incidents in the years since:

> [D]emagogy and political expediency . . . contributed to the brew. So did over-sensationalism by the press, radio, and TV, and a misguided attempt by the Department of Health, Education, and Welfare to withhold from the public for many weeks information the public was entitled to have from the beginning. Also involved were timidity and lack of leadership; a complete failure to educate the public properly on vaccines (despite all the propaganda); the constitutional unwillingness of scientists to give absolute guarantees; and many things more.
>
> The point is, though, that the mess was unnecessary. . . . The vaccination program was perhaps not the best that could have been devised. But it was not a bad program. It could have been kept on the track. All that was necessary was a little judgment

and some capacity for decisive action in the right places at the right time.

* * *

One final legacy of the polio vaccine's scandal-tinged rollout involves a little known agency of crack infectious disease specialists housed within the CDC. The Epidemic Intelligence Service (EIS) was the brainchild of a public health giant named Alexander Langmuir, who in 1951 convinced federal health officials that the country needed a team of medical doctors to conduct the first wave of investigation whenever new and disturbing disease patterns emerged.

It had been the EIS that had first traced the defective batches of polio vaccine back to Cutter Laboratories, and the speed with which it worked had been a major factor in preventing the total collapse of the polio campaign. But in the same investigation in which Langmuir's ground troops had fingered Cutter, they discovered that Wyeth Pharmaceuticals, one of the other manufacturers licensed to make the vaccine, had also produced defective batches. When confronted with the possibility that a second revelation could spark a panic that would fatally cripple the vaccine program, Langmuir concluded that the risks attributable to Wyeth were small enough that it would be better to keep quiet. Seven years later, Langmuir faced a similar decision after EIS investigators found that in very rare cases—less than one person out of every million vaccinated—the oral polio vaccine developed by Albert Sabin caused paralysis. By that point, polio had all but vanished from the United States. Given this situation, Langmuir chose to bury the data implicating Sabin's vaccine, calculating that it was better to avoid the potential reemergence of the virus than to let the public know all the facts. Whether he made the correct decision is a subject for medical ethicists and philosophers to reason through. What is beyond debate is that in the years to come, public health officials learned the disastrous consequences of failing to educate the public about the risks and realities of fighting disease.

CHAPTER 4

FLUORIDE SCARES AND SWINE FLU SCANDALS

Even taking into account the Cutter Incident, the polio vaccine was, by any objective standard, an enormous success. Within a half-decade of its introduction, cases of paralytic polio in the United States decreased by almost 90 percent. Parents were no longer afraid to let their children play outside in the summer, and the sight of a child stuck inside an iron lung became a rare occurrence. In its 1962 annual report, the New York City Health Department declared that the country was on the verge of winning mankind's centuries-old "battle against infectious disease" and would soon be ready to commence a "great era of cold war against chronic diseases for which we do not have biologic cures."

In order to reach this victory, the government's role in vaccinations had to evolve from sporadic advocacy to codified law. One of the first indications of this shift was the 1964 formation of a federal committee that would help legislate national immunization practices. Soon, state legislatures joined in the effort: In 1969, New York became the first state to enact a compulsory school vaccination law. By the mid-1970s, thirty-nine other states had followed suit.

The rapid expansion of vaccination law established freedom from disease as a national priority and not just an individual prerogative. The benefits of these laws were felt almost immediately. Before the measles vaccine was introduced in 1963, the country averaged more than 500,000 cases annually. By 1968, only 22,000 cases were reported—a decline of more than 95 percent.

The retreat of measles and other viruses rippled through nearly every aspect of American life. A 2006 study by researchers at the Harvard School of Public Health estimated that the polio vaccine alone has saved the United States more than $180 billion, a bottom-line quantification of the more than one million cases of paralytic polio and 160,000 deaths it has prevented. (To put that number in context: The population of Providence, Rhode Island, is 171,000.) The benefits of widespread vaccination have been particularly acute for women, who, no longer needing to spend weeks quarantined at home with sick children, have had greater freedom to join the workforce.

Just as the benefits of this patchwork of municipal, state, and federal regulations were becoming clear, the shortcomings of the hasty way these laws had been constructed began to be obvious as well. Physicians were devising their own health policies without repercussions: Some counseled against receiving vaccines while others declined to administer them altogether. School vaccination requirements were routinely ignored. The poor were perpetually under-immunized, either because of their lack of access to medical care, their suspicions about the motives of public health workers, or both. As a result, measles cases surged just as the disease appeared poised to disappear: In 1970, there were more than twice as many infections as there'd been two years earlier. The next year, the figure rose another 65 percent.

The persistence of measles highlighted once again the inescapable paradox of vaccines: When a disease is endemic, any potential side effects of its vaccine appear slight compared to the risks of not getting immunized. The flip side is that the more effective a given vaccine, the more the disease it targets will become an abstraction. (Who born in the last fifty years can remember seeing a case of polio?) This is a well-known catch-22 for public health officials, who constantly fear that vaccines will start to feel, like vegetables or vitamins, like something one could just as easily do without.

Too often, this quandary has led to a failure to acknowledge potential side effects. Like taking on increasing amounts of debt without a plan for how to pay it off, those promissory notes cause even greater harm in the long run: When officials gloss over concerns that prove to be legitimate, they raise the suspicions of everyone who takes for granted that public health policy is designed to do exactly what it claims—keep people healthy. Combine this drop of doubt with a press corps that's unconcerned with the particulars of science and you create a scenario in which the beliefs of conspiracy theorists start to sound a lot less bizarre.

A quintessential example of this dynamic at work is the controversies over the fluoridation of public drinking water. The first hint that fluoride might be an effective bulwark against tooth decay came in the beginning of the twentieth century, when a dentist in Colorado noticed some of his patients had small areas of discoloration on their teeth, a condition known as mottling. Over the course of more than a decade, he connected mottling to the naturally occurring presence of fluoride in drinking water. (The territorial variation in fluoride led to some colorful regional nicknames, including Texas Teeth and Colorado Brown Teeth.) He also noticed that while large quantities of fluoride produced severe mottling, smaller quantities appeared to result in stronger teeth and fewer cavities.

The next piece of evidence came from Minonk, Illinois, a small coal mining and farming community whose water was naturally fluoridated at approximately 2.5 parts per million (ppm). (To give a sense of just how minuscule that is, that means the water in Minonk was composed of approximately .00025 percent fluoride.) In the 1930s, a retrospective study found that children who'd lived in Minonk all their lives had significantly lower levels of tooth decay than children who'd moved to the area after their teeth were fully developed. Those results spurred a total of thirteen more population-wide studies in five states between the early 1930s and the mid-1940s, by which time data from an additional dozen countries also confirmed fluoride's protective effects. Finally, in 1945, the United States Public Health Service launched what was to be a decade-long field test to be conducted simultaneously in four metropolitan areas. In each one, two communities with comparable makeups would be chosen. The water supply of one of those would receive 1 ppm of fluoride, while the water supply of the control group would remain untreated.

It didn't take ten years for the results to become clear: By 1950, the children in the communities that received fluoridated water had half as much tooth decay as the children in the control communities. That year, the Health Service recommended that all public drinking water be fluoridated. In the vast majority of places, this occurred with little fuss—which was to be expected, considering there was a half-century's worth of evidence supporting the move. In almost one thousand communities across the country, however, determined anti-fluoride activists succeeded in forcing voter referendums on the issue. Fifty-nine percent of the time, they won. (Fluoride remains a preoccupation of health policy skeptics to this day: One of New Jersey's most prominent anti-vaccine activists recently urged her followers to fight for "Water Fluoridation Choice" and warned that fluoride is suspected of causing "more human cancer death, faster, than any other chemical.")

These debates likely never would have occurred had it not been for the press's willingness to parrot quack claims under the guise of reporting on citizen concerns. In this instance, anti-fluoridationists fixated on the dangers of fluorine, a poisonous gas that is among the most chemically reactive of all elements. What we know as "fluoride" does not contain fluorine gas—it's made up of some combination of sodium fluoride, sodium aluminum fluoride, and calcium fluoride. Equating the two is like accusing Joe the Plumber of being a murderous dictator because he shares a first name with Joseph Stalin, or claiming that Joseph Stalin must be a brilliant writer due to the skill of Joseph Conrad. That parallel is not as tenuous as it sounds: Sodium fluoride is used to strengthen teeth, sodium chloride is table salt, and chlorine is a poisonous gas that was used by Germany in World War I.

Another of the activists' tactics was using specious arguments that appealed to the press's love of simplicity. This practice started with their vehement rejection of being labeled as "anti-" anything under the guise of being advocates for choice. In theory, this seemed like a reasonable distinction; in reality, personal "choice" in this matter is impossible, since the treatment of public water supplies is an all-or-nothing proposition. (With vaccination, the situation is more complex: Individual choice doesn't prohibit other members of a community from being inoculated, but it does put everyone who can't be vaccinated at risk. What's more, infectious diseases, unlike tooth decay, can be deadly.) Similarly, the anti-fluoridationists' appeal for stricter safety requirements sounded like a no-brainer—who could be against those? The problem is that the assurance they sought demanded an impossible standard: Scientists can't give an ironclad guarantee that minuscule quantities of *any* substance in the universe are completely benign—all they can say is, "To the best of our knowledge there is no danger." Finally, anti-fluoridationists exhibited a flagrant disregard for the universal truism that the dose makes the poison: When they asserted that there is some hypothetical quantity of drinking water that would contain enough fluoride to be toxic to humans, they did so confident that nobody in the media would point out that that amount was fifty bathtubs' worth. Drinking that much water is physiologically impossible: Consuming as little as a tenth of a bathtub's worth at one time is enough to cause hyper-hydration—and death.

If the mechanisms through which activists enlist journalists to promote their causes are relatively straightforward, the factors that inspire those partisans in the first place oftentimes reflect a range of moral and social concerns that seemingly have nothing to do with the issue under discussion. Much as had been the case a half-century earlier with Lora Little and her followers, the battle over the fluoridation of public drinking water became, in the words of the social anthropologist Arnold Green, a "surrogate issue" for burgeoning social anxieties about modernity and the leveling effects of faceless bureaucracies in the 1950s.

Twenty years later, a similar synchronicity between a specific health issue and larger social concerns was seen in Great Britain in relation to the whole cell vaccine for pertussis. At first blush, the declining frequency of whooping cough infections in the twentieth century looks to be another of the unalloyed triumphs of modern science. In 1942, an American scientist named Pearl Kendrick combined killed, whole cell pertussis bacterium with weakened diphtheria and tetanus toxins to create the first combination DPT vaccine. Kendrick's discovery occurred during a decade in which more than 60 percent of British children caught whooping cough. Those infections resulted in more than nine thousand deaths, which was more than the number of fatalities caused by any other infectious disease. After mass DPT inoculations began in the 1950s, these numbers immediately plummeted.*

But the success of the pertussis vaccine provides another example of the disturbing complacency on the part of trailblazing vaccinologists in the first half of the twentieth century. While whole cell vaccines can be both effective and safe, their reliance on the actual contagion itself as opposed to an isolated component makes them among the most primitive of all vaccines. The fact that a whole cell pertussis vaccine was shown to be effective in 1942 should not have stopped immunologists from seeking out more nuanced mechanisms

* Like the placement of periods before or after quotation marks and conventions about how to write out the date, the trivalent diphtheria-pertussis-tetanus vaccine is referred to differently depending on which side of the Atlantic you find yourself on, with "DTP" being the preferred abbreviation in the U.K. and "DPT" the accepted one in the United States. (The main consequence of this is the acronyms available for activist groups seeking to play off of the vaccine's name.) Further confusing things, a combined vaccine with an acellular pertussis component was introduced in the 1990s; that is referred to both as the DTaP and the TDaP vaccine. For the sake of simplicity, I'll use DPT throughout unless there is a specific reason to do otherwise.

for protecting children in the decades to come. Instead, even after reports about possible complications from the vaccine began surfacing, the reaction on the part of the scientific community was a collective shrug. Hundreds of thousands of children's lives had already been saved. Even if the claims that a tiny percentage of children had suffered post-vaccination complications were true—and there has never been definitive evidence that they were—the net result was an enormous gain. Awards and accolades come to those who protected the world against a hitherto deadly disease, not those who tinker with vaccines others had already developed.

In Britain, this veneer of nonchalance was shattered in January 1974, when a study describing three dozen children purported to have suffered neurological problems following DPT vaccination was published. At the time, social unrest in the U.K. was higher than it had been at any point since the Great Depression. After decades of close to one hundred percent employment, the number of people out of work rose to above one million for the first time since the 1930s. The seemingly imminent threat of everything from nuclear war to a world without oil contributed to a nihilistic disdain for the traditional forces ruling society.

All these factors help to explain why the British press seized upon this specific report after having ignored similarly equivocal ones that had been published in years past. Newspaper articles, magazine features, and TV segments featured interviews with parents who believed their children had been harmed, regardless of the strength or paucity of the evidence. (One particularly flimsy example ran in *The Times*, which wrote about a severely handicapped man whose parents were, for the first time, claiming their son had been injured two decades earlier.) Once concerns about vaccine safety had been raised, advocacy groups like the Association of Parents of Vaccine-Damaged Children helped ensure that these issues remained in the forefront of public consciousness. Within months, vaccination rates began to fall—and whooping cough infections began to rise for the first time since World War II.

While the United States avoided this particular scare, a sense of disaffection was pervasive there as well. With the nation still reeling from more than fifty thousand deaths in Vietnam and the disgrace of President Nixon, its leaders were about to discover that the public was ill disposed to take part in a public health campaign just because they were told to do so.

On February 4, 1976, nineteen-year-old Army private David Lewis collapsed at his barracks in Fort Dix, New Jersey, shortly after completing a routine training hike. Later that day, Lewis was hospitalized with what looked like the flu. Less than twenty-four hours later, he was dead; soon, four of his fellow soldiers were hospitalized with what appeared to be similar symptoms.

When health officials completed their autopsy on Lewis's body, they were startled by what they found: Lewis had indeed died of influenza. More disturbing was their realization why a healthy teenager in excellent physical condition was felled by a disease most dangerous for infants and the aged or infirm: The strain of flu Lewis had been infected with was very similar to the one thought to have caused the 1918 flu pandemic, in which up to 5 percent of the world's population was killed. (Current estimates of the total number of fatalities that resulted from "the greatest medical holocaust in history" range from 20 million to more than 50 million.) When CDC officials realized that hundreds of other soldiers at Fort Dix had been infected, they feared they had the makings of a disaster on their hands.

For Gerald Ford, who'd become president just eighteen months earlier, the crisis was fraught with political peril. That November, he would ask voters to return to the White House the only person in the nation's history to have served as both vice president and president without being elected to either office. Thus far, the lifelong Republican's tenure as commander in chief had been defined by his unpopular pardon of his predecessor, the Democrats' historic gains in the 1974 midterm elections, and the worst inflation since the collapse of the Confederate dollar in 1864. If he presided over the deaths of tens of thousands of U.S. citizens, he'd go down as one of the biggest failures in American political history. If, on the other hand, he rushed to vaccinate the population against an epidemic that was at that point only speculative, he risked being seen as having stirred up a panic and having subjected citizens to unnecessary risks.

The unanimity of Ford's advisors made his decision as to which course of action to adopt easier to make. On March 11, CDC director David Sencer delivered a memo to the president in which he wrote, "The Administration can tolerate unnecessary health expenditures better than unnecessary death and illness." Two weeks later, Sencer took Jonas Salk, Albert Sabin, and a coterie of other leading virologists to meet with Ford in the White House. Sencer began by adding some urgency to what he'd written in his memo: With flu season only months away, it was "go or no go" time as far as a mass vaccination campaign was concerned. Salk spoke next, and he, too, unequivocally endorsed a mass vaccination campaign. (In a later interview, Salk acknowledged that he was considering other, long-term benefits as well: "I certainly thought of it as a great opportunity to fill part of the 'immunity gap' [between antigens in the environment and the antibodies of a given population]. We should close the gap whenever we can.") At the end of the meeting, Ford asked the doctors in the room whether any of them had reservations about moving forward, or whether any of them thought a nationwide effort was an overreaction. None replied. He then announced that he would remain in his office for the next ten minutes if anyone wanted to express doubts to him privately. When no dissent emerged, Ford walked directly to the White House's press room and live on national television urged "each and every American to receive an inoculation this fall." It was, he said, a "subject of vast importance to all Americans."

The logistical and safety issues raised by Ford's mass vaccination campaign were as numerous as they'd been in 1955, but unlike the polio trials, the country was not united in an effort to combat what was being referred to as "swine flu." There were also legal concerns that hadn't been present in the 1950s. The announcement on June 21 that no major side effects had been identified during the truncated CDC tests did nothing to mollify the leaders of the American insurance industry, who said that covering drug companies against liability from adverse reactions to such a hastily developed and widely circulated shot was not "feasible . . . at virtually any price."

Fears of vaccine-induced torts were not unfounded: In the years since the Gottsdanker decision, the legal barriers needed to successfully sue drug companies had declined even further. The decision with the biggest impact had come just two years earlier, when a jury ordered Wyeth Pharmaceuticals to pay $200,000 to the family of Anita Reyes, an eight-month-old who'd contracted paralytic polio two weeks after being vaccinated. In that case, Wyeth had gone to trial with a seemingly impregnable two-pronged defense: The chance that Anita Reyes had been infected by its vaccine was two thousand times less likely than the chance that she'd been infected by a naturally occurring strain of the virus, and the vaccine Anita Reyes received was packaged, like all of Wyeth's vaccines, with an insert that listed its risks and potential complications. Despite those facts, the jury ruled that because the vaccine's warning label was not shown to the Reyes family at the health clinic where Anita received the shot, the drug company itself was liable. If that standard were consistently applied, there was essentially nothing pharmaceutical companies could do to limit their liability. In the months following the Reyes ruling, half of the companies producing vaccines in the United States announced their withdrawal from the market.

Months after Ford's on-air announcement, Joseph Stetler, president of the Pharmaceutical Manufacturers Association, testified before Congress that the only way the drug companies would distribute the vaccine they'd been stockpiling was if they were indemnified against litigation arising from purported vaccine injuries. Otherwise, he said, the liability they faced was impossibly high, "particularly if you are talking about a nationwide immunization program." Shortly after the Reyes verdict, an article had appeared in *Pediatrics* in which the author, a noted vaccine expert, wrote, "Society—not the manufacturer, the physician, or the patient—should support those who suffer the adverse consequences of our laws." This argument was essentially the same one that drug companies were now adopting: By establishing herd immunity, vaccinations provided for the greater good; therefore, any unintended consequences should be borne by the country as a whole.

Congress was unmoved—at least until that summer, when a false alarm about a swine flu outbreak in Pennsylvania spurred it into action. On August 12, Ford signed legislation that transferred responsibility for vaccine injuries from manufacturers to the federal government. In doing so, he not only restarted the stalled swine flu campaign, he established the basis for the federally funded vaccine compensation program that exists today.

On October 1, the government's massive immunization campaign officially began when the sick and elderly in Boston and Indianapolis were injected with doses of the vaccine. The first month of the program proved uneventful, with no reports of significant complications or further deaths from swine flu. Then, in late November, the news turned. The much feared flu pandemic hadn't materialized—David Lewis was still the only recorded fatality—but there *were* reports of people developing Guillain-Barré syndrome soon after being immunized. By the end of December, five hundred cases of Guillain-Barré had been identified among the forty million people who'd received the swine flu vaccine, which was about seven times higher than what would normally have been expected—a figure that was certainly significant enough to raise questions but was nowhere near enough to prove causality. (Establishing a causal connection is especially difficult when dealing with uncommon conditions such as Guillain-Barré, which is so rare—it occurs at an incidence of about 1 per 100,000 members of a population—that its pathology remains a mystery.)

With Ford having already lost that November's presidential election to Jimmy Carter and the prospect of a large-scale outbreak looking increasingly less likely, there was no reason or incentive to continue the immunization campaign. On December 16, it was called off. By that time, Gerald Ford's fumbling effort to vaccinate everyone in the country against a threat that never materialized was widely viewed as one more example of the federal government's incompetence, its engagement in nefarious conspiracies, or both. "Any program conceived by politicians and administered by scientists comes to us doubly plagued," columnist Richard Cohen wrote at the time in *The Washington Post*. Few were inclined to disagree.

In the coming years, the consequences of losing the public's trust became all too apparent. By 1977, uptake of the pertussis vaccine in Great Britain had fallen to just over 30 percent. That year saw the first of three successive whooping cough epidemics. By the time they subsided, tens of thousands of children had been infected—and dozens, most of them infants, had died.

CHAPTER 5

"VACCINE ROULETTE"

On April 19, 1982, WRC-TV, the local NBC affiliate in Washington, D.C., aired a special titled "DPT: Vaccine Roulette." "For more than a year we have been investigating the P, the pertussis part of the vaccine," Lea Thompson, the story's on-air reporter, told viewers at the onset of the show, which focused on claims of vaccine-induced brain damage, mental retardation, and permanent neurological damage. "What we have found are serious questions about the safety and effectiveness of the shot. The overriding policy of the medical establishment has been to aggressively promote the use of the vaccine—but it has been anything but aggressive in dealing with the consequences." "Vaccine Roulette," Thompson said, would try to rectify that: "Our objective in the next hour is to provide enough information so that there can be an informed discussion about this very important subject. It affects every single family in America."

The most powerful segments of the show were the interviews with parents who described how their children had been left in near-comatose states after receiving a vaccine that was mandatory for public school children in the vast majority of states. Arresting visuals drove the point home: In one sequence, an image of a slack-jawed girl splayed on her family's couch was followed by one of a boy with an off-center smile and corroded teeth who seemed incapable of focusing his eyes. When healthy children were shown, they served as portentous signs of future tragedies: Another segment started with footage of a blond boy with the left sleeve of his red-and-white-striped shirt rolled up to expose his arm. After focusing on a close-up of a syringe being prepared for injection, the camera cut back to show the child squirming anxiously in his mother's lap. When the needle first broke the skin of the boy's biceps, his mother winced; after his gasp turned into shrill screams, the mother's look of unease became one of anguish and horror.

Thompson did not rely solely on discomfiting images and despairing personal anecdotes to make her point. She cited one study that found that adverse reactions to the DPT vaccine could be as high as one in seven hundred. An impressive roster of experts testified that the vaccine's dangers were being swept aside by unnamed powers-that-be. Bobby Young, who was identified as a former vaccine researcher at the Food and Drug Administration (FDA) and the University of Maryland, said that a shocking number of children were "rendered a vegetable" by the DPT vaccine, and that many others suffered from less extreme neurological damage. High-pitched crying after receiving the vaccine, Young said, "may be indicative of brain damage in the recipient child." Robert Mendelsohn, who was described as the former head of pediatrics at the University of Illinois School of Medicine, told Thompson that the pertussis vaccine was "the poorest and most dangerous vaccine that we now have." The most unsettling testimony came from Gordon Stewart, a doctor who was identified as a member of the British government's committee on the safety of medicines. The DPT vaccine, he said, was a "crude brew" that protected against a disease that wasn't even dangerous to begin with: "Whooping cough has not been a killing disease for a very long time." Later in the show, Stewart articulated the anxieties of every parent who'd ever worried about the injections being given to his or her children:

> We start off with healthy infants, and we pop 'em not once but three or four times. . . . My greatest fear is that very few of them escape some kind of a neurological damage because of this. . . . I mean, if the child isn't frankly rendered a vegetable and yet has a fever—and a very large fraction have the fever from it, also a large fraction have the screaming syndrome, which is surely an irritation of the central nervous system—you add all of this up, how many infants [are damaged]? And how can you prove that they haven't been—or that they have been? All of them are vaccinated.

"Vaccine Roulette" was the first time the American public was confronted with many of these questions. "[DPT] was without a doubt our most reactogenic [vaccine]," says Paul Offit, the chief of the Division of Infectious Diseases and the director of the Vaccine Education Center at the Children's Hospital of Philadelphia. "It's the only whole cell, whole bacterial vaccine we've ever used in our nation's history." Doctors had known for years that the DPT vaccine could cause seizures, high fevers, and fainting. (My younger sister ran an extremely high fever after her first DPT injection, which she received in the late 1970s.) There was not, however, any evidence that it caused the type of permanent damage Thompson's show warned of.

Within days of the initial airing of "Vaccine Roulette," it was clear that it would have an impact that extended far beyond its initial Washington broadcast area. With little to no independent reporting, national media outlets repeated the story. A UPI wire dispatch was typical of the focus and tenor of many of the follow-ups. "Hundreds of American children may be left brain damaged and retarded each year by a common vaccination that many states require, a year-long investigation concluded," it began. Acknowledgment of a dissenting viewpoint was essentially limited to the following: "The Department of Health and Human Services had no immediate comment." In one of several stories it published on the broadcast, *The Washington Post* gave Thompson a platform to snipe at her critics: "Lea Thompson,

Channel 4's respected investigative reporter/producer who is responsible for the program, says she did not intend to panic parents into rejecting the shot, only to spotlight facts about its dangers, facts she believes have been 'suppressed by the medical establishment.' "

After months of accolades, it came as no surprise when Thompson won an Emmy for her work. At the awards ceremony, she was serenaded to the stage with the jazz standard "Satin Doll" ("She's nobody's fool/So I'm playing it cool as can be"). During her acceptance speech, she gave her own evaluation of the long-term effects of her work: "I think babies will be saved because of this special."

In the midst of all these hosannas, Thompson's journalistic peers neglected to ferret out the reality of the situation. One of the only publications to do the minimal amount of reporting necessary to research Thompson's accusations was the *Journal of the American Medical Association.* What it found was a dispatch rife with mistakes and misrepresentations. There were quotes that sources said were taken out of context: Edward Mortimer, who'd previously headed up the committee in charge of the American Academy of Pediatrics' book of guidelines on childhood infectious diseases, said that over the course of a five-hour interview, Thompson had asked him the same question "repeatedly in slightly different ways, apparently to develop or obtain an answer that fitted with the general tone of the program." There was a reliance on inaccurate statistics: The one-in-seven-hundred ratio of adverse reactions to total vaccinations, which Thompson attributed to a study done by researchers at UCLA, was 25 percent higher than the figure the study actually reported. There were warped interpretations of independent research: The conclusions Thompson drew from the report were "totally distorted," according to James Cherry, the paper's lead author and the chief of pediatric infectious diseases at UCLA. (As an example, Cherry noted that Thompson neglected to report that not one of the more than six thousand children in his study had shown *any* lasting reactions to the vaccine, and that there had not been a single case in which there was evidence of even temporary neurological damage.) There were unfounded claims so laced with conditionals and so lacking in specificity as to be essentially meaningless: What precisely was the "screaming syndrome" that was "surely an irritation of the central nervous system"? What was the basis for the assertion that high-pitched crying "may be indicative of brain damage"? What, for that matter, is the difference between high-pitched and regular crying?

Some of the most troubling errors involved the extent to which Thompson misstated the titles, history, and affiliations of the sources she relied upon to make her case—errors that were all the more egregious because they were also among the easiest to verify. According to his former colleagues, Bobby Young, who had died by the time the program aired, had never researched the pertussis vaccine. Robert Mendelsohn, a family practitioner in Evanston, Illinois, had never been the head of pediatrics at the University of Illinois School of Medicine—although he was a well-known and outspoken opponent of *all* vaccines. Then there was Gordon Stewart, a medical professor at the University of Glasgow who believed that the positive impact of antibiotics and medical interventions had been greatly exaggerated. By the mid-1970s, Stewart's reputation within anti-vaccine communities was such that he'd become a magnet for parents convinced their children were vaccine-injured, and in 1977, he published a paper in which he cited many of those children as proof that the DPT vaccine caused brain damage. By the time he appeared on Thompson's show, he'd been for years the go-to guy for reporters looking for pithy quotes about the dangers of vaccines.

Thompson reported none of this history to her viewers, nor did she tell them of Stewart's repeated run-ins with his colleagues or his reputation for operating on the scientific fringe. When she highlighted the conclusions of a study Stewart had published, she failed to mention that it relied on data that had been collected by the Association of Parents of Vaccine-Damaged Children or that it was used in several lawsuits filed by parents seeking monetary damages for purported vaccine injuries. She also neglected to report that Stewart's anti-vaccine manifesto had contributed to the explosion of whooping cough hospitalizations in Great Britain in the late 1970s. (Stewart questioned whether these outbreaks had ever occurred.) Thompson's presentation of Stewart as someone who was able to offer objective commentary on the DPT vaccine would be akin to a newscaster acting as if Sarah Palin could function as an impartial evaluator of Barack Obama's job performance.

When asked about these problems, most of which were based on indisputable facts and not personal opinions, Thompson fell back on a tactic that had been used by the "irregular physicians" in their battles with the AMA all the way back in the 1850s: She accused her critics of trying to protect their turf. "I think much of the potshotting has come from people who have been burned," she said. "We hit a lot of raw nerves. . . . Many doctors are miffed because they have to talk to their patients now." She even found a way to lay blame for omissions in her reporting at the feet of those establishment critics: When asked why she had not devoted any of her broadcast to the ravages of whooping cough, she pointed a finger at the CDC, the FDA, and more than a dozen other medical institutions, all of which, she said, had failed to provide her with a single frame of usable footage of children sick with pertussis—a fact that was due to the success of the vaccine in the first place. Of course, if Thompson had truly felt an obligation to offer a different perspective, she could have gone to the U.K. and spoken with parents whose children had died of the disease. The problem with that, she said, was that despite dedicating a year's worth of resources to the project, her station didn't have the budget to send a reporter to London.

Thompson and her employer's role in the controversy they helped ignite did not end with the broadcast itself. In the days after its initial airing, WRC-TV helped organize a burgeoning anti-vaccine advocacy movement by providing callers with the phone numbers of other people who'd already contacted the station looking for information about negative reactions to childhood shots. Within two weeks, a group of parents had organized a face-to-face meeting, and those ad hoc get-togethers eventually coalesced into a group that called itself Dissatisfied Parents Together—DPT for short.

One of these parents was Barbara Loe Fisher, a former PR professional who'd become a full-time housewife after she'd given birth to her son, Chris, four years earlier. When "Vaccine Roulette" aired, it had been more than a year since Chris had started displaying symptoms of what would eventually be diagnosed as a range of developmental disorders. Many years later, Fisher would describe the transformation of her "happy-go-lucky little boy" into one who was "listless and emotionally fragile, crying at the slightest frustration as if his heart would break." When Chris's behavior first started to change, Fisher said, she repeatedly pressed her pediatrician for an explanation and sought any steps she could take to transform her son into the perfect child she remembered him being.

It wasn't until she saw Thompson's broadcast that the pieces fell into place. The reactions that Thompson described—convulsions, loss of affect, permanent brain damage—were, Fisher realized, identical to those experienced by her son. Suddenly, Fisher remembered in meticulous detail what had happened one day eighteen months earlier, when Chris had received the final dose of his DPT vaccine:

> When we got home, Chris seemed quieter than usual. Several hours later I walked into his bedroom to find him sitting in a rocking chair staring straight ahead as if he couldn't see me standing in the doorway. His face was white and his lips slightly blue, and when I called out his name, his eyes rolled back in his head, his head fell to his shoulder and it was like he had suddenly fallen asleep sitting up. I tried, but could not wake him. When I picked him up, he was like a dead weight and I carried him to his bed, where he stayed without moving for more than six hours, through dinnertime, until I called my Mom, who told me to immediately try to wake him, which I finally did with great difficulty. But he didn't know where he was, could not speak

> coherently and couldn't walk. I had to carry him to the bathroom and he fell asleep again in my arms and then slept for twelve more hours.

It's an incredibly moving story, and one that Fisher has told to congressional panels, federal committees, and state legislatures, and at national press conferences, for more than twenty-five years. In all that time, she's almost never been questioned about the specifics of her narrative—and there are parts that, if nothing else, certainly are confounding.* Fisher, as she told a government vaccine safety committee in 2001, is "the daughter of a nurse, the granddaughter of a doctor, and a former writer at a teaching hospital" who viewed herself as "an especially well-educated woman when it came to science and medicine." How was it that her only response to finding her unresponsive son displaying symptoms associated with heart attacks, strokes, and suffocation was to carry him to bed and leave him alone for six more hours? And if Chris's reaction to his fourth DPT shot was so severe that it transformed an ebullient boy into a sluggish shell of his former self, why had he been fine after receiving the first three doses?

Shortly after the formation of Dissatisfied Parents Together, Fisher founded the National Vaccine Information Center. Since then,

---

* I tried to interview Fisher several times over the course of more than a year. She explained in an e-mail that her ultimate refusal was due to an article about vaccines and the anti-vaccine movement that appeared in the November 2009 issue of *Wired* in which Paul Offit was quoted as saying, "She lies. Barbara Loe Fisher inflames people against me. And wrongly. I'm in this for the same reason she is. I care about kids." In her e-mail to me, Fisher wrote, "As you know, character assassination in order to demonize and marginalize individuals or groups belonging to a minority in society is a classic form of propaganda. Historically, it has been used as a component in overt and covert political disinformation campaigns designed to neutralize individuals and groups dissenting from the majority opinion." Then, citing the fact that *Wired* shares a corporate parent with *Vanity Fair*, where I am a contributing editor, she wrote, "I hope you understand why I cannot subject myself to another opportunity to have my position mischaracterized and my reputation further damaged by foolishly trusting again and cooperating with a journalist I don't know, who works for Condé Nast." In December 2009, Fisher sued Offit, Condé Nast, and the author of the *Wired* story for libel. The case was summarily dismissed.

she's played an essential role in organizing a movement that has targeted the press, politicians, and the public in equal measures. The result has been a steady erosion of vaccine requirements and a steady increase in the percent of the population skeptical of vaccine efficacy. Today, every state save for Mississippi and West Virginia has religious exemption laws on the books. Eighteen states have philosophical exemption laws, which allow parents to cite nothing more than personal beliefs as a reason to limit their children's vaccinations. In 2010, *Pediatrics* released a study that reported that despite all evidence to the contrary, 25 percent of parents believe that vaccines can cause developmental disorders in healthy children. "The genie is not going back in the bottle," another leader of the anti-vaccination movement wrote in a congratulatory essay posted on one of the movement's most popular Web sites. "The fear of vaccines is going to rise. Our community is only gaining strength."

CHAPTER 6

Autism's Evolving Identities

In 1943, Leo Kanner, an Austrian doctor and the author of the first English-language textbook on child psychiatry, published "Autistic Disturbances of Affective Contact." The paper, which reported on Kanner's work with eight boys and three girls, was the first to use the term "autism" to describe children who, in addition to being developmentally disabled, were incapable of "form[ing] the usual biologically provided affective contact with people." (The neo-Latin word *autismus* had been coined in 1910 by Swiss doctor Eugen Bleuler as a way to depict the "morbid self-admiration" of schizophrenics who were stuck in their interior worlds. Its root is the Greek word *autos*, or "self.") At the time, Kanner, who'd immigrated to the United States in 1924, was widely regarded as the world's foremost expert on childhood emotional development. In addition to their inability to form normal human attachments, Kanner said, autistics exhibited an extreme lack of empathy and a tendency to become unnaturally absorbed in routine tasks.

Kanner's ultimate goal was to use his classification as a tool to develop more effective treatments—but before he could do that, he had to determine the disorder's root causes. In the early stages of his research, Kanner looked to the families of autistic children in his search for factors that made them unique. It didn't take long for him to identify significant ways in which the mothers of his test subjects stood apart from his perception of women in the rest of the population: They were intellectually assertive, professionally ambitious, and emotionally restrained. Perhaps, he speculated, the symptoms of children with a biological predisposition to autism were exacerbated when they were raised in environments in which they didn't receive enough affection. Six years later, in Kanner's second major paper on the subject, he further developed this line of thinking: The reason autistic children found such comfort in solitude was that they'd been victimized by a "genuine lack of maternal warmth," which he likened to being "kept neatly in refrigerators which did not defrost." Kanner's metaphor, which conjured up images of children incubating in wombs of physically and psychologically frigid women, would remain the dominant trope through which autism would be understood for more than thirty years.

Beginning in the late 1950s, the "refrigerator mother" theory was expanded on and promulgated by another Austrian émigré, the celebrated University of Chicago professor Bruno Bettelheim. In 1959, Bettelheim published a paper titled "Joey, a 'Mechanical Boy,' " in which he described a child "devoid of all that we see as essentially human and childish, as if he did not move his arms or legs but had extensors that were shifted by gears."

> How had Joey become a human machine? . . . A colicky baby, he was kept on a rigid four-hour feeding schedule, and was not touched unless necessary and was never cuddled or played with. The mother, preoccupied with herself, usually left Joey alone in the crib or playpen during the day. . . . Joey's mother impressed us with a fey quality that expressed her insecurity, her detachment from the world and her low physical vitality. We were especially struck by her total indifference as she talked about Joey. This seemed much more remarkable than the actual mistakes she made in handling him.

Bettelheim's status as a pioneering child psychologist, his academic bona fides, and his media savvy gave his opinions even more weight than those of Kanner. Throughout the 1960s, he used increasingly vituperative language when promoting his belief that psychological abuse, not hardwired physiology, caused autism. (This dichotomy is an example of the nature/nurture debate that has been a constant source of tension for students of human behavior: Are distinct aspects of our personalities the result of our innate biology or the unique circumstances of our individual lives—or some combination of the two?) Bettelheim, a Jew who escaped the Nazis in the years before World War II, went so far as to compare the households in which autistic children were raised to concentration camps and refrigerator mothers to Nazi guards. By the time his book *The Empty Fortress: Infantile Autism and the Birth of the Self* was published in 1967, he was widely considered the world's foremost expert on the disease.

From today's perspective, Bettelheim's theories are as simplistic as they are offensive. Even before the ethicality of Bettelheim's research on emotionally disturbed children was called into question, his analytical and methodological errors should have been obvious to his colleagues.* As counterintuitive as it may seem, this is precisely why his work remains so instructive: The readiness with which Bettelheim's theories were embraced illustrates how what are thought of as indisputable, evidence-based conclusions are influenced by prevailing social and cultural currents. Post–World War II America was a place of rapidly changing gender roles. As is the case after every war, returning veterans found the transition back to society to be a difficult one. In their absence, women had worked in factories, provided for their families, even formed professional baseball leagues—all activities that were previously the exclusive domain of men. Surely,

* By the 1980s, the majority of Bettelheim's conclusions had been called into question. After he committed suicide in 1990, dozens of former patients of the residential treatment program he ran at the University of Chicago claimed they'd been physically and mentally abused while in his care.

women couldn't succeed in traditionally male roles while *also* fulfilling their customary responsibilities, such as raising children and supporting their husbands. An internalization of those social anxieties, coupled with a widespread acceptance of the notion that behavioral disturbances stemmed from childhood emotional traumas, simultaneously penetrated Bettelheim's research and informed the public's reaction to his work.

An appreciation of the interconnectedness of scientific discourse and societal constructs not only makes clear why early hypotheses about the roots of autism were so readily accepted, it also explains why those hypotheses ring so false today. Over the last fifty years, having women in positions of power and mothers of young children in the workplace has gone from seeming abnormal to being accepted as commonplace; advances in genetics, epidemiology, and neurobiology have caused scientists to examine psychological phenomena through an increasingly mechanistic lens; and the paternalistic attitude of self-appointed authorities has come to be viewed with skepticism, if not outright hostility. The assumptions that guide what we think of as provable, when we decide something has been proven, and whom we consider capable of proving it, all change over time. This awareness—that knowledge is never absolute and notions of what we can conceive of are fluid—is the underpinning of the modern-day understanding of science.

Throughout the first half of the twentieth century, the ways in which psychological and mental illnesses were viewed were defined almost exclusively by Sigmund Freud, Carl Jung, and their intellectual descendants. That began to change in the 1950s, as the steady rise of treatments targeting biological mechanisms—ranging from prefrontal lobotomies to psychotropic medications—began to weaken the intellectual stranglehold of these psychoanalytic pioneers. Psychiatry began to move away from a fixation on *why* a patient had a given disorder (e.g., autism was the product of poor mothering) and toward diagnoses and treatments based on defined behaviors that respond to specific remedies.

In the United States, this shift was illustrated by the 1952 release of the American Psychiatric Association's (APA) *Diagnostic and Statistical Manual of Mental Disorders*, commonly referred to as the DSM. Ever since its initial publication, the DSM has determined how medical professionals in much of the world categorize psychiatric illnesses. The implications of these classifications extend far beyond mere semantics, with tiny changes dictating what is considered diseased and what is thought of as normal. This, in turn, can have an enormous effect on the day-to-day realities of people's lives: Insurance companies use the DSM to determine which treatments they will cover; pharmaceutical companies look to the DSM when deciding what types of drugs to develop; lawmakers refer to the DSM when making health policy; even judges and juries have pointed to DSM classifications to justify rulings and verdicts. To be sure, this process is not unidirectional: Just as minor alterations in the DSM can have far-reaching implications in the real world, shifts in societal mores can result in significant changes to the DSM. One of the best-known examples of this shift occurred in 1973, when the gay rights movement successfully pressured the APA to remove homosexuality from the "sexual deviations" that were listed in the seventh printing of the second edition of the manual, known as the DSM-II. There have also been a number of instances in which behavior that threatened the social order was pathologized in the DSM. Jonathan Metzl writes about one of the most astonishing of these in his book *The Protest Psychosis*, which details how the category of schizophrenia was expanded to include "threats to authority" in the 1960s. The result was a mushrooming of African-American men being institutionalized after receiving diagnoses of the disorder.

The evolution of the DSM's handling of autism is another example of the two-way interplay between science and society.* In the manu-

---

* The notion that scientific progress can be understood only within specific social contexts was articulated most famously by Thomas Kuhn in his 1962 book, *The Structure of Scientific Revolutions*. In it, Kuhn argued that scientific advances are not merely

al's first edition, the only mention of autism came under the heading of "schizophrenic reaction, childhood type," which listed "psychotic reactions in children, manifesting primarily autism" as the disorder's primary symptoms. In the DSM-II, which was published in 1968, the characteristics of childhood schizophrenia were expanded to include "autistic, atypical, and withdrawn behavior; failure to develop identity separate from the mother's; and general unevenness, gross immaturity, and inadequacy in development." It wasn't until 1980, with the publication of the DSM-III, that "infantile autism" was listed in a distinct category. Seven years later that diagnosis was changed to "autistic disorder," and in 1994 it was folded into the newly expanded class of "pervasive developmental disorders," which included Asperger's disorder, childhood disintegrative disorder, Rett disorder, and the loosely defined pervasive developmental disorder, not otherwise specified (PDD-NOS).

One reason a clinical designation of "autism" has gone from being virtually nonexistent to broadly inclusive is the movement toward identifying a widening number of behaviors as being indicative of a psychiatric disorder: Where the DSM-II weighed in at just over one hundred pages, the DSM-IV ran to more than nine hundred. (This expansion has been mirrored by a dramatically increased interest in psychiatric specialization, as seen in the rise in the number of members of the American Academy of Child and Adolescent Psychiatry from 400 in 1970 to 1,200 in 1980 to 3,300 in 1990 to 7,400 in 2000.) Another, equally important, factor was the growth of a parent-based advocacy movement led by a psychologist named Bernard Rimland, whose autistic son, Mark, was born in 1956. At the time, awareness of

---

the result of the steady accumulation of more sophisticated data; instead, scientific thought undergoes "paradigm shifts" in which accepted "ways of knowing" are replaced by new methods of understanding the world. Kuhn's prototypical illustration of a paradigm shift at work occurred in sixteenth-century, pre-Enlightenment Europe: It wasn't until the all-enveloping primacy of the Catholic Church was challenged by a growing belief in empiricism and experimentation that Ptolemy's earth-centric theory of the universe was overturned by Copernicus's heliocentric one, in which the earth is just one of the planets orbiting the sun.

the disorder was so scant that Rimland had to diagnose Mark himself on the basis of information he gleaned from some old college textbooks. The lack of general knowledge—or even interest—in the disorder led Rimland to seek out every scientific paper, every case study, every research proposal and draft article and Ph.D. dissertation on that subject that he could find.

In 1964, at the precise time that the medical community was embracing Bettelheim's theories, Rimland published *Infantile Autism: The Syndrome and Its Implications for a Neural Theory of Behavior*, a piece of scholarship so impressive that Leo Kanner himself wrote the foreword. (With that contribution, Kanner signaled his ultimate refutation of his earlier behavior-based model of the disease.) At the time, Rimland's work was largely ignored—except, that is, by other parents, who were the people he was most interested in reaching anyway. In 1965, Rimland founded the Autism Society of America, which was in all likelihood the first mainstream organization that regarded a diagnosis of autism as anything other than a cause for despair and a source of shame. The Autism Research Institute (ARI), which Rimland set up two years later, was an even more ambitious undertaking: In creating an institution that would plan, collect, and interpret medical research on its own, Rimland was setting out to completely circumvent a medical hierarchy that he felt had turned its back on autistic children and their families.

By the end of the twentieth century, the movement Rimland set in motion had helped bring about a dramatic shift in attitudes about the disorder.* Instead of shipping children off to all-purpose institutions for the insane or retarded, parents of autistic children were organizing rallies and using their political clout to guarantee access to

---

* One of the most significant factors in this regard was *Rain Man*, the 1988 blockbuster film starring Dustin Hoffman as an autistic, card-counting savant named Raymond Babbitt and Tom Cruise as his self-absorbed younger brother. The film was a critical and commercial juggernaut: It was the highest-grossing movie of the year and won Academy Awards for Best Picture, Best Director, Best Actor (for Hoffman), and Best Original Screenplay. Rimland served as an advisor on the movie, and Hoffman's character was based in part on Rimland's son.

a wide array of resources. Lobbying efforts in states including Colorado, Indiana, Maryland, Massachusetts, and Wisconsin resulted in waivers that allowed middle-class families to receive Medicaid assistance to help pay for expensive treatments. These efforts, in turn, resulted in still more diagnoses: As the anthropologist Roy Richard Grinker explains in *Unstrange Minds*, a book prompted by his discovery that his daughter was autistic, "Part of the reason [children are classified as autistic] is that clinicians are more likely to give a child a diagnosis that he or she thinks will help the child receive the best services or school placement than a diagnosis that conforms to the DSM but will not facilitate the best form of intervention." Grinker goes on to quote Judy Rapoport, the chief of child psychiatry at the federal government's National Institute of Mental Health (NIMH): "I'll call a kid a zebra if it will get him the educational services I think he needs."

The ongoing elasticity of "autism" has meant there are huge variations in who would have been designated as autistic at any given time: One study comparing the DSM-III and the DSM III-R found an increase of almost one hundred percent in the clinical diagnoses of the same group of 194 children. The most remarkable example of the arbitrary nature of diagnostic requirements came in 1994, when the mistaken inclusion of draft text in the published version of the DSM-IV resulted in the official adoption of a much less restrictive list of criteria than intended, as "impairment in social interaction *and* in verbal or nonverbal communication skills" was changed to read "impairment of reciprocal social interaction *or* verbal and nonverbal communication skills, *or* when stereotyped behavior, interests and activities are present." One study found that 75 percent of children for whom a diagnosis of a pervasive developmental disorder was ruled out according to the intended standards would have been identified as having a PDD under the published guidelines.

That does not mean that the rise in the number of children with autism spectrum disorders, from an average of 1 in 1,000 in the 1980s to 1 in 110 today, is due entirely (or even primarily) to what scientists refer to as "diagnostic creep." In the past several years alone, researchers have identified an ever-growing number of risk factors, including advanced maternal age (women who give birth after they've turned forty are between 50 and 77 percent more likely to have an autistic child as women in their twenties), paternal age as related to maternal age (the risk of autism appears to rise when women in their twenties have children with men over forty), proximity to families with autistic children (parents' awareness of the disorder increases the chance their child will receive a similar diagnosis), and the growing use of a type of muscle relaxant used during pregnancy to treat asthma and prevent early labor. There's also the strong likelihood that there are other environmental factors that have yet to be identified.

These nuances are close to impossible for reporters to convey and for audiences to absorb in a short news report, which explains in part why every parent you meet in the park knows that autism rates have been ballooning but so few are aware of the complex range of potential causes influencing the rise. It's also hard to explain what you don't understand (or have time to learn): Over the past twenty years, there's been an industry-wide bloodletting in the news media that has led to the jettisoning of science reporters—and in a growing number of cases, of entire science sections. From 1989 through 2005, the number of newspapers with weekly science sections fell from ninety-five to around thirty-five, and that figure has fallen even more precipitously since then. In 2008, CNN got rid of its *entire* science, space, and technology unit. It's no wonder that in a 2009 survey, 44 percent of health care journalists said their news organizations "frequently or sometimes" ran stories that relied on news releases "without substantial additional reporting or contacting independent sources." More than 90 percent said budgetary pressures were hurting the quality of health care reporting.

The end result when generalists are commandeered to cover subjects that require specialized knowledge can be simultaneously amusing and infuriating. Take a 2009 piece about advances in functional magnetic resonance imaging that ran in the *New York Times*–owned *International Herald Tribune*. Unlike regular MRIs, which use a powerful magnetic field to get images of your insides, fMRIs measure blood flow in the brain in an effort to determine which areas are activated by specific thoughts, reactions, or emotions. One area of research has involved efforts to match distinct words and phrases with identifiable brain patterns. While some early research has shown promising results, this is not a very practical way to find out what someone is trying to tell you: For a patient to receive an fMRI, he needs to go to a hospital or high-tech research facility, where he'll be placed inside a magnetized cube that's roughly seven feet high by seven feet wide. He'll also need to be attended to by technicians with extensive training.

Contrast those realities with the first three paragraphs of the *Herald Tribune* story, which was titled "Watch What You Think":

> When the police stopped me for running a red light recently, I was thinking, "Don't you cops have anything better to do?" But the words that came out of my mouth were a lot more guarded, something like, "Sorry, I thought it was green." Sometimes it's good to play the dumb foreigner.
>
> The policewoman, a tough lady smoking a cigarette, glared at me. Was she reading my mind? No, I guess not, because she only gave me a warning. But beware, in a few years she might actually carry a device that can do that.
>
> Research is rapidly advancing to allow thought-decoding through brain-scan technology, and it scares me to death. I don't want anyone else in my head, and certainly not the police.

As Vaughan Bell, a clinical and research psychologist at Kings College, London, put it, "It's a masterpiece of superficial reading of the scientific evidence and interpreting it in the most unrealistic and panicky way possible."

The cumulative impact of this type of journalism is unquestionably negative: It affects the public's sense of whether scientists are reliable and research money is being spent wisely, and it creates unrealistic expectations for the future. That said, a story like "Watch What You Think" is unlikely to have any *immediate* impact. (No one who should have gotten an fMRI was frightened out of it, and studies involving fMRIs didn't lose funding.) But misguided, ill-informed, and cavalier coverage of science and medicine is not always so benign: It influences how hundreds of millions of research dollars are spent, it sucks up the time and energy of public health officials already stretched thin, and it bestows credibility on people's delusions and fantasies, with occasionally calamitous results.

CHAPTER 7

Help! There Are Fibers Growing Out of My Eyeballs!

In 2007, a sixty-year-old woman confronted a doctor at New York Presbyterian Hospital with stacks of paper, each covered with what she said were tissue samples that had fallen off her skin. These samples were meticulously labeled (one read "white stuff from buttocks") and, the woman said, were the result of an infestation of an organism that was causing her to scratch herself until she bled. The woman had all the symptoms of delusional parasitosis, a well-documented psychological disorder in which patients are convinced that parasites have burrowed underneath their skin.* Indeed, this woman had predicted that her doctor would give her that diagnosis—but, she said, she was absolutely certain that she was suffering from a real physical disease and not a psychological disorder. She knew this, she explained, because she'd been in touch via the Internet with thousands of other people suffering from the exact same thing.

* Scabies, a highly contagious disease in which parasitic mites cause intense itching, rashes, and infections, is an example of a nondelusional parasitosis.

The patient was referring to Morgellons syndrome, a disease that was invented in 2002 by Mary Leitao after doctors refused to acknowledge that flesh-eating bugs were the cause of the bloody sores covering the undersides of her two-year-old son's lips. (At least one doctor she consulted speculated that Leitao could be suffering Münchausen syndrome by proxy, a disorder that involves deliberately harming another person—usually a child—in order to gain attention.) Leitao, whose husband was an internist at a Pennsylvania hospital and who had once worked as a lab technician in a Boston hospital, later said she was always certain that her son was suffering from some unknown disease. After using a RadioShack microscope to examine tissue samples taken from her son's scabs, she claimed to have discovered microscopic fibers that were pushing through his skin. When the medical community continued to ignore her claims, she founded the Morgellons Research Foundation. (The name "Morgellons" is taken from a seventeenth-century monograph describing children in southern France who had "harsh hairs on their backs.") For the first four years of its existence, Leitao's foundation received little attention outside the growing online community of people convinced they were also suffering from the disease.

That all changed in May 2006, when Leitao hired a professional communications consultant to launch a national PR blitz. By midsummer, NBC's *Today* show and CNN's *Paula Zahn Now* had aired pieces on Morgellons. On ABC's *Good Morning America,* Cynthia McFadden anchored a segment that focused on "one of thousands of patients across the country" who suffer from "a mysterious illness called Morgellons disease" that causes "burning and itching," "open sores," and "string like fibers literally coming out of your skin." Several weeks later, a *Time* magazine article on Morgellons opened with a description of Gregory Smith, a Georgia pediatrician who described looking into the mirror one night only to see a white fiber "burrow down into his eye." The fiber, Smith said, glowed deep blue when examined under a black light. The article continued:

> "Yes ma'am, I was a little bit distraught," recalls Smith, 58, who says he can no longer work because his mysterious ailment has also robbed him of his memory and neurological function. "I tried to grab ahold of it with tweezers and it would not come out. It was quite painful, so I threw up my hands and went to the Emergency Room with my wife."
>
> For Smith, and some 4,000 people across the nation who claim to suffer from similar symptoms, it's the reaching out that has been problematic. The disease, called Morgellons after a reference in a 1674 medical paper, isn't officially recognized by the medical community.

The burst of coverage was so striking there were rumors that Morgellons was the product of a covert marketing campaign to support the big-screen adaptation of the Philip K. Dick novel *A Scanner Darkly*, which starred Keanu Reeves and Winona Ryder. (Reeves played an undercover police agent who goes insane when his drug addiction causes the two hemispheres of his brain to "compete" with each other.) One pop culture blog wrote, "Suspicions were aroused when it was noted that the Wikipedia entry on Morgellons was first created only in February, and linked to the website Morgellons.org. This website seems to put a large emphasis on media coverage of the disease, and claims that a 'national news broadcast' will occur in June or July regarding the illness. Coincidentally—or not, as the case may be—the release date for the film is July 7."

It wasn't just fans of surreal sci-fi classics who took note: In the coming year, the number of hits from a Google search of Morgellons went from 15,400 to 56,500, and the number of families registered with Leitao's foundation went from 3,300 to 11,000. Press coverage before that spring's media blitz was sparse: A Nexis search of dates prior to May 1, 2006, turns up a total of nine newspaper stories that cite Morgellons, seven of which involve Texas State Medical Board hearings against a doctor named George Schwartz for improperly prescribing high doses of narcotics and amphetamines.*

Even taking into consideration that claims of glow-in-the-dark filaments growing out of people's eyeballs are like catnip to journalists, the utter lack of independent reporting that went into the coverage of Morgellons is astounding. Many stories—like those in *Time* and on CNN—failed to quote a single person who was not a patient that believed Morgellons was a legitimate disease. Leitao, who studied biology in college, was often referred to as a "scientist" or "biologist" (or both, as was the case on *Good Morning America*). Virtually every time a "mainstream" expert was quoted, it was one of four people connected to the Morgellons Research Foundation: an Oklahoma State University associate professor named Randy Wymore, an Oklahoma State University assistant professor named Rhonda Casey, San Francisco's Union Square Medical Associates' medical director, Raphael Stricker, and Ginger Savely, a nurse practitioner in Stricker's practice. Inevitably, any relevant context about their work was omitted—which is not to say it wasn't readily available: That March, Savely had left Texas when the doctor she'd been working with was threatened with the loss of his medical license if he continued to sign off on her practice of prescribing a year or more's worth of antibiotics to patients who claimed to be suffering from Lyme disease or Morgellons. Wymore, whose last published study as a lead author had come in 1997, was on the advisory board of Leitao's charity and ran OSU's Center for Health Sciences for the Investigation of Morgellons Disease.

---

* Schwartz began treating Morgellons patients after hearing about the disease on *Coast to Coast*, a nationally syndicated AM radio show with a heavy emphasis on UFOs and conspiracy theories. In 2005, an article in *The Santa Fe New Mexican* described Schwartz's treatment of both Morgellons and heroin addiction, which often took place in a room at a local Red Roof Inn. One of Schwartz's patients told the *New Mexican* that she couldn't remember many details from the three days she spent in the motel under Schwartz's care because Morgellons made her suffer from "brain fog." In 2006, Schwartz's license to treat patients was suspended; two years later, he agreed to "not practice medicine or seek an active license to practice medicine anywhere in the United States, now or in the future."

He claimed to have "physical evidence" proving that Morgellons was an actual condition, but as yet had not offered any of this evidence to his peers to review. Casey, meanwhile, had never been the chief investigator of a published study. On her official OSU staff page, she included the following among her professional accomplishments: "Invited to speak to the Gifted and Talented Class for Bixby High School. The subject of the 1 hour-long lecture was Education for Success, and included audience interaction."

Then there was Stricker, who ran a business that advertised itself as having six specialties: "VIAGRA Treatment" (initial consultation $400, individual pills $15 thereafter), "Weight Loss Clinic" ("All you have to do is to come by our offices to pick up your first two weeks of medications. . . . It has been said that these medications are nearly 'magical' in the way they work, and for many people it truly seems that way"), "Fertility Clinic" ("The exact mechanism by which [our treatment] works is still unknown"), "Hyperbaric Oxygen Therapy," "Immunodeficiency/AIDS Clinic," and "Lyme Disease." Because Stricker's protocols were rarely covered by insurance, he demanded full payment by his patients at the time of service.

In the months after Leitao's media blitz, doctors across the country received an influx of patients convinced that they, too, suffered from Morgellons. Many of these claims were dismissed out of hand, but there were sympathetic practitioners who ran sophisticated tests in an effort to figure out what was going on. Time after time, those tests came back with similar results: Patients didn't have any fibers in their festering sores, or the fibers they did have were readily identifiable as coming from clothes or household fabrics. Common psychological ailments also cropped up repeatedly: Some Morgellons patients were found to have paranoid delusions that caused them to literally peel the skin off their bodies, while others' conditions were either caused or worsened by drug addiction or severe stress. All the while, coverage of the condition continued unabated: In October 2006, Stricker was the sole non-Morgellons patient who unconditionally recognized the disease in a *New York Times* article that claimed,

"Doctors themselves are divided over whether Morgellons is a medical or a psychiatric illness." That was true only in the sense that scientists are divided over evolution: There was a tiny group operating on the fringe who held one view, while the vast majority, and the vast majority of evidence, lined up on the other side.

While news outlets were remiss in not enumerating the major caveats that attended Morgellons, a growing body of research suggests that merely reporting on the supposed condition, regardless of the context, might have been all that was needed to give it legitimacy. In a 2007 study, researchers at the University of Michigan presented subjects with a pamphlet titled "Flu Vaccine: Facts and Myths." The flyer listed common perceptions about the vaccine and labeled each as either true or false (e.g., "The side effects are worse than the flu: FALSE"). After studying the sheet, subjects were quizzed on its contents. In the first round of questioning, conducted only a few minutes after the pamphlet was taken away, people generally fared well separating the myths from the facts. But as more time elapsed, subjects were less and less able to correctly identify the myths, although they were still able to remember the contents of the pamphlet itself—and, since they'd been told that the information came from a credible source, they began to assume that *all* the statements they'd heard were likely true.

On January 20, 2008, the *Washington Post*'s weekend magazine ran a 7,200-word cover story on Morgellons disease by Metro reporter Brigid Schulte. The story began with a lengthy description of Sue Laws, a suburban housewife whose life was shattered one afternoon in October 2004 when she was overcome by a sensation akin to thousands of bees simultaneously stinging her and burrowing under her skin. By the time Laws's husband responded to her screams, Schulte wrote, countless numbers of microscopic red fibers were poking out of her back. The pain was at its worst at night: The bugs, Laws said, were more active in the dark. In an effort to diminish her agony, she began sleeping with the lights on. She also started showering for hours on end, coating her skin with baby powder, soaking in bathtubs full of vinegar, and washing her sheets in ammonia. Eventually, Laws's hair began to fall out and her teeth began to rot—and still, the shimmering red, black, and blue fibers kept popping out of her skin.

According to the *Post*, the medical establishment didn't offer Laws any help; when doctors did acknowledge her pain, they inevitably said it was the result of a mental illness. "In time, Sue, 51, came across a condition like her own, and joined 11,036 others from the United States and around the world who, as of earlier this month, had registered on a Web site as sufferers of what they say is a strange new debilitating illness."

It wasn't until sixteen paragraphs later that the story acknowledged that contemporaneous records kept by Laws's longtime personal physician documented her drinking up to thirty cups of coffee and smoking three packs of cigarettes a day. (Laws told the *Post* her doctor was exaggerating by approximately 600 percent.) The story did not describe how nicotine, in addition to raising metabolism, releases epinephrine—also known as adrenaline—or how caffeine prolongs the effects of epinephrine in the system. It also did not inform readers that one of four recognized caffeine-induced psychiatric illnesses is caffeine intoxication, which is caused by "recent consumption of caffeine, usually in excess of 250 mg (e.g., more than 2–3 cups of brewed coffee)," or that one symptom of severe caffeine withdrawal is psychomotor agitation, which can include pulling one's own skin and biting one's lips until they bleed. Finally, while the *Post* article did tell readers that the onset of Laws's Morgellons coincided with the diagnosis of her twenty-year-old son with a fatal brain tumor, it didn't tell them that psychomotor agitation is also a well-documented symptom of severe depression.

In the coming pages, Schulte described more "Morgellons sufferers," many of whom had "lost their jobs, their homes, their spouses and even had their children taken away because of the disease." There was Pam Winkler, a Bel Air housewife who said Morgellons had exhausted her to the extent that she turned to cocaine and the narcolepsy medication Provigil just to stay awake. (Both her husband and her doctors were of the opinion that Winkler's cocaine addiction *preceded* her self-diagnosis and that she was just looking for attention. Winkler's response? "You know me. I'm a shallow person. I'm vain. Do you think I'm doing this to get attention? If I wanted attention, I wouldn't look this skanky. I'd get boobs.") There was the former Miss Virginia who spent eight hours a day picking fibers out of her skull, the grandmother who refused to see her children for fear of infecting them, and the patients with worms coming out of their eyeballs and flies emerging from their lungs.

On their own, each one of those stories might sound outlandish, but, Schulte implied, who was she to pass judgment? For that matter, why should anyone have faith in the so-called authorities who claimed the disease didn't exist? After all, Schulte wrote, "Western medicine has been guilty of closed-mindedness in the past. There is even a name for it: the Semmelweis Reflex, the immediate dismissal of new scientific information without thought or examination. It was named for a 19th-century Hungarian physician who was roundly vilified by his colleagues when he asserted that the often-fatal childbed fever could be wiped out if doctors washed their hands in a chlorine solution. He was right."*

---

* This rhetorical sleight-of-hand, which is sometimes referred to as the Semmelweis Stratagem, is a variation on the Galileo Gambit, whereby someone whose work is debunked argues that the fact that Galileo's work was also debunked proves he is actually correct. Semmelweis is frequently invoked by anti-vaccinationists. While explaining to me why the "mainstream" ignored Andrew Wakefield, one of his supporters and financial backers said, "We know there's never been a time in history when people, in a short period of time, have looked at the data and thought, 'Oh, we *do* have a problem here,' or even, 'We might have a problem here—maybe you *should* wash your hands after you dissect bodies and before you go deliver babies.' God forbid you should wash your hands!"

Two days after her article ran, Schulte took part in an online chat with readers. "I went into this article with a completely open mind," she said. "I reported what I found. . . . I think it's pretty clear there are still questions out there that need to be looked at and answered." When one reader asked Schulte why she'd left out details such as Savely leaving Texas after her treatment methods were questioned by the Texas Board of Nursing, Schulte replied, "That is a whole separate, controversial topic that I touched upon in the story but would have taken it in a different direction to explore in more detail." (In fact, while Schulte wrote that Savely's clinical approach was controversial, there was no mention of her problems with Texas medical authorities.)

By that time, the affliction Mary Leitao had created in response to what she viewed as mistreatment by medical authorities had been officially recognized as a potential public health crisis. Just as reporters had been quick to respond to the Morgellons Research Foundation's bait, politicians jumped at the opportunity to support a cause that had single-minded, passionate advocates on one side and no obvious or organized opposition on the other. More than three dozen members of Congress, including Hillary Clinton, Barack Obama, and John McCain, urged the CDC to investigate Morgellons. Some went further: California senator Barbara Boxer wrote a total of seven letters to the agency, while Iowa's Tom Harkin inserted language into a Health and Human Services bill that called for "this important research" to be undertaken "as quickly as possible."

The CDC, whose budget is determined by Congress and the president, had no choice but to acquiesce, and for perhaps the first time in its history it agreed to fund a widescale study into a condition that doctors and scientists overwhelmingly agreed didn't exist. In a statement announcing a study that will cost millions of dollars to complete, the agency said that "an increased number of inquiries from the public, health care providers, public health officials, Congress, and the media" led to its agreeing to look into "an unexplained skin condition." To this day, there has not been a single independently verified case of Morgellons syndrome.*

* The CDC's "Unexplained Dermopathy Project" began in January 2008 with an epidemiologic study conducted by—and this is a mouthful—the National Center for Zoonotic, Vector-borne, and Enteric Diseases' (NCZVED) Division of Parasitic Diseases (DPD). In September 2009, the DPD put together an external review panel to review the study's findings and advise about the best course of action. As of October 2010, the panel had not released its final report.

Part Two

CHAPTER 8

Enter Andrew Wakefield

The 1990s was not a good decade to be a public health official in the U.K. In September 1992, British officials ordered the withdrawal of the two most widely circulated brands of the MMR vaccine after the mumps component was shown to cause a mild form of meningitis in a tiny fraction of recipients (estimates ranged from one in six thousand to one in eleven thousand). The fallout from that decision provided yet another illustration of the government's impotence to counter media-fueled scares, as nothing tempered the panic that followed the announcement—not the fact that the symptoms of the vaccine-induced meningitis were limited to bouts of fever, vomiting, and "general malaise"; or that one in four hundred of those infected naturally with mumps came down with a much more severe form of meningitis; or that in 1988, the year the single-dose MMR vaccine was introduced in Britain, there were 86,000 cases of measles and six deaths, and in 1991 there were ten thousand cases and no deaths. Within twenty-four hours of the announcement, the government, according to a story in *The Daily Mail*, was facing "a welter of legal demands from parents who claim their children were damaged" by the vaccine.

Less than two years later, before the repercussions of the mumps vaccine scare had fully receded, Britain found itself facing a potential measles epidemic. In the first half of 1994, there were close to nine thousand reported cases of measles infections, which was twice as many as there'd been the year before. The most alarming thing about that spike was that many of those infections were in older children, for whom the disease was likely to be more severe. With epidemics in Great Britain historically occurring every five to seven years, the World Health Organization warned of an outbreak that had the potential to infect hundreds of thousands of children and cause thousands of cases of blindness, deafness, brain damage, and death. That July, as measles infections continued to rise, Kenneth Calman, the government's chief medical officer, announced that a £20 million effort to vaccinate seven million schoolchildren would begin in the fall. (Officials, mindful of the mumps vaccine scare, decided to include only measles and rubella in the shots distributed as part of the campaign.) Well aware of the need for positive publicity, Calman dedicated a good portion of that money to television and print advertisements that would highlight the dangers of measles infections. Calman also sent a letter to every doctor in the country, outlining the details of the campaign.

Those efforts proved to be for naught. As soon as vaccinations began in September, the program was attacked by parents complaining that they hadn't been kept sufficiently informed about the program. The media, while quick to cover any unease—one typical headline read, "Why Another Needle, Mummy?"—was much slower to examine the accuracy of speculative concerns. In the coming weeks, the news only got worse. Ampleforth and Stonyhurst, two of Britain's most prestigious Catholic schools, announced they were boycotting the campaign because the vaccine had been developed in part with tissue from aborted fetuses; days later, Muslim leaders said that they were opposed to the shot as well. In London, a group of homeopaths set up something called the Disease Contacts Network, through which parents could pay to receive notification of outbreaks so that they could deliberately facilitate the infection of their children. "Then," *The Independent* reported, "all they have to do is party. The theory is that naturally acquired disease confers immunity for life, and that if caught when children are between about five and nine, the illness itself is likely to be less severe." The fact that this theory was total rubbish went unmentioned.

The government's campaign also spurred the growth of a citizen-led anti-vaccine movement similar to the one that had taken root in the United States a dozen years earlier. Jackie Fletcher, a suburban housewife who lived about twenty miles outside Manchester, quickly established herself as the movement's de facto spokesperson. In many ways, Fletcher's story resembled that of her American equivalent, Barbara Loe Fisher: According to what Fletcher told reporters, her son, Robert, had been a "healthy, one-year-old boy" until he received an MMR shot a year and a half earlier. Soon after, Fletcher said, he began having the seizures that eventually led to diagnoses of everything from encephalitis and epilepsy to speech and learning disabilities. Fletcher first became convinced of the connection between her son's illnesses and vaccines when, during one of Robert's frequent trips to the hospital, she met another mother who said her son had also gotten sick not long after receiving his MMR jab. There was no way, Fletcher decided, that those similarities could be mere coincidence. Working with her Community Health Council, she placed ads in local papers in an effort to locate other parents who believed their children had also been injured by vaccines. Several dozen responded, and in December 1992, Fletcher founded Justice, Awareness, and Basic Support—JABS, for short—which she described as a group dedicated to protecting Britain's children.

To accomplish this, Fletcher embarked on a strategy that combined public campaigning with legal maneuvering. When Britain's Legal Aid Board approved funding for investigating MMR injury claims, Fletcher was ready with a list of hundreds of families willing to consider taking part in a class action lawsuit against vaccine makers. Soon, she was working with a solicitor named Richard Barr as he coordinated families' cases and fed stories to the press. From the outset, Barr claimed that the government's withdrawal of some versions of the mumps vaccine was evidence that the other two components of the MMR vaccine were dangerous as well: "[O]ut of 170 cases [Barr] has details of, most showed side-effects 14 days or sooner after the MMR vaccination," *The Economist* wrote in its October 29, 1994, issue. "That is too soon to have been caused by the mumps virus but plausibly the result of the measles component." (Unlike many publications, *The Economist* made clear that "little published data supports this case.")

For more than a year, Fletcher and Barr used the media to keep the specter of a massive lawsuit alive. In late 1995, Fletcher told reporters that JABS had assembled a list of dozens of children who had fallen ill as a result of the previous year's "government mass-vaccination campaign." According to one typically credulous account, many of the families considering legal action had children who'd "contracted debilitating, crippling and, in some cases, life-threatening diseases." (The coverage was at times unintentionally humorous: The headline on a *Daily Mirror* story read, "My Son Went Bald After His Measles Jab; Mirror Health on How a Simple Injection Has Changed a Young Boy's Life Forever; Doctors Say His Hair May Never Return.") At the time, it wasn't clear how these threatened vaccine-related lawsuits would proceed: Since 1979, the only outlet families in the U.K. had had to recover damages was the Vaccine Damage Pay Unit, where the burden of proof was high and compensation was capped at £30,000. In late 1996, Barr let the press in on his legal strategy: He was going to use the Consumer Protection Act, a European Union law that established no-fault liability, to sue the pharmaceutical companies directly.

Barr and Fletcher both knew that would not be an easy feat. As Fletcher's husband, John, told *The Guardian*, "[The pharmaceutical companies] say prove it and argue that it was coincidental." In order to establish cause and effect, the Fletchers would need more than stirring anecdotes—they'd need something that looked like scientific proof. As it turned out, they'd already begun collaborating with a researcher who would provide them with that very thing.

• • •

For more than twenty years, Andrew Wakefield has published work in which he's claimed to have uncovered new ways of looking at old problems. This pattern began in the late 1980s, when Wakefield, a Canadian-trained surgeon then in his early thirties, challenged the conventional thinking about a debilitating and incurable inflammatory bowel disease (IBD) called Crohn's disease. Already, Wakefield was a seeming contradiction of brash self-confidence and unrealized potential. He was a former amateur rugby player who'd earned his doctorate when he was only twenty-five years old—but according to medical databases, he'd published a total of just three research papers during the first decade of his career.

In focusing on Crohn's, Wakefield had chosen a disease whose precise etiology has never been established, although then, as now, the vast majority of evidence indicated that it is an autoimmune disorder that causes the body to incorrectly identify partially digested food as invasive matter. When the immune system attacks the perceived threat, white blood cells build up on the walls of the small intestine, resulting in painful, and occasionally deadly, inflammations. This way of understanding Crohn's, Wakefield speculated, was all wrong: The disease's trademark inflammations were not the pathological features of an immune system malfunction, they were the result of clogged blood vessels in the wall of the gut. If true, Wakefield's theory stood to completely change the way Crohn's was treated—but, as would be the case repeatedly in the years to come, Wakefield's data did not justify his conclusions. "[It was] an interesting idea," Thomas MacDonald, the dean for research at the Barts and London School of Medicine, said later. "But just wrong."

Such appraisals did not prompt Wakefield to reexamine or revise his theory; if anything, they seemed only to reinforce his belief that he was correct and his critics were mistaken. In the early 1990s, building on his earlier work, Wakefield began to focus on measles as a likely cause of the IBD-causing stomach inflammations. In later years, Wakefield would say that the genesis of that insight was neither lab research nor collaboration with infectious disease specialists: His "eureka" moment occurred one winter night when, while flipping through an old virology textbook, he happened upon a description of how the measles virus had, on occasion, been shown to cause ulcers. "You could have been reading an account of Crohn's disease," he said. "It was very exciting."

Those years also marked the point in his career during which the focus of Wakefield's work increasingly converged with public health concerns. According to an investigative journalist named Brian Deer, it was around that time that Wakefield began contacting high-level officials at the British Department of Health, requesting face-to-face meetings to discuss financial backing for his work. His entreaties seemed to go beyond a sober recitation of the value of his research: In an October 1992 letter to David Salisbury, who at the time was the head of the British vaccine program, Wakefield wrote, "My concern is that although measles, and in particular the vaccine, may have no association with Crohn's disease whatsoever, what will be picked up by the press is the apparent association between the increasing incidence of disease and the vaccine." It was an odd point for a researcher to make. Instead of arguing the merits of his work, Wakefield seemed to be warning about the possible public fallout of his conclusions.

By the beginning of 1993, Wakefield, having failed to obtain government funding, was fully immersed in the work that would dominate the rest of his life. That spring, he and several co-authors submitted a paper to *The Journal of Medical Virology* that claimed to have found evidence that "the measles virus is capable of causing persistent infection of the intestine" and that Crohn's disease "may be caused by a . . . response to this virus." That study created such a furor that Britain's Medical Research Council (MRC), which oversees and promotes health-related research, convened a panel to examine the data that Wakefield had relied on—and found that it had such significant problems as to render the study's conclusions all but meaningless.

Wakefield's next attention-getting paper was published in 1995, not long after the government's measles vaccination campaign: Now, he was "examin[ing] the impact of measles vaccination" on "the rising incidence of inflammatory bowel disease"—and in doing so, fulfilling the prophecy he'd made to Salisbury three years earlier. By that point, the nascent concerns about Wakefield's early research were appearing with increasing frequency. In the years to come, teams in Japan and the United States would try, and fail, to replicate the 1995 paper's results. That was not a surprise: The shortcomings of Wakefield's research were so serious that they all but negated his results. In one instance, the chemical solutions he'd used to identify the presence of measles in intestinal tissue were not specific for the virus, which meant that positive reactions could have signified almost anything. In another, the results that Wakefield claimed were indicative of the presence of measles actually stemmed from contamination by a clone of the virus that had been provided to Wakefield's lab for use as a positive control. Even if his data had been more reliable, his conclusions were oftentimes contra-logical: If measles infections did, in fact, cause Crohn's in some subset of the population, why had there been an increase in Crohn's diagnoses over the previous two decades, a period during which measles infection rates had plummeted?

By 1997, the emerging consensus about Wakefield was, as Thomas MacDonald explained later, that he often neglected to carry out "important things that a credible, decent scientist would feel duty bound to do." But if Wakefield had an increasingly checkered reputation within the insulated worlds of medicine and academia, he was a rising star in the public sphere, where his skill at public relations endeared him to a London press corps willing to hype results first and check reliability later, if at all.

CHAPTER 9

THE *LANCET* PAPER

On February 26, 1998, journalists from London's dailies were invited to a press conference at the Royal Free Hospital School of Medicine. The occasion was the publication of a dense academic paper in *The Lancet*, the august medical journal that is one of the most cited scientific publications in the world. The paper's title didn't make it an obvious candidate for articles in the next day's broadsheets; even the most creative headline writers would have trouble coming up with a pithy leader for a piece about "Ileal-Lymphoid-Nodular Hyperplasia, Non-Specific Colitis, and Pervasive Developmental Disorder in Children." But the Royal Free's PR team gave hints that this was no ordinary paper: They'd put together a twenty-minute promotional video for the occasion and assembled a panel of five of the hospital's researchers to address the report's implications. Andrew Wakefield, the paper's lead author and its "senior scientific investigator," was the star of both.

It didn't take long to figure out what all the fuss was about. Contrary to the paper's title, the main thrust of the press conference was not the possible connection between intestinal and developmental disorders—it was Wakefield's supposition that the MMR vaccine, which had been used in the United States since the early 1970s and in Great Britain for the previous decade, could very well be responsible for the dramatic rise in rates of autism. In order to support this theory, Wakefield piggybacked on his claim that he'd found the measles virus in the intestinal tracts of IBD patients—a claim that had already been discredited by other studies. Now he said he'd come up with a potential biological pathway that linked the MMR vaccine with IBD and autism. Some children, he speculated, had immune systems that, for some unknown reason, were unable to handle the combination of the three vaccines at once. As a result, the measles component of the vaccine took root in the lining of the small intestine, causing a "leaky gut." The next step in Wakefield's hypothesis was dependent on a widely discredited "opioid excess" theory of autism, which drew parallels between autistic children and doped-up lab rats: After the opioid peptides that are naturally produced during digestion escaped through the gut's newly porous walls, Wakefield argued, they breached the blood-brain barrier and overwhelmed developing children's brains. The result was autism.*

Knowing that the paper's findings would be controversial from the start, the five experts who addressed the media had agreed beforehand that regardless of their individual interpretations, they'd deliver one overarching message: Further research needed to be done before any conclusions could be drawn, and in the meantime, children should continue to receive the MMR vaccine. Once the tape recorders began to roll, however, Wakefield went dramatically off-script: "With the debate over MMR that has started," he said, neatly eliding over the fact that he was, at that very moment, the person responsible for igniting the debate, "I cannot support the continued use of the three vaccines given together. We need to know what the role of gut

---

* The opioid excess theory was the brainchild of an Estonian-born psychologist named Jaak Panksepp, who, in the late 1970s, said he was struck by the behavioral similarities between morphine-addicted rats and autistic children, including "a lack of crying during infancy, a failure to cling to parents, and a generally low desire for social companionship." Later in his career, Panksepp's work on rodents led him to conclude that "boys born to mothers using opiates during the second trimester would be expected to have a higher incidence of homosexuality than the offspring of nonaddicted women."

inflammation is in autism. . . . My concerns are that one more case of this is too many and that we put children at no greater risk if we dissociated those vaccines into three, but we may be averting the possibility of this problem."

Almost immediately, the press conference descended into near chaos. Even if Wakefield's study had been more comprehensive and his data more robust, it was virtually unprecedented for a research scientist to advocate wholesale changes to health policy. After stressing that the MMR vaccine had been given to millions of children around the world and had saved untold numbers of lives, Arie Zuckerman, the dean of the Royal Free Hospital's medical school, became so agitated he began banging on the lectern. "If this were to precipitate a scare that reduced the rate of immunization," he said, "children will start dying from measles."

Zuckerman's frustration was understandable. As scientists around the world already knew, there were ample reasons to view Wakefield's latest effort skeptically. After an initial peer review raised questions about the quality of Wakefield's research and the soundness of his reasoning, Richard Horton, the editor of *The Lancet*, demanded the paper be rewritten in such a way that made clear the speculative nature of the work and slapped an "Early Report" label above the title and on the header of each page. Horton also took the even more unusual step of asking Robert Chen and Frank DeStefano, two American vaccine specialists at the CDC, to prepare an evaluation of Wakefield's paper that would appear in print. "Usually, when they publish a commentary, it's to extol the study, or show how it's advanced the field," DeStefano says. That was obviously not the case here. When he first read the paper, DeStefano says, his reaction was, "There really didn't seem to be that much there. It was kind of like, Why were they publishing the article?" *

* By the time the journal hit the stands, it looked as if Horton had been wondering the same thing. Only two years had passed since, at age thirty-three, he'd become the youngest editor in *The Lancet*'s 173-year history. During that time, he'd established a reputation for shaking things up; already, some of his professional peers were

You did not need to have advanced scientific training to share DeStefano's confusion; in fact, knowledge of basic principles of experimentation were all that were needed to understand why Wakefield's work was so unimpressive. Broadly speaking, there are three ways scientists collect data to test new theories. The best possible method is through a randomized clinical trial, in which researchers take a sampling of a population and arbitrarily test their hypothesis on one half while leaving the other half untouched. This is preferred because by engineering the test population, it is possible to control for other potentially mitigating factors.*

Unfortunately, randomized clinical trials are not always feasible. Sometimes this is for ethical reasons: You can't determine what quantity of a given substance is toxic in humans by administering doses that could prove fatal. Other times, it's because of logistical problems: It's impossible to test the lifelong effects of living in a given environment by transplanting half the population. The second-best option is a case control study, where investigators analyze a group that has been naturally subjected to the issue under examination. There are a number of reasons case control studies provide less definitive proof than randomized trials. Many of those reasons stem from their ret-

---

grumbling that his approach was more appropriate for buzzy, mass-market magazines than a staid research journal. In this instance, however, Horton tried to play it safe: *The Lancet* conspicuously declined to participate in the Royal Free press conference, a decision a spokesman said was made because it wanted to "avoid being alarmist," but which some at the time interpreted as Horton's desire to distance himself from a media maelstrom that he realized was inevitable. In his published commentary on the imbroglio, Horton simultaneously defended his decision to publish the paper and distanced himself from Wakefield's publicity campaign: "Reported adverse comments about the safety of MMR vaccination were made at [the Royal Free's] press conference," he wrote. "By contrast, the views expressed in the paper are unambiguously clear: 'we did not prove an association between measles, mumps, and rubella vaccine and the syndrome described.' "

* There are several ways to conduct randomized trials. The ideal one is through a double blind test, in which neither the researcher nor the test subjects know who is being studied and who is a control. Double blind trials protect against the placebo effect on the part of the subject and observer bias (or wishful thinking) on the part of the researcher.

rospective nature, which means the population under review hasn't been randomly determined and there is no way to control for other factors that may have influenced the result.

The least convincing data comes from the type of study Wakefield had conducted: a simple case series. These are oftentimes nothing more than an interesting phenomenon someone happened to notice. Generally speaking, case series are starting points for new hypotheses—and most of the time, what at first blush looks to be significant is nothing more than the result of the random nature of the universe.

To understand why a case series is a tenuous place to hang your hat, take the example of gender. Even though the population as a whole is split almost evenly between boys and girls, there are numerous examples of individual families with all boys or all girls, and there are many more examples of families where a series of children born in a row are of the same sex. Now, imagine that an alien is sent to earth to learn about what types of human offspring are born to parents living in different states. The first couple he meets, Alexander Baldwin and Carolyn Newcomb of New York, has four children: Alec, Daniel, William, and Stephen. Based on that set of data—which is the equivalent of a single case series—the alien would assume that all earthling children born in New York were boys (and that they all were actors with a penchant for appearing in the tabloids). If the second couple the alien meets was George W. Bush and Laura Bush, he'd assume that all children born in Texas were, just like Barbara and Jenna Bush, twin girls. Those conclusions would, of course, be incorrect—but the only way to realize that would be to collect more data.*

As Chen and DeStefano demonstrated with their 839-word evis-

* A real-world example of a theory proven to be true after all three types of studies were performed is the beneficial effects of fluoridation. First was the case series: A dentist noticed that patients with mottled teeth seemed to have less tooth decay. Next was the case control study: Children who'd spent their whole lives in an area with naturally fluoridated water were compared to those who'd moved there after their teeth had fully developed. Finally came the randomized clinical trial: A relatively homogeneous population was divided in two, with only one half receiving fluoridated water.

ceration, the shortcomings of the *Lancet* paper went far beyond the limitations of the way its data had been collected. There were serious problems with its entire premise: Hundreds of millions of children had received the measles vaccine since it was first introduced and the vast majority of them had no chronic bowel or behavioral problems; the "syndrome" Wakefield purported to have discovered was already well documented and was nonspecific to patients with autism; autism was a well-known condition long before the MMR vaccine became available; and despite hypothesizing that the MMR vaccine led to IBD led to autism, in most of the cases Wakefield cited, the behavioral changes *preceded* the bowel problems. There were methodological problems, the most glaring of which was selection bias: The parents who came to Wakefield, who was not a pediatrician and had never been clinically trained to work with children, did so because he was known as someone interested in connecting the MMR vaccine with inflammatory bowel disease. There were concerns about the reliability of the paper's data: Wakefield was dependent on parents' post-facto recollections about the temporal connection between vaccination and onset of their children's symptoms, and in the three years since Wakefield first reported finding the measles virus in patients with IBD, "other investigators using more sensitive and specific assays, have not been able to reproduce these findings." Finally, and most damningly, there was "no report of detection of vaccine viruses in the bowel, brain, or any other tissue of the patients in Wakefield's series." The entire report wasn't built on a house of cards—there weren't any cards to begin with. Several years later, when the dean of research at the London School of Medicine called the study "probably the worst paper that's ever been published in the history of [*The Lancet*]," he was merely acknowledging what many scientists had thought all along.

In the coming days, as his study came under increasingly intense fire, Wakefield took advantage of the fact that public opinion wars are not won through the use of dry, academic language. While his detractors were explaining that "the most striking and consistent endoscopic feature, lymphoid nodular hyperplasia in the terminal ileum, is not unusual in children," Wakefield was trotting out what had become a standard response of vaccine denialists to accusations of unreliable or inaccurate data: He condemned his critics for caring more about their standing in the scientific community than about sick children. "[Advocating the discontinuation of the MMR vaccine] is a moral issue for me," Wakefield told *The Independent*, going on to describe the frantic phone calls he was fielding from desperate parents. "These are the people to whom we are answerable," he said. And Wakefield had answers: Parents should insist their children receive separate measles, mumps, and rubella vaccines, and that there be a minimum of a year's break between each one. The fact that every child who delayed vaccination would be at risk was beside the point, as was the lack of availability of single-dose vaccines—if the pharmaceutical companies cared more about patients than profits, Wakefield implied, they'd figure out what to do. In the meantime, Wakefield's office set up a dedicated phone line to field calls from frantic parents. When they called, they were offered a fact sheet Wakefield had put together and were told to direct any further questions to their doctors. Even when Wakefield was called on to defend his work in the March 21 issue of *The Lancet*, he relied on emotion instead of scientific rigor to buttress his conclusions:

> Our publication in *The Lancet* and the ensuing reaction throws into sharp relief the rift that can exist between clinical medicine and public health. Clinicians' duties are to their patients, and the clinical researcher's obligation is to test hypotheses of disease pathogenesis on the basis of the story as it is presented to him by the patient or the patient's parent. Clearly, this is not the remit of public-health medicine. The approach of the clinical scientists should reflect the first and most important lesson learnt as a medical student—to listen to the patient or the patient's parent, and they will tell you the answer.
>
> Accordingly, we have now investigated 48 children with de-

> velopmental disorder in whom the parents said "my child has a problem with his/her bowels which I believe is related to their autism". Hitherto, this claim had been rejected by health professionals with little or no attempt to investigate the problem. The parents were right. They have helped us to identify a new inflammatory bowel disease that seems to be associated with their child's developmental disorder. This is a lesson in humility that, as doctors, we ignore at our peril. In many cases, the parents associated onset of behavioural symptoms in their child with MMR vaccine. Were we to ignore this because it challenged the public-health dogma on MMR vaccine safety?

It wasn't hard to predict that the London press would find the lone-wolf doctor narrative more appealing than the one told in reams of data. Typical of the ensuing coverage was a *Guardian* feature on Karen Prosser, the mother of a severely autistic child. "Prosser knew something was not quite right with her second baby, Ryan, when he was 15 months old," the piece began. By the time Ryan was two, he'd lost the ability to speak. From the start, Prosser suspected her son's developmental difficulties were related to his severe constipation. Her doctor, she said, told her she was being ridiculous.

> Then Mrs Ryan [Prosser] saw an article about the work Dr Wakefield and colleagues had been doing at the Royal Free into possible links between the MMR vaccine and inflammatory bowel disease. . . .
>
> "I can say I believe 100 percent it is the bowel that has caused his autism," said Mrs Prosser. Ryan, four today, is a different child as long as he takes his medicines to keep the bowel disease at bay. He can now talk. The tantrums have stopped. He will look at people now and wants to be friends. Every day he learns new words.
>
> Mrs Prosser believes the MMR vaccination set off the bowel disease which in turn caused Ryan's autism. "Something caused

the bowel to do this," she said. "Bowels don't just create inflammation. Something happened."

Throughout that spring, the controversy raged on. In March, a group of thirty-seven leading health experts, ranging from pediatricians to psychiatrists, epidemiologists to gastroenterologists, and immunologists to virologists, was convened for a review of Wakefield's paper. It, too, found his research lacking, his theories "biologically implausible," and his conclusions essentially worthless. In the group's final report, which was sent to every doctor in the country, it noted, "It would be surprising if the link had not been noted in other countries with good diagnostic facilities for autism where MMR has been widely given for many years."

In the coming weeks, *The Lancet*'s correspondence pages would be filled with further indictments.* Ten co-signers from the Institute of Child Health at University College London Medical School said they were "surprised and concerned that *The Lancet* published [a] paper . . . yet provided no sound scientific evidence"; three more from the Barnsley Health Authority's Department of Public Health Medicine suggested it might be time for research publications to "to carry health warnings" so people can be "adequately appraised about the strength and quality of evidence presented."

The substance of virtually all those letters was ignored by the media, which treated the evolving story as more of a horse race between competing claims: "Ban Three-in-One Jab, Urge Doctors After New Fears" and "Doctors Link Autism to MMR Vaccination" one day, "Research Rejects Autism Link with Vaccine" and "Triple Jab Is Safe, Says Medical Chief" the next. In those cases, at least, reporters

* One of the few voices of support for Wakefield came from Barbara Loe Fisher, who wrote, "It is perhaps understandable that health officials are tempted to discredit innovative clinical research into the biological mechanism of vaccine-associated health problems when they have steadfastly refused to conduct this kind of basic science research themselves. . . . US public health officials will not accept any independent thinking or scientific investigation into vaccine-associated health problems that does not carry their imprimatur."

and editors could claim the precise scientific arguments were over their, or their readers', heads. The same could not be said about the contents of a letter that was sent to *The Lancet* by a doctor in the Wiltshire Health Authority's Department of Public Health named Andrew Rouse. It arrived at *The Lancet*'s offices on March 4, just six days after Wakefield's press conference. An edited version ran in the journal's May 2 issue:

> After reading Andrew Wakefield and colleagues' article I did a simple Internet search and quickly found the *Society for the Autistically Handicapped*. I downloaded a 48-page fact sheet produced for the society by Dawbarns, a firm of solicitors in King's Lynn.
>
> It seems likely then that some of the children investigated by Wakefield *et al* came to attention because of the activities of this society; and information from parents referred in this way would suffer from recall bias. It is a pity that Wakefield *et al* do not identify the manner in which the 12 children investigated were referred (eg, from local general practitioners, self-referral via parents, or secondary/tertiary or international referral). Furthermore, if some children were referred, directly or indirectly, because of the activities of the Society for the Autistically Handicapped, Wakefield should have declared his cooperation with that organisation.

In the form that it appeared, Rouse's letter was rather oblique. What was the Society for the Autistically Handicapped, and why did Rouse think it "likely" that the group had referred some of the children in Wakefield's study? How was Dawbarns, the law firm at which Richard Barr worked, involved? The first public hints of the answer came from Wakefield himself, whose written response included a denial of "litigation bias" despite his having agreed to "evaluate a small number of these children on behalf of the Legal Aid Board." An even fuller picture emerged on the Web address Rouse cited, which featured a copy of the Dawbarns pamphlet, which recommended to parents "who believed [their] child has been damaged" that they "seek proper compensation in the courts." "We are working with Dr Andrew Wakefield of the Royal Free Hospital London," it read. "He is investigating this condition. . . . If your child has developed persistent stomach problems (including pains, constipation or diarrhea) following the vaccination, ask us for a factsheet from Dr Wakefield."

Still more pieces of the puzzle were available to anyone who did the type of "simple Internet search" Rouse had performed. For instance, an online search for "Dawbarns" and "Royal Free" would have turned up an article that had appeared fifteen months earlier: On November 27, 1996, *The Independent* ran a 1,300-word story about Rosemary Kessick, a forty-two-year-old former business manager whose son was one of more than three hundred children whom Dawbarns had assembled for its suit. The boy was also, according to the article, "one of 10 children taking part in a pilot study at the Royal Free Hospital in London, which is investigating possible links between the measles vaccine with the bowel disorder Crohn's Disease, and with autism. The study is being organised by Norfolk solicitors Dawbarns, one of two firms awarded a contract by the Legal Aid Board in 1994 to co-ordinate claims resulting from the MMR vaccine." It turned out that the parents of some of the children Wakefield had used to launch a worldwide vaccine scare were planning on suing the manufacturers of the MMR vaccine, that Wakefield's lab had been paid £50,000 to collect data for the suit, and that the initial research proposal for the Wakefield-led, Dawbarns-funded study had stated its goal was to "establish the causal link between the administration of the vaccines and the conditions outlined." *

None of this was reported at the time, nor was the public informed that nine months before his *Lancet* paper was published, Wakefield had filed a patent for "a new vaccine for elimination of MMR and measles virus and to a pharmaceutical or therapeutic

* In total, Wakefield received £435,643 for his involvement in the ultimately unsuccessful, publicly funded lawsuits filed by parents against the drug companies GlaxoSmithKline, Aventis Pasteur, and Merck.

composition for the treatment of IBD . . . and regressive behavioural disease." Instead, London's newspaper readers were treated to headlines such as "Measles Turned My Son into an Autistic Child" and "I Want Justice for My James—Study Boosts Mum's Battle." It would not be until years later that a freelance investigative reporter uncovered the truth behind Wakefield's work—and by that point the damage had already been done.

CHAPTER 10

Thimerosal and the Mystery of Minamata's Dancing Cats

On the morning of January 28, 1928, just as he had most mornings for the previous two weeks, George Thomson began his day at the diphtheria vaccination clinic in Bundaberg, a small town on the Burnett River on the eastern coast of Australia. Thomson was Bundaberg's general practitioner, and he doubled as the town's chief public health official. Although diphtheria—a highly contagious respiratory infection that had long been one of the leading childhood killers in the region—had been in decline ever since the discovery of an effective antitoxin serum thirty years earlier, Thomson knew even a handful of cases carried the risk of a deadly outbreak. Hoping to forestall any potential epidemics, he'd implemented a voluntary vaccination program targeted at school-age children, who were among those at the greatest risk for the disease.

The program had begun smoothly enough. Using little more than a set of hypodermic needles and a large glass bottle of diphtheria vaccine, Thomson had vaccinated almost two dozen children in the first week alone. Twenty-one children showed up for their shots on the

28th, some for their first in the series and some for their second. With each patient Thomson performed the same routine: He swabbed the injection site with antiseptic, removed a syringe from a tray of saline solution, pierced the rubber top of his vaccine bottle with a needle, and administered the shot.

That day, however, something went horribly wrong. By early afternoon, eighteen of the twenty-one children Thomson had injected were convulsing, vomiting up "offensive green bile," and showing other signs indicative of a sudden, aggressive infection. Less than two days later, twelve of those children were dead. An ensuing Royal Commission investigation discovered that Thomson's bottle of vaccine, which he stored each night in a cupboard in his office, had at some point become contaminated with staphylococci bacteria. The children he'd injected on January 28 had contracted a particularly virulent form of staph infection, and the sickest had died of toxic shock. In an effort to protect his hometown, George Thomson had caused one of the most concentrated vaccine catastrophes on record.

As word of the Bundaberg tragedy spread, public health officials around the world began looking for ways to use sterilizing preservatives in multi-dose vials of vaccine. By the early 1930s they had settled on a recently synthesized ethylmercury compound called thimerosal. It seemed ideally suited to the role: It was inexpensive to manufacture, side-effect-free, and lethal to a broad range of microbes without undermining the potency of the vaccine it was mixed with.

Mercury's unusual qualities—it is one of only six heavy metals that are liquid at or near room temperature—and its shimmering, silvery appearance have made it a source of fascination for millenia. In 210 B.C., Qin Shi Huang, the first ruler of unified China, died after consuming a mixture of mercury and jade he hoped would grant him eternal life. The ancient Romans, none the wiser, often used it as a cosmetic applied directly to their faces.

In the nineteenth century, a mercury compound called calomel was used to treat everything from tuberculosis and parasites to toothaches and constipation. By the 1950s, in addition to its role as a vaccine preservative, mercury was a common ingredient in any number of over-the-counter medicines. It has been used in amalgam dental fillings and topical antiseptics, and, for a period, was an ingredient in teething powders and de-wormers for small children. Even as its uses changed and multiplied, scientists had little codified knowledge of how mercury interacted with the human body. It wasn't until 1956 that they had the chance to systematically assess its effects.

That spring, a five-year-old girl was admitted to a hospital in Minamata, the largest city on Kyushu, the southernmost of the four main islands of Japan. The suddenness with which she had fallen ill and the severity of her condition were shocking: In a span of just several days, she went from appearing perfectly healthy to having trouble speaking or using her hands to suffering from crippling seizures. Forty-eight hours later, the girl's younger sister began exhibiting similar symptoms; her hospitalization was followed by those of other children in the neighborhood. Something was clearly wreaking havoc on these children's central nervous systems—but nobody had any clue what it was.

The more local officials investigated, the more it seemed as if Minamata had been cursed with an otherworldly plague. Residents spoke of a "dancing cat disease," where previously healthy cats began to spastically convulse and then drop dead. On more than one occasion, floating fish corpses had freckled the local port. Birds dropped from the sky without warning. By October, this mysterious new contagion had stricken forty Minamata residents, fourteen of whom had died, which made for a mortality rate several times higher than that of most cancers—and still, investigators remained stumped. It wasn't until the following year that a common denominator emerged: All the victims had come from families that subsisted in large part on locally caught seafood. Even with that clue, it took another two years before officials determined that the cause of "Minamata disease" was toxic waste from a local factory that was poisoning the region's fish with an organomercurial compound called methylmercury.

When scientists studied the effects of methylmercury on humans, what they found was sobering: Not only were large quantities of mercury more dangerous than previously assumed, but they were particularly dangerous for unborn children. There were, however, obvious limitations to what could be learned. The level of mercury consumed by residents of Minamata was so massive—the city's typical resident had forty-seven times more methylmercury in his hair than the average person elsewhere, and some of the city's inhabitants had close to two hundred times more than normal—that while the data demonstrated the consequences of severe mercury poisoning, it was less useful in helping to determine what levels of mercury could be consumed safely.

In the coming decades, epidemiological studies of both man-made catastrophes and naturally occurring phenomena further clarified the danger mercury presented to humans. In 1971, nearly five hundred Iraqis died when a shipment of grain that had been treated with a methylmercury-based insecticide was stolen and sold as food. (Labels marked "DANGER" and "POISON" that were stamped on the grain sacks were printed only in English; the grain was intended for use as seed and never meant to be eaten.) Spurred in part by the Iraqi tragedy, teams of scientists conducted field studies in the Faroe Islands, off the coast of Northern Europe, where islanders' periodic consumption of whale meat led to bursts of extreme methylmercury exposure; and in the Seychelles, an archipelago nation in the Indian Ocean, where residents' seafood-intensive diets resulted in prolonged, low-level mercury intake. In the Seychelles, researchers from the University of Rochester found no difference in children whose mothers had consumed a significant amount of methylmercury during pregnancy; in the Faroe Islands study, there was some evidence of slight cognitive defects when children were tested at seven years of age.

All the while, the number of childhood vaccines recommended

by the federal health authorities rose, from around six in the 1960s to eight in the 1980s to eleven by the end of the 1990s. Variations of a handful of those vaccines used thimerosal as a preservative. As is often the case with sweeping public policies that are implemented over many years, the officials in charge of the vaccine program tended to ignore the bigger picture: When new vaccines were under consideration, they were evaluated on their own merits, with little thought given to the collective effect of piling one thimerosal-containing shot atop another. This was partially due to the assumption that because these vaccines were so widely used, any possible side effects would already have come to light. Vaccinologists also knew that ethylmercury—the form present in vaccines—was less toxic than methylmercury, which caused the tragedies in Minamata and Iraq and was the subject of the studies in the Seychelles and the Faroe Islands.*

Regardless of the relative harmlessness of ethylmercury, there was a troubling self-assuredness that underlay health officials' blasé attitude to thimerosal's potential risks, which echoed the complacency regarding the whole cell pertussis vaccine in the 1970s. Pharmaceutical manufacturers and immunologists at times acted as if the fact that vaccines had saved more lives than any other single invention rid them of the responsibility of ensuring that the vaccines already on the market were as safe as possible.

The trigger that finally led to a systematic examination of thimerosal came about almost by accident, as a by-product of broader concerns about mercury in the environment. In the fall of 1997, a Democratic congressman from New Jersey named Frank Pallone introduced the Mercury Environmental Risk and Comprehensive Utilization and Reduction Initiative (or MERCURI). In addition to monitoring mercury emissions from coal-fired power plants, Pallone's bill would regulate waste from household items such as batter-

* A good parallel of the difference between ethylmercury and methylmercury is that of ethyl alcohol, or ethanol, and methyl alcohol, or methanol: Two shots of the former give you a buzz; two shots of the latter are lethal.

ies and thermometers and require the military to curtail its use of the heavy metal. MERCURI never passed, but Pallone did insert a provision into the FDA Modernization Act of 1997 that called for a federal report on all mercury-containing food and drugs within two years.

The opportunity to review already licensed drugs was a rare one for the FDA. Unlike the Environmental Protection Agency (EPA), which uses newly developed diagnostic tools to reevaluate pesticides every ten years, the FDA typically operates on a "one and done" model: A drug deemed safe when it's introduced to the market is generally considered safe forever unless evidence to the contrary emerges. Thimerosal began to be used as a preservative in vaccines in the 1930s, and while it did receive an FDA safety review in 1976, the scientific understanding of toxicity had improved considerably since then. Use of the preservative had never set off any precipitous alarms, but health officials who spoke candidly admitted that they didn't know what the ultimate risks were.

The task of assessing mercury in vaccines fell to the FDA's Center for Biologics Evaluation and Research (CBER). In December 1998, it asked all vaccine manufacturers to provide detailed information regarding their use of thimerosal. By the following April, the agency had determined the precise concentration of thimerosal in each of the vaccines that included it as an ingredient. Then it narrowed this list down to those that were recommended for infants, which left the two versions of the hepatitis B vaccine and some formulations of the DPT and Hib vaccines. After running the numbers, the agency realized that an infant whose parents adhered to the recommended vaccine schedule could receive anywhere from 0 micrograms (μg) to 62.5 μg of ethylmercury at each of his two-month, four-month, and six-month checkups, for a maximum combined total of 187.5 μg.

Unfortunately, the lack of federal guidelines for ethylmercury consumption made this figure almost impossible to interpret. The confusion was heightened by the fact that there wasn't even a clearly defined level above which methylmercury exposure was cause for concern, with three separate federal agencies each setting distinct benchmarks tailored to their individual mandates: The FDA's was the amount that was unsafe for prolonged consumption in food, the EPA's was the amount that would trigger an investigation of mercury levels in the environment, and the Agency for Toxic Substances and Disease Registry's was the amount that could cause illness in the most sensitive members of a population. What's more, none of those figures was for methylmercury injected directly into the body. It was clear that 187.5 micrograms of injected ethylmercury was higher than the FDA's limit for ingested methylmercury—but what, exactly, did that mean?

Confronted with this mishmash of incomplete science and inconsistent regulations, a consensus emerged that until further studies were conducted, the best course of action was to remove thimerosal from pediatric vaccines, both as a safety precaution and to guard against the country's entire vaccination program being called into doubt. There was considerably less consensus on how to go about this, with opinions ranging from the immediate suspension of all thimerosal-containing childhood vaccines to maintaining the status quo while quietly phasing in thimerosal-free vaccines in the months and years to come. In an effort to break this deadlock, the FDA solicited input from outside experts. Included among those was Neal Halsey, the director of the Institute for Vaccine Safety at Johns Hopkins University. The issue had a special resonance for Halsey: He was wrapping up a four-year stint as chairman of the AAP's committee on infectious diseases, and over the previous decade he'd been one of the strongest advocates for increasing the number of recommended vaccines to eleven for children under two years old. Now he learned that during that period the amount of thimerosal infants could potentially receive had as much as tripled. Halsey's first reaction, he says, was one of disbelief. "There was no safety data looking at whether that was safe or whether there was evidence of long-term outcomes," he says. "No one had done six- or seven-year follow-ups on those children, or even thought to do it. That's the situation we were in."

From the outset, Halsey was convinced there was only one way to proceed: In a series of meetings and conference calls from June 25 through June 30 that included at various times officials from the AAP, the CDC, the EPA, and the FDA; independent toxicologists; and representatives from vaccine manufacturers, Halsey argued forcefully for a public statement supporting the use of thimerosal-free vaccines for all infants younger than six months old. The concerns about the preservative would be public knowledge by the end of July at the latest—that was when the FDA was planning on sending letters about the issue to all vaccine manufacturers and to members of Congress—and, Halsey argued, from a PR standpoint alone they'd be well served by getting ahead of the news.

In retrospect, what is most stunning about the events that followed is the rapidity with which they occurred. Thimerosal had been an additive in vaccines for more than half a century. The FDA Modernization Act had given CBER two full years to conduct its review. And then, in the course of a handful of days in the summer of 1999, an ad hoc jumble of policymakers, doctors, scientists, and industry representatives held a series of frenzied meetings in locations chosen specifically because they allowed for a circumvention of public disclosure laws. Established protocols were thrown out the window; experts who had never been a part of public policy discussions threatened to go public if their recommendations were ignored; ultimatums were given and then withdrawn.

When those initial meetings failed to produce a consensus, the debate spilled over into a series of frantic phone calls and hastily written draft statements over the July 4 holiday weekend. At the last minute, Surgeon General David Satcher stepped in to broker a compromise over the particularly sticky issue of the thimerosal-containing hepatitis B vaccine, which was given at birth. (The AAP wanted to postpone the shot, while the CDC was more concerned about the consequences of delaying it.) Ultimately, Satcher decided the AAP's stance was the one most likely to result in a unified front.

On July 7, only two weeks after many of the parties involved had first learned about the issue, the two organizations released coordinated statements. The CDC's read as follows (the added emphases are mine):

> The Food and Drug Administration (FDA) Modernization Act of 1997 called for the FDA to review and assess the risk of all mercury-containing food and drugs. . . . There is a significant safety margin incorporated into all the acceptable mercury exposure limits. Furthermore, *there are no data or evidence of any harm* caused by the level of exposure that some children may have encountered in following the existing immunization schedule. Infants and children who have received thimerosal-containing vaccines do not need to be tested for mercury exposure.
>
> The recognition that some children could be exposed to a cumulative level of mercury over the first 6 months of life that exceeds one of the federal guidelines on methyl mercury now requires *a weighing of two different types of risks* when vaccinating infants. On the one hand, there is the known serious risk of diseases and deaths caused by failure to immunize our infants against vaccine-preventable infectious diseases; on the other, *there is the unknown and probably much smaller risk, if any,* of neurodevelopmental effects posed by exposure to thimerosal. The large risks of not vaccinating children far outweigh *the unknown and probably much smaller risk, if any,* of cumulative exposure to thimerosal-containing vaccines over the first 6 months of life.
>
> Nevertheless, *because any potential risk is of concern,* the Public Health Service (PHS), the American Academy of Pediatrics (AAP), and vaccine manufacturers agree that thimerosal-containing vaccines should be removed as soon as possible. Similar conclusions were reached this year in a meeting attended by European regulatory agencies, European vaccine manufacturers, and FDA, which examined the use of thimerosal-containing vaccines produced or sold in European countries.

The AAP's statement echoed many of the same points, but it did so in language designed to be more colloquial. "Parents should not worry about the safety of vaccines," it began soothingly. "The current levels of thimerosal will not hurt children, but reducing those levels will make safe vaccines even safer. While our current immunization strategies are safe, we have an opportunity to increase the margin of safety."

In the following days, the most palpable response was confusion. Pediatricians were given no guidance on how to explain to worried parents that a small, hypothetical risk was overwhelmingly outweighed by the consequences of catching potentially deadly diseases like pertussis or diphtheria. While the CDC had advised hospitals to suspend hepatitis B vaccines for newborns except in cases where the mother was hepatitis B positive, many hospitals suspended the shot altogether, either because they were unsure of how to interpret the new recommendations or because they decided what was best on their own. Regardless of the reason, the consequences included an untold number of infections and at least one death, of a three-month-old who died of acute liver failure brought about by hepatitis B.

But it was the seeming ambiguity and not the alternately pedantic and convoluted wording of both organizations' statements that proved to have the most negative consequences in the years to come. When officials at the CDC wrote that there was "no data or evidence of any harm," they meant to offer the strongest reassurance they could, given that they could never say definitively that thimerosal had *no* side effects—there was always the possibility that some piece of new evidence might emerge in the future. But to a public not well versed in the language of science the line read like an equivocation. Even worse was the AAP's phrase "make safe vaccines even safer," which came off as institutional doublespeak of the worst kind.

Of course, it's unlikely that even the most mellifluously worded statements would have prompted the media to provide nuanced explanations or include necessary context. Barbara Loe Fisher, who was oftentimes identified in the press as a seemingly impartial "vaccine expert" and as the head of a parent advocacy group that fought for children's health, was relied on by many reporters in need of a quick reaction to the move. (One *Los Angeles Times* article helpfully listed her organization's contact information for those interested in learning more about the subject.) Fisher wasn't the only vaccine opponent who passed herself off as impartial: A *Denver Post* story quoted at length from a "forceful statement" distributed by the "Association of American Surgeons" that read, "Federal vaccine policy results in the violation of informed consent, and is based on incomplete studies of efficacy and potential adverse effects of the vaccines." The organization the *Post* was referring to—its actual name is the Association of American Physicians and Surgeons (AAPS)—is an extreme-right-wing group that openly derides "evidence based medicine" and has accused the AMA of emulating Mussolini's National Fascist Party.

The overall tenor of the media's coverage only fueled the conviction of those doctors and health officials who believed it had been a mistake to conduct scientific deliberations in public, especially before all the evidence was in. The most outspoken proponent of this view was Paul Offit, the Philadelphia-based pediatrician and vaccinologist. (Earlier that year, Offit had coauthored *Vaccines: What Every Parent Should Know*, the first of many books he'd write on the subject.) In support of his argument, Offit pointed to the quandary the CDC found itself in after "Vaccine Roulette" had linked the DPT vaccine with neurological damage. "At the time, there was no evidence disproving [the program's charges]," Offit says. "Companies couldn't say, 'Look, here are the epidemiological studies, look at the children who did or did not receive the vaccine,' in order to show that the incidence of seizures and mental retardation was the same in both groups." By the time enough evidence had been collected to show that the fears were baseless, there was no way to undo the damage that had been done.

Now, Offit said, a similar situation was unfolding: Reporters were skimming over the differences between ethylmercury and methylmercury and news segments were leaving out the risks of curtailing vaccinations in favor of a more sensational narrative. It would have been better, he argued, to have simultaneously developed thimerosal-free vaccines and conducted studies into ethylmercury's effects on children. If the preservative was shown to be harmless in the quantities that had been used, the news would barely register in the public consciousness. (There's not a lot of pizzazz in a headline that reads, "Chemical Removed from Vaccines Years Ago Turns Out to Have Been Safe After All.") If it was shown to have been potentially injurious, the government could truthfully say it had taken action as soon as it realized there was a potential problem.

Neal Halsey was among those who thought that Offit was being dangerously naive. "A congressman would have been the person who stood up there and said, 'Look, here's a problem and nobody's doing anything about it,' " Halsey says. "What would the impact be on our public image if the public health authorities and the AAP, if we had learned about this and said, 'Don't do anything'? What would the public's perception have been of those agencies and organizations responsible for recommending vaccines to children? We would have suffered an enormous credibility gap. . . . That's where Paul Offit doesn't get it. He simply doesn't understand."

In the months to come, the debate spilled into public view: That fall, Halsey published an editorial in the *Journal of the American Medical Association* defending the AAP and CDC. Offit responded with a letter charging that the groups' recommendations "might cause some children to miss vaccines that they need." Stanley Plotkin, a colleague of Offit's who'd helped develop the rubella vaccine in the 1960s, wrote a response to Halsey's piece that was even more to the point: "The editorial by Dr. Halsey presents a positive picture of the events surrounding the new recommendations for vaccines containing thimerosal; in my view, what happened was nothing less than a public health disaster."

• • •

Even with the benefit of hindsight, it's impossible to say whether Halsey's or Offit's position had more merit. On the one hand, the events of the past decade give credence to Offit's fears about the ways in which the news would be received. On the other hand, a Republican congressman from Indiana named Dan Burton began holding hearings on the country's immunization policies later that summer, and it seems likely he would have grabbed hold of the issue irrespective of what the AAP and the CDC had recommended. Ever since Burton was first elected to the House in 1982, he has shown a flair for controversial statements and actions that invariably resulted in minor media frenzies. In 1997, he staged a backyard demonstration in which he shot a pumpkin in an effort to prove that Clinton associate Vince Foster had been murdered by White House hit men—a demonstration that prompted Calvin Trillin to write, "He failed to realize that in humans other than himself what's inside the head bears no resemblance whatsoever to what's inside a pumpkin." * The following year, Burton's release of falsified prison transcripts of former associate attorney general Webster Hubbell prompted Republican leader Newt Gingrich to shout, "I'm embarrassed for you. I'm embarrassed for myself, and I'm embarrassed for the conference at the circus that went on at your committee." Burton also had a long and antagonistic history with medical authorities and an uneasy relationship to scientific facts: He was so terrified of catching AIDS that he brought his own scissors when visiting the House barber and refused to eat soup in public.

When Burton began committee hearings on August 3, he an-

---

* There have been conflicting reports on what Burton sacrificed as part of his reenactment. The pumpkin remains the most referenced victim—in addition to Trillin's witticism, it prompted a Democratic House of Representatives staffer to show up at a Burton-led hearing wearing a pumpkin suit and a button that read "Don't shoot"—but watermelons have also been cited. More recently, in June 2009, a *Washington Post* reporter wrote that a former Burton aide swore the cantaloupe was Burton's melon of choice. As the *Post* noted, that may have been the most logical choice, since the cantaloupe "is much closer to the size of a human head than either a watermelon or a pumpkin."

nounced that his interest in vaccines had been prompted by the experience of his granddaughter, who he said had been rushed to the hospital after receiving the hepatitis B vaccine, and his autistic grandson, who Burton claimed had been fine before he received multiple shots in one day. Up to that point, there'd been virtually no mention of autism in any of the stories about the thimerosal controversy; in fact, there hadn't even been any studies linking *methylmercury* to autism. "There has never been strong or reasonable scientific evidence to point to a link between thimerosal and autism," says Saad Omer, a vaccine specialist who teaches global disease epidemiology at the Johns Hopkins Bloomberg School of Public Health. "Even the people who advocated removing [thimerosal from vaccines], the Public Health Service, when they did that, they were not talking about autism." That was about to change. As Burton vowed that day, "We are going to be beating on this issue as long as I am chairman of this committee."

That Burton's approach to addressing concerns about vaccines and the vaccine industry was overwrought shouldn't obscure the fact that vaccine manufacturers had not behaved ethically in regards to thimerosal. In 1991, Maurice Hilleman, who was at the time working as an advisor to the pharmaceutical company Merck, warned in a confidential memo that there had not been sufficient safety studies done on thimerosal. Hilleman was a giant in the vaccine world: He developed more vaccines than any other person in history, including those for chickenpox, hepatitis A, hepatitis B, Hib, measles, meningitis, mumps, and pneumonia. In his memo, which was not made public until it was leaked by an unidentified whistle-blower in 2005, Hilleman wrote, "It appears essentially impossible, based on current information, to ascertain whether thimerosal in vaccines constitutes or does not constitute a significant addition to the normal daily input of mercury." He went on to say that while his own opinion was that thimerosal did not pose any risks, his calculations showed that the amount of ethylmercury children stood to receive from vaccines exceeded several established safety guidelines for methylmercury.

Hilleman also pointed out that he was not the first person to have reached this conclusion: The governments of Sweden, Denmark, and Norway had all recently banned the use of thimerosal in vaccines in an effort to protect against a loss of public confidence.

Merck never told health officials about Hilleman's memo, and somehow the Scandinavian governments' concerns were not noticed in America. As a result, the chance to forestall both Paul Offit's and Neal Halsey's worst fears from being realized was squandered. "If only we had done this in 1992," Halsey says, "we wouldn't have faced this problem in 1999."

CHAPTER 11

The Mercury Moms

Lyn Horne and Tommy Redwood met in Birmingham, Alabama, in 1986. She was a thirty-year-old nursing student raising two kids from her first marriage; he was a twenty-nine-year-old medical school student at the University of Mississippi. The couple got married the following year, and after Tommy finished his coursework they moved to a suburb outside Atlanta.

Six years after the Redwoods' wedding, Lyn learned she was pregnant with her third child. She's described that moment as the fulfillment of a dream: Finally, her husband would "have a child who would call him dad instead of Tommy." Will Redwood's birth was not an easy one—Lyn had to undergo an emergency cesarean section after the doctors realized Will was coming out feet first—but Lyn remembers that first year of Will's life as a joyous one.

Things began to change shortly after Will turned one. Suddenly, he was constantly sick, and each illness seemed as if it was worse than what had come before. By the time he was a year and a half old, Will had had strep throat, a rarity in children under the age of two; rotavirus, which causes severe diarrhea and in the mid-1990s resulted in the hospitalization of tens of thousands of American children each year; and an upper respiratory infection so serious it required hospitalization. Around the same time, he began acting increasingly withdrawn.

By Will's third birthday, relative strangers had begun hinting to Lyn that her son's difficulties might be indicative of a learning disorder—perhaps even autism. That, Lyn thought, was ridiculous.

Lyn did acknowledge that, at the very least, Will needed more guidance and attention than he was getting in a regular nursery school class, and in the fall of 1997 he began attending an intensive program for developmentally disabled children. That November, Lyn met with a group of specialists at Will's school, including classroom instructors, a speech and language pathologist, the director of special education services, and the school's principal. One of them asked Lyn to describe her goals for Will. "I want him to know his name," she said. "That is my first goal." Then she was asked if there was anything else. "By the time he starts kindergarten," she said, "I want him to be indistinguishable from his peers." As the journalist David Kirby wrote in a book that Redwood helped to assemble, "Her son did not have an official diagnosis, and she was going to cling to her faith that his dire condition might only be temporary."

It was not to be. By the time Will started kindergarten, Lyn had accepted that the child who was going to fulfill her and Tommy's life together was on the autism spectrum. She'd also decided that she was going to do more than dedicate herself to making sure Will received the best care possible. She was going to determine exactly what had happened to her son and figure out what she could do to fix it.

In 1979, four years after they'd met as freshmen at Harvard, Sallie McConnell and Tom Bernard got married and moved to downtown Manhattan. The early 1980s might have been a bleak economic time for the country as a whole, but the future looked bright for the young couple: Tom landed at the powerhouse Wall Street investment bank Salomon Brothers and Sallie found a job as an advertising executive. Before she'd turned thirty, Sallie had struck out on her own and started a market research company based in suburban New Jersey.

In 1987, the Bernards learned their life was about to change forever: Sallie was pregnant with triplets. In preparation, they moved to Summit, a New Jersey suburb whose name reflected the aspirations of its residents. That September, five weeks before her due date, Sallie went into labor. Fred arrived first: He weighed a healthy five pounds, twelve ounces. Jamie was next. He was a full pound less than his brother; still, the doctors predicted he'd need to stay in the hospital for only a couple of extra days.

Bill, however, was a different story altogether. At just barely three pounds, he fell between what's referred to as Very Low Birth Weight (less than three pounds, four ounces) and Extremely Low Birth Rate (less than two pounds, three ounces). Whatever the official designation, his size put him at risk for a wide range of immediate complications and lifelong health problems. For four weeks after Fred and Jamie were discharged, Bill remained under constant watch at Lenox Hill Hospital's Neonatal Intensive Care Unit on Manhattan's Upper East Side.

Almost as soon as Bill arrived home, his parents noticed he lagged behind his brothers in almost every developmental milestone. At first, the differences could be measured in weeks, but over time they became longer and longer. In early 1990, a couple of months after the triplets' second birthdays, Sallie and Tom scheduled Bill for a full battery of neurological tests. Those tests didn't lead to a definitive diagnosis; the closest they got was a doctor's scribbled notation that Bill should be watched for "autistic-like tendencies." It would take another two years before Bill was determined to have a pervasive developmental disorder, not otherwise specified, which, the Bernards learned, was a fancy medical term for someone on the high-functioning end of the autism spectrum. On the afternoon they got the news, Sallie announced to Tom that she viewed their son's diagnosis as a positive development: "Now we have something to work with. Now we can form a plan of attack."

In the coming months, numerous experts told Sallie that autism was genetic, that there was no way to treat it, and that there was virtually no chance of Bill ever recovering. At first, this litany merely frustrated her, but as time went on, it began to make her mad. She was sick of doctors talking down to her and dismissing her out of hand. Maybe, she thought, their arrogance was just a way of covering up their own ignorance. The more she thought about it, the more she realized that a lot of what she was being told didn't make any sense: If autism was a genetic disease, how was it that Fred and Jamie were perfectly fine? And how could genetics explain the fact that there seemed to be more and more children diagnosed with autism every day? When she realized that the number of students in the school Bill attended for developmentally disabled children had tripled in less than two years, Sallie decided it was time she started researching the situation on her own. What she found astonished her.

By the mid-1990s, more than three decades had passed since Bernard Rimland had embarked on his own quest after his son, Mark, began exhibiting signs of autism. The challenges facing Rimland in the early 1960s were significantly different from the ones confronting Bernard and Redwood. Much of that change was due to Rimland himself, whose work at the Autism Research Institute had helped to redefine the disorder and had broken a trail for subsequent generations of DIY parent-researchers. By the time Will Redwood and Bill Bernard were diagnosed as being on the autism spectrum, Rimland's notion that autism was primarily a neurological condition was almost universally accepted.

That did not mean that Rimland felt his work was finished: Despite his success at changing people's perceptions of the disease, he remained frustrated that it was not getting the attention of innovative researchers or doctors. Although a few studies had established a putatively genetic basis for the disorder, its ultimate pathology remained a mystery. When parents asked their doctors what they could do to help their children, much of the time they were answered with silence. Once again, Rimland decided it would be easier to create a whole new infrastructure than to try to find a way to work within the mainstream, and in 1995, he launched Defeat Autism Now!, an organization that would collaborate with those medical professionals who were "turning their backs on the medical establishment and using the DAN! approach."

This time around, Rimland was no longer a solitary voice in the woods. In 1994, Eric and Karen London, a psychiatrist and corporate attorney, respectively, living in Princeton, New Jersey, with their autistic son, had become so frustrated with the lack of resources available to them that they launched the National Alliance for Autism Research (NAAR). Meanwhile, about 120 miles up the coast from Rimland's San Diego headquarters, Hollywood producer Jonathan Shestack and his wife, television art director Portia Iversen, were starting their own nonprofit, which they called Cure Autism Now (CAN). Like DAN!, CAN and NAAR aimed to be more than traditional advocacy organizations: They wanted to shape the direction and scope of biomedical research on autism spectrum disorders by conceiving of, planning, and funding projects on their own.

Instead of reading through old textbooks and hunting for journal articles as Rimland had done, this second generation of autism advocates used the Internet to access cutting-edge research: Thanks to service providers like America Online and search engines like Alta Vista, information that had previously been available only to the select few was in wide circulation. When she started CAN, Iversen was so inept at using a computer that she had to pay someone to come to her house and help her download files. At the time, the only noteworthy repository for scientific research papers available to laypeople was a collection of medical literature maintained by the National Institutes of Health's (NIH) National Library of Medicine.* Then,

* The database's official name—the Internet's Grateful Med—belied the degree of expertise needed to navigate its contents. Its instruction page told users to start by entering a search term, after which they were directed to "click on the 'Find MeSH/Meta Terms' button located at the top of the page. IGM then displays the Metathesaurus Browser Screen. The IGM algorithm displays the MeSH terms that are

seemingly overnight, Iversen said, "all these incredible medical databases suddenly became free." Without any prior experience or knowledge of biology or genetics, the onetime art director had access to enough information to allow her to analyze unstable regions of the human chromosome and compare the human genome to that of fruit flies.

There are some arenas in which the democratic assessment of information makes sense—it's a good bet that if a new Stephen King book has fifty thousand five-star ratings on Amazon, fans of *Cujo* and *Carrie* aren't going to be disappointed with it—but epidemiology, which uses enormous amounts of data to analyze health and disease on a population-wide level, is not one of them. Determining cause and effect can be difficult even when conducting laboratory research, and epidemiologists don't have the luxury of setting up controlled experiments with a minimum of variables: They need to factor in the uncontrollable actions and unpredictable behaviors of anywhere from dozens to millions of individuals. It's no wonder that the ambient statistical noise can prove deafening even to professionals who have spent decades processing information on such a large scale. That's exponentially truer for lay practitioners untrained in statistical analysis.

One well-known example of the pitfalls of amateur epidemiology occurred on Long Island in the early 1990s. In March 1992, fifty-year-old Lorraine Pace was diagnosed with breast cancer. This came as a shock—not, Pace said, because a cancer diagnosis is always upsetting, but because she thought her healthy lifestyle should have protected her from the disease. Even more alarming to Pace was her sudden awareness of the large number of cancer patients in the area: Eventually, she counted a total of twenty people in her neighborhood in West Islip who'd also been diagnosed with cancer in the

---

the nearest to the search terms (the Ovid 'mapping' function provides quite similar results). Click on the best MeSH term." And so on.

past several years alone. This might have been an informal sampling, but it left Pace convinced that an unidentified toxin was stalking her community.

Frustrated with what she saw as a lack of official concern, Pace took matters into her own hands and founded the West Islip Breast Cancer Coalition. Day after day for months, the organization's members would meet in Pace's living room to add new data points to a giant, color-coded map: yellow dots for homes with malignant breast tumors, pink for benign tumors, and blue for no tumors at all. After analyzing its data, the group announced that cancer rates in the area were 20 percent higher than the state average.

The media and local politicians alike jumped on the story, much to the dismay of scientists, who knew that epidemiological studies that start with desired outcomes in mind are almost by definition worthless.* In this case, there were a number of obviously mitigating factors that hadn't been considered by Pace's group: West Islip has a high proportion of wealthy, white women with better access to health care than most of the country, which likely meant that slow-growing tumors that would have gone unnoticed in other communities were being identified; West Islip women tended to defer childbearing to later in life, which was known to increase the risk of breast cancer; and the citizens of West Islip tended to live longer than average, and cancer rates climb dramatically with age.

Ultimately, Pace's crusade spawned a controversy that raged for nearly a decade at a cost to taxpayers of more than $30 million. When, in 2002, an exhaustive federal study found that breast cancer rates in Long Island were in fact barely distinguishable from those in the rest of the country, the news received a fraction of the attention the initial scare had caused. Even if that hadn't been the case—even if there *had* been a higher-than-average rate of breast cancer in the

* The inherent problem here is the tendency, conscious or otherwise, to define your population in such a way as to achieve a predetermined result—as occurs when Democrats and Republicans carve out ever-more serpentine congressional districts in order to protect their incumbents.

area—the most likely explanation would have been that Long Island found itself in a random eddy of disease rather than the victim of a hidden carcinogen.*

If large-scale pattern recognition is hard to practice in your neighborhood, it's nearly impossible to conduct over the Internet. Even when you know that an online community selects for a certain type of person—say, politically minded liberals or ardent conspiracy theorists—sustained encounters with a small group of like-minded people almost inevitably lead to the conclusion that everyone thinks the way you do. (This phenomenon is addressed in more depth in Chapter 16.) In the summer of 1999, the CDC's and AAP's joint statements on thimerosal served as a catalyst for parents like Lyn Redwood who up until that point had not considered that vaccines might have played a role in their children's conditions. "I started reading about mercury online, and I was awestruck by the symptoms being similar to my son's," Redwood says. "They said no testing was necessary and there was no evidence of harm. How could they know that if they hadn't looked? I felt this huge sense of urgency to share my concerns, and the records I had on my son and a dozen other children whose stories were exactly the same." Armed with only the fervency of her beliefs, Redwood started an online mailing list devoted to mercury and autism. "By word of mouth, people heard about it, so I had ten, and then one hundred, and then a thousand, and then four thou-

* In order to understand why this is true, it's important to keep in mind the role randomness plays in all areas of life. For example: Both of the boys who lived on my block and were in my school year when I was growing up were named Seth. This does not mean that one hundred percent of the male population is named Seth, or that one hundred percent of the population in my grade in school were named Seth, or even that boys who lived in my town and were in my grade were at an elevated risk for being named Seth. This demonstrates one of the major hurdles in proving causality: In order to demonstrate that the apparent "Seth cluster" on my block was not just the result of random noise, you'd need to explain *how* and *why* it occurred and not simply demonstrate that it *had* occurred.

sand people as part of this list, comparing notes about their children's development [and] mercury. It in essence created a community and a movement." Along with Sallie Bernard, one of the most stalwart of her geographically diverse allies was Liz Birt, a Chicago attorney with an autistic son named Matthew. Together, they formed the core of the Mercury Moms, an informal coalition of parents that would influence health policy and vaccination practices around the globe for years to come.

One of the group's first coups was its courting of a Florida woman named Danielle Sarkine, who was the mother of an autistic son named Christian—and, more importantly, the daughter of Dan Burton. It didn't take long for Sarkine to agree to pressure her father to hold congressional hearings focusing specifically on thimerosal. She was confident she'd be successful: Burton already believed it was more likely than not that his grandson's autism was the result of an "injury" he'd suffered after being injected with multiple vaccines at once.

By the time Burton's House Government Reform Committee began its hearings on thimerosal in the spring of 2000, he had become as committed a partisan as anyone. In his opening remarks, Burton described Christian as a "beautiful and tall" baby who "was outgoing and talkative [and] enjoyed company and going places." His parents, Burton said, were convinced he was going to star in the NBA—at least, that is, until he got vaccinated. Almost immediately, Burton said, Christian began "banging his head against the wall, screaming and hollering and waving his hands. . . . [H]e had those shots, and our lives changed and his life changed." Burton also praised the spirit of parents who looked outside mainstream medicine for answers to their children's conditions.

While no specific legislation came out of Burton's hearings, they served as yet another potent illustration of the connective power of the online world. In her prepared remarks, a Louisiana housewife named Shelley Reynolds told of how she'd used the Internet to make contact with "thousands and thousands of parents from across the country," all of whom had identical stories to her own:

> Child is normal, child gets a vaccine, child disappears within days or weeks into the abyss of autism. If you doubt me, I invite you to attend the Hear Their Silence rally on the [National] Mall, where our Open Your Eyes project will be displayed. View the thousands of pictures we have collected and realize that forty-seven percent of those who participated believe that vaccines contributed to the development of their child's autism. Parents like me are relying on you. . . . I know my children. I know what happened to my son. As far as I am concerned, the needle that silently slipped into my baby's leg that day became the shot heard round the world.

* * *

It was 1993 when the son of Albert Enayati, an Iranian Jew whose family had immigrated to the United States in the 1970s, was diagnosed with autism. At first, Albert was so despondent he spent his nights weeping. By the end of the decade, he'd become a dedicated activist who led the New Jersey chapter of CAN. It was through that network that he first met Sallie Bernard, and the two bonded over their anger about the presence of thimerosal in vaccines. In 1999, shortly after he was laid off from the multinational drug manufacturer Pfizer, Enayati dedicated himself to collecting research that confirmed his suspicions. In 2000, just as Burton began his hearings, Bernard, Birt, Enayati, and Redwood founded the Coalition for Sensible Action for Ending Mercury-Induced Neurological Disorders, or SafeMinds. One of the group's first goals was to publicize the results of the research Enayati and the others had been conducting. Bernard, the Harvard-educated marketing consultant, was adamant that the best way for the group to disseminate its message was to get its work published in a medical journal. In order to be taken seriously, she explained, they had to speak "their language—scientists aren't going to read anything unless it's written in scientific jargon." That summer, the leaders of SafeMinds completed "Autism: A Novel Form of Mercury Poisoning." Its summary read:

> A review of medical literature and US government data suggests that: (i) many cases of idiopathic autism are induced by early mercury exposure from thimerosal; (ii) this type of autism represents an unrecognized mercurial syndrome; and (iii) genetic and non-genetic factors establish a predisposition whereby thimerosal's adverse effects occur only in some children.

While the result successfully aped the abstruse style unique to academia—one passage read, "These findings may be linked with HgP because: (a) Hg preferentially binds to sulfhydryl molecules (-SH) such as cysteine and GSH, thereby impairing various cellular functions; and (b) mercury can irreversibly block the sulfate transporter NaSi cotransporter NaSi-1, present in kidneys and intestines, thus reducing sulfate absorption"—the team of parents was unable to convince a respected academic journal that its work deserved an audience. That did not mean the paper would not appear in print: In December 2000, it was accepted by *Medical Hypotheses. The New England Journal of Medicine* this was not: *Medical Hypotheses* proudly eschewed peer review, a process it said disapprovingly "can oblige authors to distort their true views to satisfy referees." In the "Aims and Scopes" section of its guidelines to writers, the journal emphasized that it had no desire to "predict whether ideas and facts are 'true' "—in fact, it was eager to print "even probably untrue papers" so long as they spurred discussion.

"Autism: A Novel Form of Mercury Poisoning" was published in the spring of 2001. The crux of the paper's argument could be found in a table comparing ninety-four traits that it claimed were characteristic of both autism and mercury poisoning. The dozens of citations—"ASD [autism spectrum disorder] references in bold; HgP [mercury] references in italics"—may have looked impressive, but the list was little more than a meaningless grab bag of attributes. A number of the "symptoms" the authors had ascribed to autism—temper tantrums, unprovoked crying, grimacing, sleep difficulties, rashes, itching, diarrhea, vomiting—were, as any parent could attest, also "symptoms" of infancy and early childhood. In places, the traits listed to illustrate similar symptomologies were actually contradictory: The "physical disturbances" heading included both diarrhea and constipation, "unusual behaviors" listed both agitation and staring spells, and "psychiatric disturbances" included both mood swings and flat affect. When it came to specifics, some symptoms were compared seemingly for no reason other than that they both affected the same set of behaviors: repetitive behavior and impaired speech for autism, loss of muscle control and slurred speech for mercury poisoning. Other times the reactions the paper listed were directly opposite to each other: increased sensitivity to sound was cited as a common symptom of autism, mild to profound hearing loss for mercury poisoning.

The rest of the paper was marked by the conspicuous omission of anything that might contradict the authors' thesis in any way. Rather than address the known differences between ethylmercury and methylmercury—differences that were one of the main points used by those who argued that thimerosal was safe—the paper acted as if no such difference existed. (The only time either "ethyl" or "methyl" appeared was in the final sentence of the second paragraph: "The mercury in vaccines derives from thimerosal (TMS), a preservative which is 49.6% ethyl-mercury (eHg).") There was no acknowledgment of the total absence of reports of autism in every previous study of either ethyl- or methylmercury poisoning, nor was there an acknowledgment that the one published study that had looked for elevated mercury levels in children with autism had not found any such evidence. Finally, and most significantly, there was the paper's failure to provide even a single example of "detectable levels" of mercury in an autistic child. The paper's authors tried to explain this omission by referring back to the unproven conjecture that had launched the project in the first place: "parental reports of autistic children with elevated Hg."

All in all, the *Medical Hypotheses* paper, like almost all efforts undertaken for the sole purpose of proving what someone already believes to be true, was a tutorial in bad science. Instead of amassing evidence to prove a conjecture, it used the conjecture itself as evidence. The absence of supporting data was cited as further proof of the authors' detractors' failings: If tradition-bound doctors and researchers had given proper credence to personal anecdotes and untested observations, they would already have identified the subset of children that was uniquely susceptible to this "novel" form of poisoning. In the final evaluation, the paper's title may have been more accurate than its authors intended: Autism was a "novel" form of mercury poisoning only in that it was entirely fictional.

CHAPTER 12

The Simpsonwood Conference and the Speed of Light: A Brief History of Science

The Salk vaccine trials in the 1950s and the subsequent eradication of polio in the United States proved that with enough money, resources, and willpower, government agencies and private businesses could join forces to produce massive change. In the decades since, no comparably proactive effort has been made to safeguard public health. Instead, it has taken a controversy or a catastrophe to precipitate most advances, from the monitoring of drug safety to the implementation of reporting and compensation systems for vaccine-related injuries.

For example, it wasn't until 1955, in the wake of the Cutter Incident, that the National Institutes of Health established the Division of Biologics Standards, and it wasn't until 1962, a year after the sedative thalidomide was shown to cause birth defects when used by pregnant women, that Congress passed a law that required drug manufacturers to prove their products were safe before bringing them to market.

In 1974, when the parents of Anita Reyes were awarded $200,000 in their lawsuit against Wyeth Pharmaceuticals, sixteen years had passed since the verdict in *Gottsdanker v. Cutter* had shown the need to clarify culpability questions relating to vaccines—and still there was no nationwide policy in place governing drug manufacturers' ultimate liability.

The lack of comprehensive regulations meant that all the parties involved, from patients and pediatricians to drug manufacturers and the federal government, were unsure of their rights and obligations under existing immunization mandates. The American Academy of Pediatrics was among the first groups to recognize this as a disaster in the making, and in the 1970s its members began to lobby for a uniform vaccine compensation system. Without such a mechanism, they warned, there was an ever-present risk that manufacturers' liability fears could result in a drop in supplies and a spike in prices. During the swine flu scare in 1976, Congress had enacted a stopgap measure only after it became clear there was no other way to guarantee the production of enough vaccine to protect the country. Since then, lawmakers had continually professed concern about the need for a solution to the liability issue, and just as continually had found reasons to delay taking action.

Throughout it all, vaccine-related tort claims continued to rise, from twenty-four in 1980 to 150 in 1985, a five-year span during which total damages awarded to plaintiffs topped $3.5 billion. By then, the AAP's warnings had proven to be prescient, as the number of companies willing to distribute vaccines in the United States fell from several dozen in the mid-1960s to just four. The panic that followed the 1982 broadcast of "Vaccine Roulette" prompted new promises of reform, but every time a deal appeared to be imminent, discussions between the various parties broke down. It wasn't until 1986, when the departure of one of the two remaining suppliers of the DPT vaccine prompted the CDC to warn that the country was on the verge of running out of an adequate stockpile, that a compromise was finally reached. Just hours before adjourning for the year, Congress passed the National Childhood Vaccine Injury Act (NCVIA), which aimed to create a simple, transparent process for dealing with injury claims.

The central feature of the NCVIA was a uniform compensation program that would be funded by a surtax on each dose of vaccine. The program would be implemented by a group of federal judges designated as Special Masters, who would preside over a "Vaccine Court" that existed outside the traditional legal system.* The law also established the Vaccine Adverse Event Reporting System (VAERS), which would funnel reports of the thousands of vaccine injuries claimed to have occurred each year to a single, centralized database managed by the CDC and the FDA.

VAERS was, without question, a major step forward in monitoring the efficacy of vaccines already on the market. Still, VAERS's voluntary, post-facto reporting system meant that it wasn't until *after* concerns about vaccine safety had been raised that investigations got under way. By always playing catch-up, the federal government left an opening for reporters like Lea Thompson, self-styled renegades like Andrew Wakefield, and impassioned activists like the Mercury Moms to raise some legitimately troubling questions—and when they did, they felt it was their right to decide which answers were satisfactory. In each one of those situations, the underlying concerns were valid ones: Could the DPT vaccine cause permanent harm? Could the MMR vaccine cause gastrointestinal problems or developmental disorders? Could thimerosal cause neurological problems? Without satisfactory answers, it was impossible to stop the spread of the most outlandish strains of vaccine skepticism from the fringes to the mainstream.

---

* Typically, the term *Special Master* refers to someone who has been granted the authority to carry out a course of action designated by a court. In those instances, Special Masters do not need to be judges. The National Childhood Vaccine Injury Act of 1986 created a distinct category of Special Masters, who are appointed by the United States Court of Federal Claims to rule on vaccine-related claims.

• • •

If there was one thing citizens, scientists, and safety monitors agreed on in the wake of the 1999 announcements about thimerosal, it was the need to analyze the data that was already available. An obvious first step was a comprehensive review by the Institute of Medicine (IOM), an independent organization devoted to "providing objective and straightforward answers to difficult questions" related to health and science policy. (The scientists who serve on IOM committees all do so on a volunteer basis.) From the outset, the IOM knew it needed to address the concerns of those most suspicious of vaccines if it wanted its work to have any validity. As a result, this report, unlike the four vaccine safety reviews the institute had conducted between 1988 and 1994, was mandated to address "the significance of the issue in a broader societal context" in addition to answering questions about scientific plausibility. As the process evolved, the committee's efforts led to invitations to Sallie Bernard, Barbara Loe Fisher, and Lyn Redwood to participate in planning meetings and conference calls that would typically have been off-limits to members of the public.

An awareness of the snowballing anxiety surrounding vaccines had the exact opposite effect on a group of scientists who were asked by the CDC to review a massive collection of HMO patient records called the Vaccine Safety Datalink (VSD). The aim of that effort was to get vaccine experts to reach a consensus on what, if any, further research should be done to determine the possible side effects of the vaccines that had been administered over the previous decade. From the outset, the group's members sought to conduct their work outside the public eye, mindful of previous instances in which complex science had been poorly translated for a general audience: Inevitably, it seemed, data was misinterpreted, preliminary results were assumed to be definitive, tiny sample sizes were blown out of proportion, and the mere act of conducting an experiment was taken as an admission that a given hypothesis was correct.

The first round of studies of the VSD data was completed in early

2000, and that June, a total of fifty-one specialists assembled for a closed-door, two-day meeting. The gathering took place at Simpsonwood, a United Methodist Church conference center on 227 acres of woodlands located about an hour's drive north of Atlanta. Included among the dozens of participants were pediatricians, infectious disease specialists, and immunotoxicologists. Robert Chen and Frank DeStefano, the two CDC scientists who'd critiqued Andrew Wakefield's 1998 paper for *The Lancet*, were there, as was John Clements, the top vaccine specialist at the WHO. Scientists from Aventis Pasteur, Merck, Wyeth, and SmithKline Beecham, the four pharmaceutical companies that continued to distribute vaccines in the United States, were also in attendance.

On the morning of Wednesday, June 7, Walter Orenstein, the head of the CDC's immunization program, called the meeting to order. "We who work with vaccines take vaccine safety very seriously," he said. "Vaccines are generally given to healthy children and I think the public has, deservedly so, very high expectations for vaccine safety as well as the effectiveness of vaccination programs." The seriousness of the concerns about a "possible dose-response effect" linking thimerosal in vaccines with "certain neurologic diagnoses" necessitated "a careful scientific review of the data." The rest of the day, Orenstein said, would be given over to presentations relating to the VSD data, while the following day's session would be devoted to free-ranging debate of the various recommendations of how to proceed.

The pivotal presentation of the conference was given by Thomas Verstraeten, a CDC epidemiologist who'd led a team that examined the cumulative level of ethylmercury exposure in 110,000 children when they were one month, two months, three months, six months, and one year old. Verstraeten opened his talk with a quip about how the invariably ambiguous results produced by studies that merged toxicology with environmental health made it "very hard to prove anything." Because of that, he said, his study was the one "that nobody thought we should do."

Even a cursory look at the potentially confounding factors that

Verstraeten had to take into consideration showed the depth of the challenge he'd faced. Since there was no definitive data about ethylmercury's half-life (the time it takes for half the amount of a given substance to be eliminated from the bloodstream), Verstraeten used previously established half-life figures for methylmercury in order to estimate the aggregate amount of thimerosal in infants' bodies at any given point.* Because of the privacy concerns inherent with using HMO patient records, he'd been unable to question (or even identify) the people whose data he was using—which meant he had no way of accounting for possible selection bias on the part of parents (it was possible that "the same parents that bring their children for vaccination would be the same parents that bring their children for assessment of potential developmental disorders"), or doctors (there was "a potential that certain health care providers use more hepatitis B [vaccine] at birth and would also be more likely to diagnose some of the outcomes"), or both.

There was also the indiscriminate list of potential outcomes Verstraeten had to consider: In an effort to provide as definitive an analysis as possible, he looked at all of the "neurologic and renal disorders" he could "not exclude" as being linked to mercury, regardless of whether it was from direct poisoning, secondary exposure, or external contact. He also considered conditions like autism that had *not* been linked to mercury poisoning but that had captured the public's imagination. The final scope of his study encompassed everything from barely noticeable irritations to potentially deadly afflictions.

Not surprisingly, Verstraeten's results were all over the map. Rates of kidney disorders, night terrors, and cerebral palsy declined as the total amount of thimerosal went up, rates for developmental disorders in premature babies stayed the same, and rates for speech delays, autism spectrum disorders, and attention deficit disorders went up. In every case, however, the figures were within a statistical margin

* As it turns out, ethylmercury's half-life ranges from ten to twenty days in children and is as short as seven days in infants, while methylmercury's half-life is around seventy days.

of error. Still, Verstraeten said, at the very least he thought the hypothesis that thimerosal was contributing to a range of childhood disorders was "plausible."

That alone was the most persuasive argument for further research to be done, and by the end of the conference's first day, there was a general agreement that any concerns about fear mongering were superseded by the need for a comprehensive round of new experiments. Robert Brent, a developmental biologist and practicing pediatrician at the Nemours/Alfred I. duPont Hospital for Children in Delaware, and one of the conference's most outspoken attendees, encapsulated this view:

> Even if we put the vaccine in single vials and put no preservatives tomorrow, we still want the answer to this question, because remember, epidemiological studies sometimes give us answers to problems that we didn't even know in the first place. Maybe from all this research we will come up with an answer for what causes learning disabilities, attention deficit disorders and other information. So I am very enthusiastic about pursuing the data and the research for solving this problem.

Brent's opinion in favor of moving forward carried special weight: As someone who'd testified as an expert witness in three vaccine-related lawsuits, he had personal experience with the dangers of "junk science." "It is amazing who you can find to come and testify that such and such is due to a measles vaccine," he said. "They are horrendous. But the fact is those scientists are out here in the United States."

It didn't take long for the Simpsonwood participants' premonitions about the misinterpretation of the VSD data to prove to be prescient. When, just two weeks after the conference, Verstraeten gave a boiled-down version of his report to an open meeting of the CDC's Advisory Committee on Immunization Practices, Lyn Redwood was among those in attendance. Despite Verstraeten's warnings that "it is difficult to interpret the crude results," Redwood drove away convinced the issue had been settled once and for all. "My God," she said to herself, "this really did happen. Our worst fears were true. I guess we're not so crazy after all."

Of all the issues that arise after a vaccine is put into widespread use, one of the trickiest is figuring out how to investigate newly raised safety concerns. In order to test whether a vaccine carries a specific risk, researchers need a representative control group to forgo the vaccine in question—an ethically dicey proposition when you're talking about a drug with potentially lifesaving benefits. The difficulty of putting together a useful post-licensure study increases with the rarity of the condition to be investigated. Take, for example, a hypothetical mosquito bite vaccine suspected of causing blindness in 50 percent of recipients. In that scenario, an analysis of a few dozen subjects might be sufficient to indicate whether there was actual cause for alarm.

Now consider the situation as it relates to autism. With both MMR and thimerosal, the hypothesis was that some component or combination of vaccines served as a trigger in a small subset of children who were genetically predisposed to a specific variant of a disorder with inconsistently applied diagnostic criteria and a broad range of date of onset. In order to reach any reliable conclusions, you'd need to study hundreds of thousands of children over a period of five or more years. Typically, such a challenge would be tackled with retrospective studies, but in this case, even that was tricky: In the United States, mandatory vaccination laws limited the pool of unvaccinated controls, inconsistent record keeping compromised whatever data was available, and VAERS's passive design meant that some injuries would be underreported, while others—especially those that received a lot of attention in the media—would likely be overrepresented. (An extreme example of underreporting is post-vaccine measles rashes, which some researchers estimate go undocumented 99 percent of the time.) Even if every single "adverse event" were registered, there was no way to get a reliable count of the number of vaccines that had actually been administered, which made it impossible to calculate with a high degree of accuracy the frequency with which the theorized injuries would have occurred.

There was, however, one area of the world without those limitations: Scandinavia, where governments' dedication to social welfare was matched only by their fervor for record keeping. Denmark was an especially attractive option for researchers. Its Civil Registration System kept detailed records of every child born in the country; immunizations were administered through its National Board of Health; a government-run agency called the Statens Serum Institut (SSI) produced its vaccine supplies; and it did not have compulsory vaccination laws. There was even a ready-made way of comparing children who'd received thimerosal-containing vaccines with those who hadn't: Until March 1992, the three-part, whole cell pertussis vaccine produced by the SSI contained a total of 250 μg of thimerosal; from April 1992 on, the pertussis vaccine distributed in the country was thimerosal-free.

By late 2000, epidemiologists from around the world had begun sifting through millions of pages of medical records produced by several decades' worth of Danish children. Even before this work had been completed, activists were looking for ways to spin the results in their favor, regardless of the studies' outcomes. One tactic was to promote themselves as insurrectionist truth tellers doing battle with an establishment more concerned with defending its turf than with improving people's lives. The success of such a gambit depended on the public's misunderstanding of the approach scientists rely on to understand the world.

The steps of the scientific method are the same whether you're a sixth grader prepping for a science fair or a physicist proposing a new framework for the universe: Observations lead to hypotheses, hypotheses are tested by experimentation, results are analyzed, conclusions are submitted for publication, and the whole process undergoes peer review. (To be fair, for a typical twelve-year-old, "publication and peer review" usually means "write it up on a piece of poster board and appeal to your teacher for a good grade.") It's a formula so central to the very definition of science that it's easy to assume it has been accepted as gospel throughout history.

That, however, is not the case. The scientific method is actually a relatively new construct, the product of several millennia worth of arguments about the merits of purely hypothetical analysis versus the observation of the world outside ourselves. Aristotle, who believed the only way to truly understand the universe was through a set of abstract "first principles," was a proponent of the first camp; the tenth-century Persian polymath Ibn-al Haytham, whose evidence-based experiments disproved Aristotle's speculative theory of light and vision, showed the advantages of the second. During the Renaissance, Descartes' philosophy of rationalism was based on the primacy of deductive reasoning, while Galileo and Francis Bacon came down on the side of experience-based investigation. By the twentieth century, the empiricists had pretty much carried the day. Their victory was due, in large part, to the fact that unlike every metaphysical creed the world has ever known, scientific proofs rely on evidence and not on faith.

One way to understand the distinction between science and the ideologies it superseded is through the theory of falsifiability, which states that in order for a hypothesis to be a legitimate subject of inquiry, it has to have a single, corresponding null hypothesis—that is, it needs to be disprovable. (That's why "God exists" is not a legitimate scientific hypothesis: The null hypothesis—"God does not exist"—can't be proven.) There's no way to overstate how fundamental this concept is to the scientific process. Since it's impossible to prove a negative, the closest one can come to absolute proof for *any* theory is through an exhaustive, and unsuccessful, effort to prove the null hypothesis.

A good illustration of how this works is found in the Austrian philosopher Karl Popper's anecdote about seventeenth-century Europeans, who assumed that all swans were white. Since there's no way for anyone to know for certain that he's identified every swan in the universe, and since there's no way to know what the swans of the future will look like, the best anyone attempting to prove the white swan hypothesis could do was fail to prove the null hypothesis: that *not* all swans are white. In 1697, a Dutch sea captain chanced upon a black swan in Australia, at which point the null hypothesis was proven true and the white swan theory had to be discarded. This story also illustrates the necessity of framing a null hypothesis in the broadest way possible: "At least one swan is black" would *not* have been a valid null hypothesis, since it would have left room for the discovery of a red swan or a green swan or any other color of swan that would have invalidated the original hypothesis without satisfying the null hypothesis.

One of the most famous examples of the null hypothesis at work involves two men often referred to as the greatest scientists the world has ever known: Albert Einstein and Isaac Newton. While still in his mid-twenties, Einstein became obsessed with an apparent contradiction between two widely accepted theories explaining the workings of the physical universe: Newton's laws of motion and a series of equations formulated by a nineteenth-century Scottish physicist named James Clerk Maxwell. On the one hand, Newton claimed that the velocity of a body in motion remains constant in the absence of any other forces.* A corollary to the theory that the force exerted on a body is equal to its mass times its acceleration, or $F=ma$, is that the speed of a body in motion increases in direct proportion to the amount of force exerted upon it—which, in turn, means there should not be a limit to how fast a given object can travel. On the other hand, Maxwell's equations showed that electric and magnetic waves

---

* This might seem obviously incorrect: If you throw a ball, it won't continue moving at the same rate of speed forever. That is due to various forces working on objects within the earth's atmosphere, including gravity (which pulls objects downward) and friction (which slows objects in motion). The precise workings of this defy easy explanation; suffice it to say that one omnipresent source of friction is air resistance. This also helps explain why objects outside the earth's atmosphere can travel forever without losing speed—at least in theory.

traveled at a constant speed—the speed of light. In 1905, Einstein, who at the time was working as a patent clerk, began to focus on the incompatibility of these two theories: If everything in motion has a measurable mass, and if the speed of light is constant, then no amount of force can make light waves travel any faster (or slower) than they already do. In order to reconcile this contradiction, Einstein proposed that energy (E) and mass (m) were analogous concepts, and they related to each other through the speed of light in a vacuum (c), squared—a hypothesis that, if true, would mean that force does *not* equal mass times acceleration.*

This hypothesis raised some perplexing questions. If Einstein was correct, why had the laws of motion appeared to be accurate for the past 220-odd years? For that matter, were science teachers around the world lying to their students when they taught them that F=ma? The answer, as you've probably guessed, is yes—and no. Objects do stay at rest unless some external force is applied, and the more force you apply, the faster that object will travel—as long as you're not talking about really, really precise calculations relating to things that are really, really small or traveling really, really fast. Newton's laws applied to everything he could measure in *his* known universe, and they apply to everything that those of us not playing around with photons or particle accelerators can measure in ours. Instead of thinking of F=ma as being wrong, think of $E=mc^2$ as being more right.†

The story of Einstein and Newton is another example of why scientists are so inflexible in their insistence that there are no absolute certainties in the world, and why concepts we accept as true—like

---

* Proof of this theory, which Einstein developed between 1905 and 1907, didn't come until 1919, when a British astrophysicist's photograph of a solar eclipse documented the sun's gravitational pull "bending" starlight; this led to a banner headline in *The Times* (London) that proclaimed "Revolution in Science—New Theory of the Universe—Newtonian Ideas Overthrown."

† Don't worry if you're having a hard time following this oversimplified explanation of physics' most challenging problem. For most of us, understanding special relativity is a little like true love: We should consider ourselves lucky if we can grasp hold of it for even one fleeting moment.

quantum mechanics and evolution and the big bang—are referred to as theories: There's always the chance that someone, somewhere will discover a scenario in which they no longer apply.* This is actually a good thing. Scientific progress, as Carl Sagan wrote in his book *The Demon-Haunted World*, thrives on errors.

> False conclusions are drawn all the time, but they are drawn tentatively. Hypotheses are framed so they are capable of being disproved. A succession of alternative hypotheses is confronted by experiment and observation. Science gropes and staggers toward improved understanding. Proprietary feelings are of course offended when a scientific hypothesis is disproved, but such disproofs are recognized as central to the scientific enterprise.

The result of all this groping and staggering is that scientists with widely accepted theories spend their careers in a state of cautiously optimistic limbo: Regardless of how many times their work is corroborated, a single contrary result will cause it all to come crashing down. The centrality of "disproofs" also highlights the crucial importance of those final two steps in the scientific process: publication and peer review. Until the authors of a given theory have provided a detailed explanation of exactly how they got their results, they're essentially telling the rest of the world to accept their conclusions on faith—which puts them back on the side of the ideologues who define "truth" as whatever they happen to believe at the moment.

This emphasis on disproving what your colleagues had previously

---

* This also illustrates the uni-directional nature of scientific discoveries: If someone *does* find evidence that special relativity isn't always applicable—that Einstein's theory about how the universe works isn't true all of the time—we wouldn't revert to accepting Newton's theories as being correct just because the hypothesis that first superseded the laws of motion was shown to have its own shortcomings. An illogical bastardization of this misapprehension is often used by creationists as "proof" that evolution is incorrect: Since Darwin was wrong about some things, they argue, he must be wrong about everything, and if he's wrong about everything, then a biblical understanding of the world must be correct.

believed to be accurate can make listening in on scientific debates feel a little like eavesdropping on a newly divorced couple arguing over child visitation rights. The realities of the scientific method also present an uncomfortable challenge for anyone tasked with explaining to the public why this inherent open-endedness doesn't negate the high degree of certainty that accompanies widely accepted conclusions. The combination of ambiguity and authority implicit in science is hard enough to understand if you are sitting across the table from a scientist; it is an exponentially more challenging point to convey when filtered through media outlets that eschew nuance and depth in favor of attention-grabbing declarations.

CHAPTER 13

THE MEDIA AND ITS MESSAGES

On Sunday, February 3, 2002, the BBC-TV news magazine show *Panorama* aired a special report titled "MMR: Every Parent's Choice." The program's news peg was an as yet unreleased study in *Molecular Pathology* that included as one of its authors a "controversial doctor" who was, once again, loudly proclaiming that "MMR should not be used until researchers rule out the possibility the triple jab could cause autism": Andrew Wakefield. In the four years since Wakefield's *Lancet* paper had been published, only one research team had claimed to have identified the measles virus in the stomachs of children after they'd been vaccinated—and Wakefield had worked on that paper as well.

This new study was Wakefield's latest effort to prove to the world that measles—and by extension the measles component of the MMR vaccine—was responsible for a "new variant [of] inflammatory bowel disease." As part of its work, the research team used a highly sensitive technique that relied on something called a polymerase chain reaction (PCR). Because working with PCRs is so complicated, and because Wakefield and his collaborators were claiming to have detected tiny quantities of something other labs had been unable to find, it was especially important for the researchers to provide a precise explanation of how they had obtained their results. Instead, the paper left out information so basic it would have been expected to be included in a lab report written by an undergraduate. There were no details about how tissue samples were stored, nothing about the amount of time the samples remained in a freezer before being tested, no description of the quality of the material being tested, and no explanation of how results were interpreted. It appeared as if the research team hadn't employed positive controls of tissue samples known to include the measles virus or negative controls of samples without the virus, which meant there was no way for them, or anyone else, to establish a baseline for the accuracy of their results. Any other scientists interested in checking the paper's results were essentially left to guess how to go about it.

In the years to come, independent investigations would reveal that there had been good reasons for Wakefield and his co-authors to have been less than transparent about their work: The for-profit lab at which they obtained their results had been set up in consultation with Wakefield four years earlier for the purpose of testing tissue samples for MMR-related lawsuits; it was unaccredited, refused to participate in quality-control programs, and didn't always follow manufacturers' guidelines for equipment; in subsequent tests, it produced false positives at a rate that indicated massive contamination; and the tissue samples used for the study were of such poor quality they were not suitable for PCR testing in the first place. In light of this catalogue of shortcomings, the revelation that the lab's research notebooks were shown to have been altered after initial results were recorded raised questions as to whether widespread fraud was taking place as well.

It was impossible for Sarah Barclay, the BBC's on-air reporter, to have known any of this at the time. She was, however, aware of the criticism of Wakefield's earlier research—but to include details about his reputation or the validity of his previous work would have begged the question as to why the BBC was airing "Every Parent's Choice" in the first place. Instead, Barclay covered the story as if it were a political battle, with Wakefield's "serious accusations" leading to the government's "public concerted attack on work of the man they held responsible for the loss of public confidence" in its vaccine program. In Barclay's hands, this was a "war of words"—and she gave all the best lines to Wakefield and the parents who supported him. "He had the MMR and he's autistic," one mother said, her claim left unchallenged. "Overnight he had the fever, the high temperature. Literally overnight. He was never the same again. He stopped talking and his behavior was bizarre." Another mother, who was the last person quoted on the program, said all she wanted was to make sure parents were given all the "information" they needed. "I wasn't," she said. "I think other people should be given that right that my sons and myself were denied."

In contrast to her treatment of proponents of the MMR-vaccine theory, Barclay aggressively disputed statements by government officials who stressed the vaccine's well-established safety record. At one point, she pressed Britain's chief medical officer, Sir Liam Donaldson, on what guarantees he could offer concerned parents: "Just because there is no evidence now, it doesn't necessarily mean that there won't be evidence in the future. What happens if you've got it wrong?" An exasperated Donaldson tried to explain what Barclay should have been telling her viewers all along. "This isn't a situation where we're in the dark with no evidence whatsoever," he said. "But you're choosing to focus on one, or a small group, of people's claims against a wide range of other researchers who've not been able to replicate their work, who are prepared to come out publicly and sign up to unequivocal endorsements of the effectiveness and the safety profile of MMR."

Donaldson's protestations were for naught. Within days of being broadcast, "Every Parent's Choice" started a prototypical media landslide, where coverage by one outlet legitimizes the story for another one, and on and on until everyone is writing about the ginned-up controversy of the day. In this situation, in the weeks after the *Panorama* special aired, there were literally hundreds of broadcasts and print articles that ran on the subject. Later that year, when researchers from Cardiff University in Wales studied the coverage for a report on how the media affects the public's understanding of science, they found that 70 percent of stories related to MMR mentioned a link with autism, while only 11 percent reported on the vaccine's safety record. "Since most health experts were fairly clearly lined up in support of the MMR vaccine," the researchers wrote, "balance was often provided by pitching medical experts against parents, an approach facilitated by the work of parental pressure groups on this issue. This created a serious difficulty for scientists and health professionals, who are only able to propose dry generalizations against the more emotive and sympathetic figures of parents concerned for the welfare of their children." The net effect of this manufactured equivalence was that more than half the people in England went to sleep each night believing it was as likely that MMR contributed to autism as not.

Meanwhile, the results of more reputable studies kept coming in. That November, *The New England Journal of Medicine* published a report titled "A Population-Based Study of Measles, Mumps, and Rubella Vaccination and Autism." The paper, which represented the first results from the Denmark-based studies that had been launched two years earlier, tracked each one of the more than 530,000 children born from the beginning of 1991 to the end of 1998, making it by far the largest MMR-autism study to date. The results were unambiguous: After analyzing the children for a sum total of "2,129,864 person-years," the study provided still more "strong evidence against the hypothesis that MMR vaccination causes autism." Perhaps now people could stop fighting over how to interpret the data that had previously been available, the parents of autistic children could stop second-guessing their decisions to vaccinate their children, and everyone could redouble their efforts to identify the disease's causes and determine the most effective treatments.

That, of course, is not what happened; instead, the paper strengthened the bonds between those autism activists that blamed the MMR vaccine and those that blamed thimerosal. As soon as the paper was released, SafeMinds grabbed the media's attention by making provocative statements ("It's a sad day in America when injured children are denied their due process") and issuing accusatory press releases ("Vaccine Health Officials Manipulate Autism Records to Quell Rising Fears Over Mercury in Vaccines"). On November 6, the day before the *NEJM* article's official publication date, SafeMinds released an "assessment" belittling the analytical ability of the five MDs and three advanced degree recipients who'd conducted the study. The paper's results, the group wrote, "appear to support a thimerosal role in the increases in autism being reported in the study in Denmark"—an odd claim to make, considering that the MMR vaccine had never contained thimerosal and that neither "thimerosal" nor "mercury" even appeared in the paper's text. Much of the rest of the group's five-page analysis continued in this vein: "A vaccine-induced autism subset may be present at a much lower prevalence in Denmark. . . . This may indicate a co-factor effect (e.g., thimerosal) that operates to a greater degree elsewhere. . . . It is possible that MMR increases the rate of autism only when acting in conjunction with another environmental factor, such as mercury."

One reason for SafeMinds' confrontational approach to whomever it perceived as being opposed to its goals may have been an unspoken awareness that whenever the focus did not remain on its grievances, there was a chance the media would pay more attention to the fact that the group's argument was based on a single paper written by a group of understandably upset parents and published in a deliberately fringe journal. That did not mean there were not legitimate concerns about the preservative—everyone from Neal Halsey to the CDC's Tom Verstraeten had been unsettled by the haphazard way in which more and more thimerosal-containing vaccines had been added to the recommended schedule—but those concerns centered on data linking mercury with a range of distinct, well-defined neurodevelopmental problems like cerebral palsy. Equating those conditions with autism was a little like equating a broken finger with a broken toe: In both cases, a bone has been fractured, but only one outcome is likely to have had anything to do with slamming a fist into a wall.

The result of this pitched environment was that any effort to bring clarity to the debate inevitably ended up fueling the controversy. The same week that the *NEJM* piece ran, *The New York Times Magazine* featured a three-thousand-word cover story by Arthur Allen titled "The Not-So-Crackpot Autism Theory." If ever there were an opportunity for a thoughtful airing of the issues, this was it: *The New York Times Magazine* has first-rate editors and rigorous fact checkers and Allen is a talented reporter and writer with the ability to make complex issues accessible to lay audiences.

In many ways, the piece, which was packaged as a profile of Halsey, succeeded; however, the story's strengths were ultimately overshadowed by the uproar caused by three pieces of display type: The story's title, its subhead ("Reports of autism seem to be on the rise. Anxious parents have targeted vaccines as the culprit. One skeptical researcher thinks it's an issue worth investigating"), and one of its captions ("Neal Halsey says that vaccinologists have no choice but to take the thimerosal threat seriously").

Even before the story hit newsstands, Halsey was inundated by colleagues accusing him of being an apostate. The fact that he'd been so single-mindedly determined to move decisively three years earlier was one thing; now it looked as if the high-profile director of the Institute for Vaccine Safety was making claims that were contrary to the work of dozens of reputable scientists around the world. In an effort to clarify his views, Halsey sent a lengthy letter to the *Times* chastising the paper for its role in "misleading the public":

> The unfortunate use of a sensationalized title in the article published November 10, 2002 in The New York Times Magazine . . . absolutely misrepresents my opinion on this issue. Also, the caption under the photograph of me . . . is not a statement that I ever made. There is no "threat" as thimerosal has been removed from vaccines used in children. The headline, the press release issued prior to publication, and the caption are inappropriate. I do not (and never did) believe that any vaccine causes autism. . . .
>
> The sensationalized title sets an inappropriate context for

> everything in the article. Readers are led to incorrectly believe that statements in the article refer to autism. I have expressed concern about subtle learning disabilities from exposure to mercury from environmental sources and possibly from thimerosal when it was used in multiple vaccines. However, this should not have been interpreted as support for theories that vaccines cause autism, a far more severe and complex disorder. The studies of children exposed to methylmercury from maternal fish and whale consumption and the preliminary studies of children exposed to different amounts of thimerosal have not revealed any increased risk of autism.

Halsey went on to decry newspapers' "use of deceptive titles" as a means for "mislead[ing] the public." "Apparently," he wrote, "editors, not authors, write most titles. To avoid misrepresentations authors should propose titles and assume responsibility for making certain that titles do not misrepresent the opinions of individuals or information presented in the article." The *Times* didn't run Halsey's letter, which he posted in full on the institute's Web site. (Needless to say, neither did it change journalism's time-honored practice of having anonymous editors write eye-catching headlines.) In the coming days, Halsey's worst fears came true, as anti-vaccine advocates took to describing his stance in a way that helped them make their case.*

The aggressive posture adopted by advocates could not, however, stop the evidence from mounting against them. In July 2003, *Archives of Pediatrics and Adolescent Medicine* published a review of a dozen separate epidemiological reports on MMR, which collectively studied millions of children from five different countries born over a half-century. Once again, the conclusion was that there was "[n]o evidence of the emergence of an epidemic of [autism spectrum disorders] related to the MMR vaccine" and "no evidence of an associa-

* The *Times* did run a correction in which the paper acknowledged that the story's subhead "erred in saying of a possible link that Dr. Halsey 'thinks it's an issue worth investigating.' "

tion between a variant form of autism and the MMR vaccine." Then, over the course of the next four months, results from four large-scale studies of thimerosal and autism were released. In August 2003, a comparison of children in California (where the amount of thimerosal in vaccines had increased in the 1990s) with those in Denmark and Sweden (where thimerosal had been completely removed from vaccines) found that diagnoses of autism had risen by comparable rates in all three places. In September and October, two other studies of Danish children found autism rates had either increased or stayed the same after the removal of thimerosal from the country's vaccines in 1992. And in November, a detailed analysis by Tom Verstraeten of records from the Vaccine Safety Datalink found no connection between higher doses of thimerosal and increased rates of autism or attention deficit disorder. Each individual study might have been, in the words of Vanderbilt Department of Preventive Medicine chair William Schaffner, "imperfect," but together they formed "a whole mosaic of studies . . . that all add up to this theme: thimerosal is not the culprit."

By the end of 2003, even the mass media outlets that the activists relied on to disseminate their message were being subjected to the slash-and-burn tactics used in the efforts to discredit public health agencies and independent researchers. One of the more flagrant examples stemmed from a December 29 *Wall Street Journal* editorial titled "The Politics of Autism." The piece was relatively tame by the editorial page's standards, but there was no question as to where the paper stood:

> This is a story of politics and lawyers trumping science and medicine. It concerns thimerosal, a preservative that was used in vaccines for 60 years and has never been credibly linked to any health problems. Nonetheless, a small but vocal group of parents have taken to claiming that thimerosal causes autism, a brain disorder that impairs normal social interaction. The result has been an ugly legal and political spat that has spilled into Congress and is

> frightening some parents from vaccinating their children against such deadly diseases as tetanus and whooping cough. . . .
>
> Autism is a terrible disease and it's understandable that some parents would want to look for scapegoats. One lobby group, Safe Minds, has been especially active in blaming vaccines and has found a powerful ally in Indiana Republican Dan Burton, who runs the House Wellness and Human Rights Subcommittee. His family has had its own painful experience with autism.
>
> But their understandable passion shouldn't be allowed to trump undeniable evidence and damage childhood immunizations that are essential to public health. Vaccine makers stopped using thimerosal a few years ago, but the autism lawsuits threaten those companies with enough damage that their ability to supply vaccines is in jeopardy.

No sooner had the paper rolled off the press than the onslaught began. The *Journal*'s editorial board, which had once been accused of contributing to the suicide of Clinton White House staffer Vince Foster, was no stranger to bare-knuckled brawls; still, its members were staggered by the viciousness of the attacks. The writer who'd drafted the unsigned piece was threatened so convincingly that for a period of time he didn't come to the office for fear of his safety. The paper's secretaries were harassed with accusations of aiding baby killers. Before long, the *Journal* felt compelled to respond: On February 9, 2004, the board laced into their "antagonists" for using "a hornet's nest of moral intimidation" in an effort to "shut down public debate on the matter."

> In letters and e-mails we've since been accused of "fraud," a "terrorist act," and of having an "industry profit promoting agenda." We've been told we belong to a vast conspiracy—including researchers, pediatricians, corporations, health officials and politicians—devoted to poisoning their children. . . .
>
> As writers for an independent newspaper, we aren't about to

> shut up. But what worries us is that these activists are using the same tactics in an attempt to silence others with crucial roles in public health and scientific research. The campaign to silence or discredit them has already had damaging consequences. . . .
>
> None of this, we should stress, is in the interest of families struck with autism. Researchers have spent years studying the vaccine-autism link, and we hope they continue. But if the research disproves a connection—as it has up to now—the autism community needs to listen and move on. Research dollars are limited, and parents of autistic children deserve to see the money spent where it will do the most good.
>
> Autism is a terrible diagnosis, and we hope science soon gives parents the chance at a cure. But the best way to achieve that goal is through open and honest inquiry that shouldn't be stopped because of the clamoring of an intolerant few.

The unyielding approach of SafeMinds and its allies might have been effective from a bottom-line, PR perspective, but as the *Journal*'s piece suggested, it also raised a number of disturbing questions: By pushing their own agendas regardless of their veracity, were they leaching resources away from more legitimate lines of inquiry? By putting scientists, health officials, and medical professionals on notice that any questions they raised would be used against them, were they helping or hurting the cause of vaccine safety? Finally, what responsibility did they bear for declining vaccination rates—and for the reemergence of diseases once thought to have been all but eradicated?

CHAPTER 14

Mark Geier, Witness for Hire

The week of May 10, 2004, word began to leak out of Washington that the Institute of Medicine committee that had been asked to conduct a broad "immunization safety review" three years earlier had completed its final report. After hearing rumors that the CDC was "preparing [itself] for a bomb," Sallie Bernard, Liz Birt, Lyn Redwood, and some of the other SafeMinds stalwarts set to work preparing three separate statements: one if the committee recognized a link between vaccines and autism, one if it did not, and one if it split the difference.

On Tuesday, May 18, at eight A.M., the 199-page report was released. It began:

> This eighth and final report of the Immunization Safety Review Committee examines the hypothesis that vaccines, specifically the measles-mumps-rubella (MMR) vaccine and thimerosal-containing vaccines, are causally associated with autism. The committee reviewed the extant published and unpublished epidemiological studies regarding causality and studies of potential biologic mechanisms by which these immunizations might cause autism. The committee concludes that the body of epidemio-

> logical evidence favors rejection of a causal relationship between the MMR vaccine and autism. The committee also concludes that the body of epidemiological evidence favors rejection of a causal relationship between thimerosal-containing vaccines and autism. . . .
>
> In addition, the committee recommends that available funding for autism research be channeled to the most promising areas. The committee makes additional recommendations regarding surveillance and epidemiological research, clinical studies, and communication related to these vaccine safety concerns.

There was no question which of its prepared responses SafeMinds would be releasing to the media. "The IOM has not only compromised its integrity and independence but also failed the American public," their statement read. The "flawed, incomplete" report "put America's children at risk" and "violat[ed] nearly every tenet of medical science"; the only explanation for the shoddy work of the unsalaried thirteen-person committee—which included the director of Brown University's Center for Statistical Science, the co-director of Yale University's Center for the Study of Learning and Attention, and the chair of the Department of Family Medicine at the University of Washington's School of Medicine—was that its members had been "bought and paid for by the CDC." The press gave the organization plenty of opportunities to get this message out to the public: Redwood served as the report's main critic in stories by the Associated Press, *The Boston Globe*, *The Hartford Courant*, the *San Francisco Chronicle*, and the Scripps Howard News Service; Mark Blaxill, a Boston-based consultant who was also one of SafeMinds' leaders, talked to *The New York Times*, the *Orlando Sentinel*, and *The Washington Times;* and Bernard was quoted in a Reuters story. Reporters also reached out to Barbara Loe Fisher, who pointed to the breadth of the committee's work as a reason to distrust its conclusions. "This report went beyond any other report, and this is why I felt it was political," she said. It was a perfect example, she explained, of why only research done by scientists not involved with the "public health community" could be relied on to produce honest work.

Fisher's statement reinforced what the parents in SafeMinds had been unintentionally acknowledging with their reactions to any article or study that ran counter to their theory: They had never for a moment doubted that vaccines had caused their children's illnesses. Instead of trying to collaborate with the scientific community, their real goal was to get officials to admit they'd been wrong—and when they didn't, that refusal only served to confirm their suspicions of a broad, international conspiracy.

With compromise no longer an option, the war on any organization, institution, or individual that was even tangentially involved with vaccines or vaccine safety became increasingly personal. The attacks were particularly vicious toward Marie McCormick, a professor of maternal and child health at the Harvard School of Public Health, who had served as the IOM committee's chairperson.

McCormick was an unlikely bête noire for SafeMinds members and their allies. At the outset of her tenure as the committee's chair, she'd had serious reservations about what she'd perceived as the hasty way the country's vaccine program had been expanded in the 1990s. "I don't think it was well described to the general population," McCormick says. "I'm not even sure it was well described to physicians. I remember the grand rounds on hep B, thinking, You don't have the evidence for what you're talking about—that it was going to prevent STDs. The vaccine hadn't been around long enough to prove that." And unlike many of her peers, McCormick did not have a knee-jerk aversion to Andrew Wakefield's work; in fact, the ideas behind his 1998 MMR paper had intrigued her when it first came out. "I've seen what [autism] looks like, and I thought, Isn't it interesting that someone is finally taking seriously the concerns of these families about the sort of functional problems that these kids have. I also work in a developmental clinic for premature infants, and often the most distressing part of dealing with these kids are things like feeding disorders or bowel disorders. . . . I thought it was nice that someone was finally paying attention."

Despite her open-mindedness, soon after the committee was formed, SafeMinds members began to refer to McCormick as "Church Lady," a reference to the sanctimonious *Saturday Night Live* character played by Dana Carvey in the 1980s. At one point, Liz Birt was given a voodoo doll of McCormick so she could stick it "full of pins." Another time, after McCormick had told reporters that thimerosal had not been "proven to be dangerous," Lyn Redwood wrote in an e-mail, "I am out for blood here. It takes a lot to get me really pissed off, and she has done it."

Ironically, when the committee was being assembled, the IOM's strenuous effort to avoid even the appearance of a conflict of interest had led to concern on the part of the vaccine establishment that committee members wouldn't be able to understand the nuances of the issues under discussion. "Part of the criticism of the vaccine advisory panels such as the ACIP"—the CDC's Advisory Committee on Immunization Practices—"was that manufacturers, developers, people with a vested interest in vaccines were on the panel," McCormick says. As a result, the IOM decided that committee members were not allowed to have ever worked on the development or evaluation of any of the vaccines under review or to have testified on the issue of vaccine safety, nor could they have had ties to any vaccine manufacturers or any of their parent companies. The result was a panel that included pediatricians, neurologists, geneticists, public health experts, and ethicists—but no immunologists or vaccinologists. "The fear [in public health circles] was that we were setting a precedent for other vaccine advisory committees," McCormick says. There is a good chance that she was precisely the type of person the vaccine establishment had been worried about.

In the days after the IOM review was released, it became clear that, similarly to SafeMinds, many journalists had filed their stories before they'd taken the time to understand the report's contents. As a result, the news coverage overwhelmingly ignored the lopsidedness of the evidence in favor of a "he said, she said" paradigm that treated all perspectives as equally valid. In support of its conclusions, the committee cited reports written by dozens of researchers from universities and governmental agencies around the world. In contrast, the evidence that had been submitted to the committee favoring a causal connection between autism and vaccines consisted primarily of one unpublished paper by Mark Blaxill and the collected works of two men who lived together in suburban Maryland: Mark Geier and his son, David.* Unlike virtually everyone else who worked in the field, the Geiers' studies were not done in collaboration with other scientists. Instead of using cutting-edge equipment, the Geiers worked in the basement of their two-story home in what Mark Geier insisted was a "world-class lab—every bit as good as anything at the NIH." (A *New York Times* reporter who visited the Geiers' workspace in 2005 described it as "a room with cast-off, unplugged laboratory equipment, wall-to-wall carpeting and faux wood paneling." Mark Geier's training is as a geneticist; his son majored in biology in college but has not completed advanced degree work in any field.)

By the time the IOM report was published, Mark Geier had been a mainstay of the anti-vaccine world for almost twenty years. In a 2002 paper titled "The True Story of Pertussis Vaccination: A Sordid Legacy?" he showed just how closely he adhered to its preferred narrative. Like many other people, the Geiers traced America's current anti-vaccine movement to the early 1980s, when the focus of "Vaccine Roulette" on "children with severe injuries reported to be associated with whole cell pertussis vaccine" had spurred self-appointed citizen watchdogs to "band together" in order to "educate parents about the

* The committee also considered Wakefield's 1998 *Lancet* paper; however, the fact that it was "uninformative in assessing the hypothesized causal association between MMR vaccine and autism" precluded it from factoring into members' deliberations. The report did note, "[S]ubsequent studies have not been able to replicate or prove this hypothesis."

potentially harmful effects of childhood immunizations." "Vaccine Roulette" also marked a turning point in Mark Geier's life: It was then that he first learned of the team of attorneys who eventually "convinced him that he also needed to testify against the vaccine manufacturers." ("Once the ice was broken," the Geiers wrote, "other expert witnesses came forward to testify against the manufacturers.")

By the spring of 2004, Mark Geier had appeared in close to one hundred vaccine-related lawsuits. His willingness to appear on the witness stand might have bolstered his standing with activists, but it didn't help his reputation in legal or scientific circles. For more than a decade, judges had been ruling that his testimony was so unreliable as to be worthless. In a 1987 case, Geier testified that his calculations showed that a version of the DPT vaccine made by Wyeth Pharmaceuticals was four times more toxic than the next most toxic pertussis vaccine on the market. As it turned out, he was wrong by 1,200 percent; as the result of his mistake, a $15 million jury award to the family of a severely retarded girl was summarily dismissed. (A year later, Geier blithely characterized his blunder as "what we call an order of magnitude error—that is, when I did the calculation, I must have missed a zero.") In a 1991 case, a judge ruled that Geier "clearly lacks the expertise to evaluate" the conditions he was testifying about. The following year, one judge questioned Geier's "qualifications, veracity, and bona fides," while another wrote, "Were Dr. Geier an attorney, he would fall below the ethical standards for representation." In the course of two separate trials in 1993, Geier was labeled "seriously intellectually dishonest" and his testimony was determined to be "not reliable, or grounded in scientific methodology and procedure." And in 2003, after his testimony in a case before the Vaccine Court, a Special Master named Laura Millman wrote in her ruling that Geier likely did not meet the AMA's "minimum statutory requirements for qualification" as an expert witness. Millman also noted that Geier's role as "a professional witness in areas for which he has no training, expertise, and experience" appeared to contradict the AMA's ethics guidelines, which state, "The medical witness must not become an advocate or a partisan in the legal proceeding." She ended her appraisal of Geier's work with advice for plaintiffs: "Petitioners must seriously consider whether they want to proceed with a witness whose opinion on neurological diagnosis is unacceptable."

The IOM review provided still more examples for why the Geiers' papers were often regarded as substandard: Their procedures were "nontransparent," their findings were "uninterpretable," their experiments had "significant methodological limitations," and their results were so extreme as to be patently "improbable." (In one study, the Geiers claimed that 98 percent of the variation in autism rates between the mid-1980s and the mid-1990s was due to thimerosal. They didn't explain how this jibed with their simultaneous support of the theory that the MMR vaccine, which did not contain thimerosal, also caused autism.)

A close reading of the two men's work made even those assessments seem restrained. In one of the papers they'd submitted to the IOM, the Geiers had relied on the U.S. Department of Education's (DOE) tally of special education services to show that there'd been an increase in autism diagnoses in children born between 1990 and 1994 as compared to those born in 1984 and 1985. In doing their calculations, the Geiers failed to take into account that 1990 was the first year "autism" was given its own category by the DOE. They also didn't acknowledge that in 1987, the DSM had scrapped the narrowly defined category of "infantile autism" in favor of a more expansive one for an "autistic disorder." The description for the former took all of seventy-six words. It required that patients fulfill each one of the six listed characteristics to receive a diagnosis, one of which was an onset date before the age of three. The latter required 698 words to explain, and required patients to fulfill eight out of a total of sixteen total characteristics. Crucially, children no longer needed to be diagnosed before they were thirty months old; instead the age of onset could occur anytime in "infancy or early childhood."

In another paper, the Geiers referred to figures for the total number of vaccines distributed during a given time frame as if they were the total number of vaccines that had actually been administered. (To get a sense of why that's a bad idea, think of the times when you've bought cold medicine or pain relievers in preparation for a worst-case scenario. Now take a look in your medicine cabinet and see how many of those bottles you've actually used up.) They also claimed to have determined that children who'd received a version of the DPT vaccine manufactured with thimerosal had higher rates of autism than those who'd received a version of the vaccine without the preservative—but they gave no indication as to how they'd concluded which children had received which version.

In light of Laura Millman's characterization of Mark Geier as a witness for hire, there was one last factor, unrelated to the actual substance of the Geiers' work, that should have raised reporters' antennae: All eight of their studies on thimerosal and MMR had been completed since the beginning of 2003, just as the single largest vaccine-related lawsuit in history was getting under way. It also went unobserved in news circles that the millions of dollars in damages at stake in the thousands of claims pending before the Vaccine Court may have had something to do with why the IOM report had generated so much outrage.

CHAPTER 15

THE CASE OF MICHELLE CEDILLO

In 1986, when Congress established the framework for a compensation fund to simultaneously help people with vaccine-related injuries and shield manufacturers from potentially crippling lawsuits, it did so with the assumption that a streamlined, broad-reaching program would give some stability to the country's immunization efforts. In order for the program to work according to plan, plaintiffs—or petitioners, as they were referred to—had to be able to navigate the system with relative ease. To that end, the law established a specially formulated table of injuries ranging from chronic arthritis and persistent bleeding to anaphylactic shock and paralytic polio. So long as a given injury occurred within a defined period of time following the administration of a vaccine, the court would assume causality without petitioners needing to provide further evidence. This "no fault" standard was the most significant way in which the Vaccine Court functioned differently from the traditional legal system.*

* There were some exceptions to this rule, the most notable being that there could be no other immediately obvious explanation for the injury. In other words, even though anaphylactic shock was listed as a table injury for the polio vaccine, someone with a severe shellfish allergy wouldn't be granted automatic relief in the Vaccine Court just because he'd been immunized a couple of hours before eating a plate-

The trade-offs for this relaxed evidentiary standard and the court's promise of relatively quick resolutions were the program's predetermined limitations: Awards for death and "pain and suffering" were capped at $250,000; the maximum reimbursement for legal costs was $30,000; plaintiffs who accepted the court's rulings lost the right to file suit in civil court; claims had to be filed within three years of an injury's onset; and plaintiffs who rejected the court's decision faced severe restrictions on the conditions under which they could file future civil suits.*

In its first decade of operation, the program was, by almost every benchmark, a success. By 1999, when the average number of cases filed had settled to around 130 a year, immunization rates had stabilized, wholesale vaccine prices had declined, and none of the remaining vaccine manufacturers had exited the market. During that entire time, the court had addressed claims for more than a dozen different purported injuries. Autism was not one of them. It wasn't just that the disorder was not listed as a table injury—it hadn't even shown up in a significant way in the VAERS reports that served as an early warning system for vaccine reactions. That began to change after the publication of Andrew Wakefield's 1998 paper on the MMR vaccine: In 1999, a single autism case was filed. In 2000, there was one more. In 2001, the year "Autism: A Novel Form of Mercury Poisoning" came out, there were twenty-three.

Then, after Lyn Redwood teamed up with a Dallas-based law firm

---

ful of shrimp scampi. A more opaque scenario arose when a petitioner who'd displayed evidence of brain injury before being vaccinated was awarded damages after his symptoms continued to progress after vaccination. That decision was eventually reversed by a unanimous Supreme Court ruling in which David Souter explained the justices' decision to disallow awards for preexisting conditions with the phrase "One injury, one onset."

* Today, the court allows for the reimbursement of "reasonable lawyers' fees and other legal costs." There remains a $250,000 cap on "pain and suffering," but there is no upper limit on the amount of money claimants can receive to cover medical care and lost wages.

that specialized in class action lawsuits, the floodgates opened.* "Attorneys began to start focusing on it," Redwood says. She had hoped the issue could be resolved without the interference of lawyers, but, she says, litigation became the only option after "federal agencies and professional associations" refused to be "up front with the American public." With lawyers advertising in magazines and on infomercials, and autism Listservs and conferences abuzz over vaccine claims, thousands of parents began preparing their cases. The result was unlike anything the system had ever seen before: Suddenly, more autism claims were being filed each week than were usually filed for every other type of injury combined in a month or more. On July 3, 2002, Gary Golkiewicz, the Chief Special Master on the Vaccine Court, stated the obvious when he announced that the vaccine compensation program was facing an "unusual situation." Already, three hundred claims had been filed that year alleging a connection between vaccines and autism—and, Golkiewicz wrote, lawyers had informed him that they were planning on filing anywhere from three thousand to more than five thousand additional claims "during the next several months" alone. (By the end of 2004, a total of 4,321 autism cases had been filed.) In order to ensure that families could resolve their cases in a timely manner, Golkiewicz wrote, the plaintiffs and the government had agreed to address the "general causation" theories—"*i.e.*, whether the vaccinations in question can cause autism and/or similar disorders, and if so in what circumstances"—as part of a single "Omnibus Autism Proceeding." After the causation phase of the trial had been decided, the results would be "applied to the individual cases." Because of the Vaccine Court's purposely diminished burden of proof, a positive ruling in the causation phase would in all likelihood have resulted in compensation for thousands of other families with autism-related claims.

Included among the first batch of Omnibus cases was that of

* The Redwoods were never able to file a Vaccine Court claim: In 2001, more than six years had passed since Will Redwood had received the vaccines his parents believed had caused his autism.

Michelle Cedillo, a severally autistic girl from Yuma, Arizona. Michelle was born on August 30, 1994, the first child of Theresa and Michael Cedillo. "Michelle was a happy, robust baby, very loving," Theresa would testify later. "Everything about her was normal, her sleeping habits, her play habits." From the moment she returned from the hospital, Michelle was lavished with attention. "We took her everywhere," Theresa said. "We took her to lunch with us, she went to church with us, to the park, to the grocery store. You know, she was on regular outings, to visit family and family gatherings. She was happy. She ate normal. . . . She was a happy, well baby."

On December 20, 1995, when she was sixteen months old, Michelle received her first dose of MMR vaccine. A week later, she came down with a fever that approached 106 degrees. Michelle's pediatrician told the Cedillos that their daughter likely had the flu—there was a bug going around Yuma at the time—and prescribed a round of antibiotics. Michelle's diarrhea persisted for a period, but her physical symptoms seemed to have cleared up by the time of her regularly scheduled eighteen-month wellness checkup two months later. By that point, however, there was another concern: In Michelle's chart that day, her pediatrician noted that she seemed to be "talking less since ill in Jan."

Michelle's next visit to the doctor came in April 1997, when she was almost three years old. This time, her doctor told the Cedillos it was highly possible that Michelle had a cognitive disorder. The following month, Michelle saw a neurologist, who wrote in his notes, "It would appear that there was some neurological harm done at the time of the fevers. Whether this was a post-immunization phenomenon or a separate occurrence, would be very difficult to say." Two months after that, a developmental psychologist gave the Cedillos an official diagnosis: Michelle had severe autism and profound mental retardation, and at some point in the near future she'd likely need to be institutionalized.

At the time, the Cedillos had only a vague sense of what autism was. "We were completely overwhelmed and devastated by hearing that," Theresa said. Yuma is a city of fewer than 100,000 people located less than forty miles north of the Mexican border in southwestern Arizona, and it had few resources for autistic children or their families. There were no specialists who had experience with the disease and no network of families the Cedillos could turn to for support and advice. Desperate for information, Theresa Cedillo began to scour the Internet, where she came upon the story of Cindy Goldenberg, a clairvoyant who trains people to "incorporate life skills that lead to personal empowerment, shows how we can connect with a Higher Consciousness and Universal Sources, and demonstrates that we are always surrounded by a Divine Team." In the early 1990s, after learning that her son, Garrett, was autistic, Goldenberg began a multiyear quest that led her to conclude that the rubella component of the MMR vaccine had caused Garrett's illness by damaging his immune system and inflicting an "insult to the brain."

Goldenberg contacted more than four dozen physicians about her theory before she found Sudhir Gupta, an immunologist at the University of California, Irvine. Using Goldenberg's theories and Gupta's medical training, the unlikely duo proposed treating autism with a procedure called intravenous immunoglobulin (IVIG), during which the combined blood plasma of hundreds, and sometimes thousands, of donors is injected directly into the bloodstream. Because of its potential side effects, which run the gamut from fevers to meningitis, as well as the low risk of contracting a blood-borne disease, IVIG is for the most part used only to treat immune disorders and potentially lethal infections. Despite the fact that Irvine, a wealthy enclave in Southern California's Orange County, is a six-hour drive from Yuma, the Cedillos immediately called Gupta to schedule an appointment. After examining Michelle, Gupta told the Cedillos their instincts had been correct: Their daughter did have an "abnormal immune response." Unfortunately, he said, she was not a candidate for IVIG.

That setback did not deter the Cedillos; if anything, Theresa's discovery that there was an online community of families devoted to nontraditional treatments made her more determined than ever to figure out some way to help her daughter. As the months went on, Theresa connected with more parents and learned more about the dangers of vaccines. By the fall of 1998, when Theresa first heard about Andrew Wakefield's theories, the Cedillos had become convinced that Michelle's illness had been triggered by her first dose of the MMR vaccine.

That December, they filed a claim in the Vaccine Court alleging that Michelle's condition was the result of a post-vaccination brain injury. Since brain injuries (or encephalopathies) were considered table injuries for the MMR vaccine, all the Cedillos had to do to win their claim was demonstrate that the first indications of Michelle's disease occurred between five and fifteen days after she received the vaccine. It's impossible to predict what will happen in a court of law, but this seemed to be precisely the type of situation in which there was enough circumstantial evidence to make the family the beneficiary of the Vaccine Court's relaxed standards: The timing of Michelle's post-vaccination fever corresponded closely enough with her neurological deterioration that, in the eyes of the Special Masters, there likely would not have been many questions to ask.

What's more, while the Special Masters are not permitted to base their rulings on sympathy, it would have been hard for anyone to hear the Cedillos' story without being moved. In the months and years after Michelle's initial diagnosis, her condition only seemed to grow more nightmarish. In addition to her mental limitations, she suffered from a range of excruciating gastrointestinal ailments, which were all the more difficult to treat because of her inability to communicate. She alternated between bouts of extreme diarrhea and debilitating constipation. Her tendency to strike herself in the chest was, according to one doctor, potentially an attempt to attack the pain she was experiencing. There were times when Michelle's anguish grew so intense that she'd stay up for twelve or eighteen hours at a stretch, sleep for two or three hours, and then start the cycle again. During other periods, her body produced so much saliva that she'd spend hours alternately spitting into and licking her hands. Worst of all were the stretches as long as three days at a time when she'd stop eating altogether. Eventually, her malnourishment became so extreme that she developed discolorations over her entire body and had to have a feeding tube permanently installed. By that point, Michelle needed two people to attend to her at all times: one to care for her while the other prepared her medications, and both working together whenever her diaper needed to be changed. Whether or not a vaccine had actually caused Michelle's autism, the Cedillos seemed clearly not to be a family abusing the system with the hope of a payday.

By 2000, caring for Michelle had taken over the Cedillos' lives. Preparing for each of the many trips they took to the children's hospitals in San Diego and Phoenix was an ordeal in itself: First, the Cedillos shipped ahead Michelle's favorite toys; then, they packed her feeding pump, food, medications, and syringes; finally, on the advice of a physician, they sedated her to ensure she'd be able to fly without incident. As Michelle's condition deteriorated the Cedillos found that, like their daughter, they were growing more isolated from the world around them. They had taken Michelle to church with them when she was an infant—along with "baby," "mama," and "daddy," one of her first words was "Jesus"—but they stopped after it became impossible to bring her without "causing a commotion." Even something as simple as going out to lunch became more trouble than it was worth. "We tried. We made several attempts, but eventually we stopped," said Theresa. "It was too upsetting, for her, for the people in the restaurant, for us. So we stopped."

This is a common scenario even for parents with children on the milder end of the autism spectrum. "It makes people uncomfortable," says Jane Johnson, who is the managing director of the Autism Research Institute and is a member of the family whose ancestors founded the multinational pharmaceutical and medical devices company Johnson & Johnson. "Particularly for women, many of our friends are based around child rearing. . . . If you're sitting there with your neurotypical child and your friend Suzy is there with her autistic child, you're going to feel really uncomfortable when your child is running up and saying, 'Mommy, mommy, I just went down the slide'—and Suzy's child can't speak. You're going to cringe if you're a sensitive person with every word that comes out of your child's mouth, knowing how Suzy doesn't get those same experiences. I suppose they don't know what to say."

The combination of the isolation and exhaustion experienced by the Cedillos undoubtedly led to periods during which they questioned the commitment they'd made to each other in the days after Michelle's initial diagnosis: "We both agreed that we didn't ever want to . . . put her away somewhere," Theresa Cedillo said. "We both decided that we would try to concentrate on finding out what was wrong and more about what her diagnosis was and to see what form of help we could get for her." Time after time, all that love and attention seemed to be for naught, as another new therapy or experimental protocol proved to be no more effective than the ones the Cedillos had already tried.

Then, in 2001, Theresa Cedillo went to the annual Autism Research Institute/Defeat Autism Now! conference in San Diego. The experience was revelatory: She heard about children who sounded just like Michelle—and who'd been cured. "I heard several presentations by several doctors describing Michelle to a T with her—the regression, her bowel problems, how her bowel problems had persisted and what was wrong," she said. One of those doctors was Andrew Wakefield. By that point, the Cedillos had already endured four years of people telling them that they couldn't help Michelle—and that they doubted anyone else would be able to help her either. Wakefield, on the other hand, was anything but discouraging: He suggested that Michelle have a pediatric gastroenterologist perform an endoscopy, which is an invasive surgical procedure during which a fiber-optic camera is threaded up the rectum or down the mouth and through the maze of the gastrointestinal tract. "He gave me the information that I need to—medically speaking, what I needed to tell the gastroenterologist to see if he was willing to do that," Theresa said. (Many doctors are reluctant to perform endoscopies on children because the procedure puts patients at risk of punctured organs, deadly infections, and all the potential complications that accompany anesthesia.)

Less than three months later, the Cedillos sent a biopsy of Michelle's gut tissue to Unigenetics, a lab in Dublin, Ireland, that had been founded in the late 1990s in consultation with Wakefield for the purpose of testing tissue samples for U.K. MMR litigation. More recently, the lab had become popular among American families involved in vaccine-related lawsuits. "[We wanted] to determine if she had measles RNA in her colon," Theresa said later when asked why she and her husband had Michelle's tissue samples tested at a facility more than five thousand miles away from their home. "I think I found it online. . . . I can't remember where I first heard of Unigenetics, but I did hear directly from another parent to send it there."

Between the spring of 1998 and the end of 2001, the series of events that culminated in Theresa Cedillo's one-on-one encounter with Andrew Wakefield were repeated hundreds of times around the country, often with similar results. After almost every one of his speeches, a line of parents waited to tell Wakefield about their children. At times, he seemed to respond to the adulation in ways that were more reminiscent of a miracle healer than a clinician. During a conference in the fall of 1999, Liz Birt brought her five-and-a-half-year-old son, Matthew, to Wakefield's hotel room. Wakefield looked at the child and placed a hand on his stomach. "I think we could help him," he said. Two months later, the Birts were on their way to London for treatment at the hospital at which Wakefield worked. The more Wakefield's warnings about the connection between vaccines and autism were derided by mainstream organizations, the more his work took on the aura of *samizdat*—and the more he was treated as a savior incapable of doing wrong. On March 20, 1999, at a conference in California, Wakefield described for his audibly amused audience how he relied on his young son's schoolmates to obtain blood samples for his research:

> This is again my son's birthday party, so these are your perfect controls. So we lined them up—with informed parental consent of course; they all get paid £5, which doesn't translate into many dollars, I'm afraid—and they put their arms out with a cuff on and have the blood taken. It's all entirely voluntary. (Audience laughs.) And, uh, when we did this at the party two children fainted and one threw up over his mother. (More laughter.) Listen, we live in a market economy. Next year they'll want £10. (More laughter.) We have a birthday party coming up as well when I get back, so we're going—they charge me a fortune. Urine, I mean, come on—they charge me a fortune. They were all less than ten years of age.

At about the same time that Wakefield was joking about bloodletting at his son's birthday party, Vicky Debold, a nurse, was beginning to notice things in her son, Sam, that would eventually lead her to request her own consultation with Wakefield. Sam Debold was born in 1997, and for the first year of his life, Vicky says, he developed normally. After that, however, he was stricken with a series of escalating health problems. First, Vicky says, he lost interest in learning how to walk. Next came severe diarrhea. Eventually, he stopped being curious about the people around him. "He had always been very social, happy and bright-eyed, loved playing patty cake and stuff like that," Vicky says. "Then he just wasn't all that interested. When you tried to redirect him and get him involved in something other than what he was doing, which was typically just sitting on the floor lining stuff up, he would get upset."

Shortly before Sam's third birthday, a developmental pediatrician diagnosed him with autism. That his parents were already somewhat familiar with the disorder did not make the diagnosis any less painful than it had been for Michael and Theresa Cedillo: According to his doctor, Sam was so severely ill it was possible he, too, would end up in an institution. If the Debolds had any further questions about what the future held in store, the doctor said, they should consult a book titled *The World of the Autistic Child* by a psychologist named Bryna Siegel, which warned parents against maintaining a belief, "on an irrational, unconscious level, that there is a normal little person inside the autistic exterior, struggling to emerge whole." "It's the most depressing book I've ever read in my life," Vicky says. "I swear, if you're a parent and you read that you want to kill yourself. . . . The future is nothing but bleak." Even the parents Vicky met through local support groups seemed resigned to their fates. "They had older children," she says. "They were pretty—happy is not the right word. There wasn't—they were past the shock, the anger, the 'Why aren't we doing more about this' phase."

Vicky, on the other hand, was still confused and upset. As time went on, she also grew increasingly frustrated with Sam's doctor, whose counsel had been limited largely to warnings about what *not* to do: "You're an educated person," he'd told her during an appointment. "You and your husband know better than to get on the Internet." What was it, she wondered, that he didn't want her to see? Having seemingly exhausted all her other options, Vicky figured it was time to find out. Some of the first things she encountered when she looked online were the simmering controversy regarding mercury in vaccines and Wakefield's theories regarding the MMR vaccine. "I thought, This guy doesn't seem completely crazy," she says. "He's published in really good, peer-reviewed journals." Then, she says, she read that Wakefield, along with Bernard Rimland and Barbara Loe Fisher, was scheduled to speak at an upcoming Parents for Vaccine Awareness conference in Erie, Pennsylvania, close to where she lived.

Debold went to Erie to hear Wakefield's presentation, and when it was over, she approached him with the intention of telling him about her son. Before she could begin speaking, she began to sob. "Andy said, 'Look, let's calm down here,' " Vicky says. " 'First thing we need to do is, he needs to get a regular X-ray of his abdomen. He needs to be evaluated by a gastroenterologist, you need to check this and this and this, and if those things are all negative, you probably need a colonoscopy.' It was the first time any physician thought there was any merit to all of Sam's diarrhea and his weight loss. I'm sitting there going, this is so common sense. This should have been done a long time ago. Nobody paid any attention to this."

The depth of gratitude felt by many parents at Wakefield's suggestion that there might be effective treatment for their children was often enough to cause them to overlook ways in which his interests might not have been perfectly aligned with theirs. During the precise time he was becoming a hero in America, Wakefield was on the verge of disgrace in the U.K., where there was growing concern over what was perceived as his liberal attitude toward scoping children. In December 2001, after a series of disagreements with his superiors, Wakefield was asked to leave the Royal Free Hospital, where he'd been based for the previous fourteen years. "Dr Wakefield's research was no longer in line with the department of medicine's research strategy," a statement released by the hospital read. "He left the university by mutual agreement."

According to Wakefield, however, that "mutual agreement" wasn't the result of a dispute over the tactics he used to reach his conclusions—it was about the implications of his controversial results. "I have been asked to go because my research results are unpopular," he said at the time. "I did not wish to leave but I have agreed to stand down in the hope that my going will take the political pressure off my colleagues and allow them to get on with the job of looking after the many sick children we have seen." With a note of defiance in his voice, he added, "I have no intention of stopping my investigations."

For years, Wakefield had made some variation on this theme the linchpin of his defense whenever he came under scrutiny. Three years earlier, in a response to criticism of his 1998 *Lancet* paper on the MMR vaccine and autism, he'd trotted out his loyalty to the people he referred to as "desperate parents" and used them to divert attention from the questions that had been raised about his work: "We get these parents ringing up every day," he told a reporter from *The Independent* at the time. "They say, 'My child has autism and bowel problems and we believe they are linked.' You *have* to do something for them."

The genius of Wakefield's explanation was that it neutralized the charge that he was behaving unethically—if parents were eager for their children to receive endoscopies, what grounds did anyone have for criticizing the doctor who performed them?—while also making the "establishment" figures who opposed him seem patronizing in their willingness to tell parents what *not* to do despite being unable to suggest any alternatives. As the doctors and scientists with whom he'd once worked went about cutting their ties, Andrew Wakefield bound himself more tightly with each passing day to the parents on whom he'd base his future.

On January 14, 2002, two months after Theresa Cedillo met Wakefield in San Diego, the Cedillos changed their Vaccine Court encephalopathy claim to what was referred to as a "causation-in-fact" case alleging that the MMR vaccine had induced their daughter's autism. This was a huge roll of the dice, and it all but guaranteed that the litigation would be a major part of the Cedillos' lives for years to come. Had they stuck with their original table injury claim, the case would have been a straightforward one about a condition already acknowledged to be caused by vaccination in some situations. By making their case about autism, the Cedillos were creating a far more daunting and time-consuming burden of proof for themselves: Before they could even begin to argue that Michelle's autism had been caused by the MMR vaccine, they would need to prove the general form of the charge—that vaccines were capable of causing autism in *anyone*. That meant that in order for the Cedillos to receive compensation, the court would need to be convinced that the growing scientific consensus that vaccines did not cause autism was, in fact, incorrect.

The magnitude to which the Cedillos had complicated their task was highlighted in the announcement the court's Chief Special Master made that July regarding the Omnibus Autism Proceeding:

"[P]etitioners' representatives have stated that they are not prepared to present their causation case" because they needed more time "for the science to crystallize, to obtain experts, and in general to prepare their proof concerning the difficult medical and legal causation issues." This was a stunning admission: Despite encouraging thousands of families to file suit, their lawyers had absolutely no proof that the fundamental pillar on which they based their claims was correct. As a result, both sides were given sixteen months—until November 2003—to "designate experts" and begin assembling their strongest possible arguments. The deadline for submitting "expert reports with supporting authorities" would come three months after that, and decisions were expected to be handed down in the summer of 2004—which meant that the Cedillos would have to wait a minimum of two more years before they could begin to move on with their lives.

CHAPTER 16

Cognitive Biases and Availability Cascades

For the Cedillos, the decision to initiate litigation was undoubtedly the result of a number of interrelated factors, including love for their daughter, anger at the vaccines that they believed had harmed her, and the hope that they'd be able to alleviate her suffering. Conventional wisdom holds that the more emotional a decision, the less rational it is, but in his 2009 book *How We Decide*, Jonah Lehrer explains that is not always the case. By way of illustration, Lehrer describes a patient named Elliot who in 1982 had a tumor removed from an area of the brain just behind the frontal cortex. Elliot survived his surgery with his intellect intact—his IQ was exactly the same before and after he went under the knife—but he completely lost the ability to make decisions, ranging from what route to take to work to what color pen to use to write his name. Soon, one of his doctors realized there'd been another significant change in Elliot's personality: He'd seemingly lost the ability to feel.

"This was a completely unexpected discovery," Lehrer writes. "At the time, neuroscience assumed that human emotions were irrational"—which would mean that an inability to feel should make it easier to make decisions. It turns out the exact opposite is true: "When we are cut off from our feelings, the most banal decisions become impossible. A brain that can't feel can't make up its mind."

Today, there's an almost universal acceptance that what has traditionally been perceived as "rational" thought is in fact intimately connected with our emotions. This discovery has led to an explosion of interest in the cognitive biases we use to convince ourselves that the truth lies with what we feel rather than with what the evidence supports. The origins of many of these traits can be traced back to the primitive conditions in which they were selected for millennia ago. Take pattern recognition, which evolutionary biologists like to explain through fables about our ancestors: Imagine a primitive hunter-gatherer. Now imagine he sees a flicker of movement on the horizon, or hears a rustle at his feet. Maybe it was nothing—or maybe it was a lion out hunting for dinner or a snake slithering through the grass. In each of those examples, the negative repercussions of not taking an actual threat seriously will likely result in death—and the end of that particular individual's genetic line. On the other hand, the repercussions of bolting from what turns out to be the shadow of a swaying tree or the sound of a gentle breeze will likely be nothing worse than a little extra exercise. (In statistical terms, this explains why we're much more likely to make Type I errors, or false positives, than we are Type II errors, or false negatives.)

Unfortunately, evolution is a blunt tool, and a by-product of that protective instinct is a tendency to connect the dots even when there are no underlying shapes to be drawn. When our yearning to feel in control and our ability to recognize randomness are in conflict, the urge to feel in control almost always wins—as was likely the case when Lorraine Pace became convinced that there were an unusually high number of breast cancer cases in her Long Island community. (The technical name for this tendency is the clustering illusion.)

Pattern recognition and the clustering illusion are just two of literally dozens of cognitive biases that have been identified over the past several decades. Some of the others have been alluded to earlier in this book: When SafeMinds members set out to write an academic paper about a hypothesis they already believed to be true, they set themselves up for expectation bias, where a researcher's initial conjecture leads to the manipulation of data or the misinterpretation of results, and selection bias, where the meaning of data is distorted by the way in which it was collected. In addition to being a natural reaction to the experience of cognitive dissonance, the hardening conviction on the part of vaccine denialists in the face of studies that undercut their theories is an example of the anchoring effect, which occurs when we give too much weight to the past when making decisions about the future, and of irrational escalation, which is when we base how much energy we'll devote to something on our previous investment and discount new evidence indicating we were likely wrong. (Remember that the next time you refuse to turn around despite signs that you've been traveling in the wrong direction or you hold on to a stock because you're convinced you'll be able to make back the money you've already lost.) My favorite of the cognitive biases on the grounds of its name alone refers to the phenomenon of crafting a hypothesis to support your data, and in the process making it untestable. This is called the Texas sharpshooter's fallacy, named after an imaginary cowboy who confirms his skill as a marksman by shooting bullets into the side of a barn and then drawing a target around the resulting holes.

These realities of human cognition help explain why it can be so difficult to demonstrate to someone that their initial read on a situation—their instinct, their gut reaction, their *feeling*—is, in fact, wrong. They also show why two reasonable, intelligent people who disagree can be equally certain that the evidence supports their understanding of the "facts." It's at this point that confirmation bias, the granddaddy of all cognitive biases, kicks into action—which is to say, it's at the precise moment when we should be looking for reasons that we might be wrong that we begin to overvalue any indication that points to our being right. (This is part of the reason the scientific method can be so hard to grasp, and so hard to adhere to: It goes against our makeup to try to find ways to punch holes in our own arguments.) Misapprehensions about medicine are particularly vulnerable to the effects of confirmation bias, because the process by which a given intervention works is so often contra-logical: It makes no intuitive sense that rebreaking a bone would help a fracture to heal or that using chemotherapy to kill living tissue would help a person survive cancer. Now consider vaccines. Receiving a shot in a doctor's office might not activate the disgust response in the way that early inoculation methods did, but injecting a healthy child with a virus in order to protect him from a disease that has all but disappeared still feels somehow wrong. The fact that getting vaccinated is so obviously painful and that infants can't understand that you want to help and not hurt them only makes matters worse.

Confirmation bias, like all the unconscious legerdemains referred to above, is a phenomenon that occurs on an individual level. The extent to which the autism advocacy world has been taken over by anti-vaccine sentiment illustrates a whole other category of cognitive sleights-of-hand: those that arise out of group dynamics. Ten years ago, the ARI/DAN! meeting held in San Diego every fall was the only major national autism conference in the country. In 2010, in addition to ARI conventions in Baltimore, Maryland, and Long Beach, California, there was the annual AutismOne event in Chicago, Talk About Curing Autism conferences in Birmingham, Alabama; Madison, Wisconsin; and Orange County, California, and dozens of state- and county-wide meetings sponsored by groups like the National Autism Association (NAA).

Wherever and whenever these gatherings occurred, one of the dominant themes ended up being the dangers of vaccines and the venality of the medical establishment. This is not because participants arrived eager to attack the American Academy of Pediatrics or the CDC. Instead, the animus is a natural outgrowth of people with similar interests getting together in the first place: Inevitably, group members leave with views that are simultaneously more extreme and more similar to each other's than the ones they came in with. As an example, consider the current views of three people whose first exposure to an international network of parents with autistic children occurred at the 2001 ARI/DAN! conference. Vicky Debold says she was motivated to travel to San Diego by her frustration with the health care industry she worked in and had always trusted, while Theresa Cedillo was enticed by the prospect of learning about ways she might be able to help her daughter. For Jane Johnson, the convention provided a sense of emotional connectedness that had been missing from her life. "I looked out at an audience of hundreds of people, and they knew how I felt," she says. "To be surrounded by a thousand other people who know [your] pain is really powerful." That conference, Johnson says, was the first of eight consecutive ones at which she was brought to tears.

Nine years later, the opinions of all three women have coalesced around an opposition to the way vaccines are developed and administered. In a recent conversation, Johnson told me that public health officials around the world are potentially more culpable than the tobacco company executives who hid evidence that smoking is harmful: "This is the federal government giving every kid a carton of cigarettes and saying, 'Get to work.' " Vicky Debold, whose reaction to hearing Barbara Loe Fisher in May 2000 was, "You wouldn't be saying and doing the kinds of things you're doing if you had seen the kids die in the ICU that I saw," has become one of the leaders of Fisher's National Vaccine Information Center. And Theresa Cedillo lent her daughter's name to the largest vaccine-related compensation lawsuit in the world.

The process that explains this convergence of views is called an "availability cascade," a concept that was first articulated in a 1999 paper by Timur Kuran, an economics and political science professor at Duke who was then at the University of Southern California, and Cass Sunstein, who currently heads up the White House's Office of Information and Regulatory Affairs and was at the time a professor at the University of Chicago Law School. Kuran and Sunstein defined the term as a "self-reinforcing process of collective belief formation by which an expressed perception triggers a chain reaction that gives the perception increasing plausibility through its rising availability in public discourse."* Another way of saying this is that an availability cascade describes how the perception that a belief is widely held—the "availability" of that idea—can be enough to make it so. In this instance, the believability of the notion that vaccines cause autism has grown in proportion to the number of people talking about it, as opposed to the theory's actual legitimacy.

This blind-leading-the-blind phenomenon is all the more dangerous because one all-but-guaranteed effect of surrounding yourself with like-minded people is increased polarization: Think about the red-meat frenzy of the quadrennial Democratic and Republican conventions or the sign-waving mania churned up by a Tea Party rally. Or to come at it from a different angle, consider what happens when a bunch of otherwise reasonable Red Sox fans get together to watch a baseball game in a bar: It's a fair bet that nobody's going home that night feeling any warmer toward the Yankees.†

One reason this us-versus-them mentality has become so ubiquitous is that the mechanisms that have historically allowed for easy

---

* Not all of Sunstein's writing on behavioral law and economics is so academic. In the 2008 book *Nudge*, Sunstein and Richard Thaler write about external changes—nudges—that can influence human behavior. One of their examples stems from efforts to get men to stop peeing on the floor: "As all women who have ever shared a toilet with a man can attest, men can be especially spacey when it comes to their, er, aim. In the privacy of a home, that may be a mere annoyance. But, in a busy airport restroom used by throngs of travelers each day, the unpleasant effects of bad aim can add up rather quickly. Enter an ingenious economist who worked for Schiphol International Airport in Amsterdam. His idea was to etch an image of a black house fly onto the bowls of the airport's urinals, just to the left of the drain. The result: Spillage declined 80 percent. It turns out that, if you give men a target, they can't help but aim at it."

† When intergroup distrust does take hold, it becomes ever harder to undo. In a 2009 study analyzing how nonscientists view scientific claims, a team of psychologists found that at a certain point exposure to dissenting viewpoints results in "poor source credibility." (This is also referred to as a "sinister attribution error.") The scientists used the example of a student who arrives at college believing in a literal interpretation of the Bible and ends up "even more strongly committed to creationism after hearing evidence for evolution."

access to moderate positions are rapidly disappearing. Twenty years ago, it took a fair amount of effort to create an information cocoon: In 1987, nearly three-quarters of Americans tuned into a nightly news broadcast from one of the three networks, creating a sort of national common denominator for information about the world. Now that figure has fallen below one-third, as consumers abandon the presumed neutrality of the networks in favor of cable news telecasts that gratify viewers by feeding them exaggerated versions of the opinions they already hold. An even more potent force in this regard is the Internet, where it's easier than not to fall down a wormhole of self-referential and mutually reinforcing links that make it feel like the entire world thinks the way you do. The anonymity and lack of friction inherent in the online world also mean that a small number of committed activists—or even an especially zealous individual—can create the impression that a fringe viewpoint has broad support. (Kuran and Sunstein refer to these people as "availability entrepreneurs.") A 2007 study titled "Inferring the Popularity of an Opinion from Its Familiarity: A Repetitive Voice Can Sound like a Chorus" provides statistical proof of this phenomenon: The authors analyzed a series of group discussions and found that participants tended "to infer that a familiar opinion [was] a prevalent one, even when its familiarity derive[d] solely from the repeated expression of one group member."

When dealing with debates that hinge on the interpretation of facts—Are April snowstorms a sign of global warming or the opposite? Are the latest unemployment numbers a sign the Obama administration's policies are succeeding or failing?—these differences of opinion can quickly become personal: When confronted with people who see black where we see white, it's hard not to assume that there's something about them—either they're ignorant, or their emotions have the better of them, or they're lying in the service of a more base interest—that prevents them from acknowledging what to us is self-evident. Looked at through this light, the seemingly contradictory ideas that have defined the vaccine wars begin to make more sense. The inner workings of the human mind, more than any ambiguity about the facts, explain why more and more parents were rushing to file claims in Vaccine Court at precisely the same time that scientists were growing more certain that vaccines did not cause autism.

One result of these colliding interpretations of reality was a growing sense of distrust and paranoia. In the summer of 2000, Lyn Redwood became so convinced that her phone was being tapped that she hired countersurveillance teams to examine her house. (They didn't find anything.) In the months to come, Redwood's fear would be reinforced through conversations with Barbara Loe Fisher, who told Redwood of the mysterious clicks and whirrs other vaccine opponents around the country heard over their phone lines. It wasn't long, Redwood says, before national reporters grew fearful of even mentioning thimerosal in their stories. "*U.S. News & World Report* came out to talk to me about environmental toxins," she says. "When they realized mercury was from vaccines, the reporter set his pen down and said, 'I can't do this story.' . . . He just thought that it would be something that would be too controversial to include, and he had to clear it with his editor." (According to Redwood, the editor did eventually sign off on the piece and her comments about vaccines made it into print.)

The flip side of this sense of persecution was, not surprisingly, a ratcheting up of the rhetoric on both sides of the debate. In December 2003, just after the last of the four population-based thimerosal studies to be released that year was published, Mark Geier stated flatly, "This is fraud. . . . [I]t is one of the worst things ever to have happened to this United States. If a terrorist had done this, we wouldn't attack them, we'd nuke them." Eight months later, the head of the CDC's immunization program told a reporter from the *Los Angeles Times* that only "junk scientists and charlatans" who saw a "huge pot of gold at the end of the rainbow" believed in the thimerosal-autism link.

It didn't take long for this acrimony to turn into something more frightening. In the days after the publication of the Institute of Medicine report in May 2004, SafeMinds released a statement that expressed disappointment that committee members would not "have to answer later for their failures." That summer, a series of anonymous messages sent to the CDC were judged sufficiently hostile that the FBI launched an investigation. (Marie McCormick, the committee's chairperson, also received threatening letters, which resulted in Harvard University increasing security at her office.) One e-mail, which *The New York Times* obtained through an open-records request, read, "I'd like to know how you people sleep straight in bed at night knowing all the lies you tell & the lives you know full well you destroy with the poisons you push & protect with your lies." Several weeks later, the agency received an e-mail with an even more overt threat. "Forgiveness is between [the committee's members] and God," it read. "It is my job to arrange a meeting."

Part Three

CHAPTER 17

How to Turn a Lack of Evidence into *Evidence of Harm*

When David Kirby got the phone call that would change the course of his professional life, the forty-two-year-old's work history was most notable for the number of times he'd switched fields: After starting out as a wire service stringer in the mid-1980s, he'd worked for former New York City mayor David Dinkins, served as the spokesman for an AIDS charity, and even started his own PR firm. It wasn't until 1997 that he began focusing full-time on journalism, and within a year he'd become a regular contributor to one of *The New York Times*'s regional sections. Three years later, when he went back to being a full-time freelancer, he still hadn't gotten the one big break that would set him apart from the hundreds of other reporters fighting to get noticed in the media capital of the world.

Then, in November 2002, a friend in Los Angeles suggested to Kirby that he investigate the country's rising rates of autism. It wasn't a subject Kirby knew anything about. Over the years, he'd written some science pieces for the *Times*, but for the most part they were one-offs about personal health ("More Options, and Decisions, for

Men with Prostate Cancer") or topical news ("New Resistant Gonorrhea Migrating to Mainland US"). Still, Kirby was intrigued. Maybe, he thought, he could turn the idea into a feature for a well-paying women's magazine like *Redbook* or *Ladies' Home Journal.* He decided to make some calls.

One of the first people Kirby reached was a hairdresser in the San Fernando Valley. The conversation was a bit much. "She was cool," Kirby says, "but . . . she was just throwing all this stuff at me and I had no idea what she was talking about." At one point, Kirby says, the woman told him what sounded like a half-baked story about how there was mercury in vaccines. "I literally laughed," Kirby says. "I said, They don't put mercury in vaccines. That's ridiculous." By the time he hung up the phone, Kirby figured the whole thing was one more dead end.

A little more than a week later, Kirby had his television on in the background of his apartment when he overheard mention of thimerosal. He turned up the volume: A CNN anchor was talking about a provision that had been inserted surreptitiously into a just-passed 475-page homeland security bill. The amendment shielded Eli Lilly, the company that manufactured thimerosal, from vaccine-related lawsuits filed by families. "I was like, Oh, mercury is in vaccines," Kirby says. "That lady wasn't crazy."

Attention surrounding what became known as the Lilly rider was intense from the get-go, and not just among families who believed vaccines had injured their children. Lilly's ties to the Republican leadership were well documented: In the 2000 election, the company had contributed $1.28 million to Republican candidates; Mitch Daniels, the director of the White House's Office of Management and Budget, was the former president of Lilly's North American operations; and just months earlier, President Bush had appointed Lilly CEO Sidney Taurel to the Homeland Security Advisory Council. Now the appearance that someone had called in a favor sparked bipartisan outrage. Ohio's Dennis Kucinich, among the most liberal members of Congress, said the provision raised "fundamental questions about the integrity of our government," while past Republican presidential candidate and future nominee John McCain likened it to "war profiteering." Even the amendment's author seemed embarrassed: To this day, no one has acknowledged inserting it into the bill. "It was a big whodunit," Kirby says. "It went on for days. It was shaping up to be a big scandal in Washington."

In the coming weeks, Kirby approached *Vanity Fair*, *The New Yorker*, and a number of other outlets with a proposal for a story about the politics of thimerosal. No one was interested. "It was too controversial," he says. Then, one night, Kirby was taking a shower when he had an epiphany: "A voice told me to turn the story into a book," he says.

Initially, Kirby's literary agent wasn't much more enthusiastic than the various magazine editors he'd approached had been. "He said, 'You can't just write a straight-up nonfiction book about this and write about the science,' " Kirby says. " 'If you really want to tell the story well, you gotta go find people who this has happened to, parents of autistic kids who really believe the vaccines and the mercury did this. Tell their story. Tell it through their point of view. Write it like a novel.' " What the hell, Kirby figured. It was worth a shot. On Wednesday, January 8, 2003, he headed down to Washington for a rally SafeMinds had helped to organize to protest the Lilly rider.

Even without a huge number of people in attendance—the Republicans had all but promised to void the provision, which they did less than a week later—the rally gave Kirby a firsthand perspective on the intense emotions surrounding the issue. Parents were crying; some were carrying poster-board signs illustrated with pictures of their children pasted above captions like "VACCINE INJURED" or "VICTIM OF HOMELAND SECURITY." Despite having met each other only in recent years, many acted as if they were lifelong friends. The rally also served as a good illustration of the coalition's growing political clout: Dan Burton and Dennis Kucinich both addressed the crowd, as did Michigan senator Debbie Stabenow and Vermont senator Patrick Leahy. By the end of the afternoon, Kirby had introduced himself to Sallie Bernard and Liz Birt and told them he was interested in writing a book about their movement.

For the most part, they seemed receptive—but there was one problem: Lyn Redwood was working on a book of her own. In fact, she'd already put together a proposal, written a sample chapter, and even come up with a title: "Mercury Rising—The Untold Story Behind America's Learning Disability Epidemic." According to Redwood, when Kirby contacted her, he didn't launch into a sales pitch but simply asked how it was going. She had to admit things didn't look good: A number of publishers had already passed on the project. (One editor explained her reasoning in a letter: "This is such an upsetting book. . . . I wonder who would buy it.") To make matters worse, Redwood's agent had just started a six-month maternity leave. The more she spoke with Kirby, the more Redwood felt like the author could be an asset to her work. Soon, Redwood says, "We decided to combine forces and work on it together." It was not a decision she would regret. "It was the best thing that could have happened," she says. "My book would have been viewed as biased"—but Kirby would be able to present his project as an objective record of what had happened.

For the next two years, Redwood and Kirby were in daily contact, and Kirby was given access to all the data SafeMinds had collected and all the files they'd assembled—he even got printouts of "every single e-mail they'd sent each other over the years." (Kirby jokes that whenever he returned to New York from the Redwoods' home in Georgia his luggage was so full of documents that it would be over the airline's weight limit.) "He came and hung out at our house and lived down in the basement," Redwood says. "If you could see my office, I just have boxes and boxes and files and folders and documents. Just for him to ferret through all the documents and for me to explain to him all this—it was a labor of love."

By the time Kirby handed in his manuscript in 2004, it looked as if the story might have outpaced all of his and Redwood's work: The dozens of studies the Institute of Medicine committee on vaccine safety had reviewed for its recently released report had provided overwhelming evidence that there was no link between autism and vaccines—at least from a scientific perspective. Kirby's book, however, didn't engage scientists on their terms; instead, he invoked disputes over fringe research in order to move the discussion into the realm of public opinion, where the conversation would hinge on impressions gleaned from sound bites, pull quotes, and drive-time radio segments. His most brilliant tactical maneuver in this regard was his choice of a title: *Evidence of Harm—Mercury in Vaccines and the Autism Epidemic: A Medical Controversy.* It was, of course, a reference to the CDC's 1999 statement that had been meant to reassure the public about the presence of thimerosal in vaccines: "There are no data or evidence of any harm." The line had been problematic even then, and now Kirby made it sound anew like a frightening equivocation. In his introduction, dated less than four months after the IOM committee's analysis was released, Kirby took full advantage of this confusion. He granted that the report "favor[ed] rejection of a causal relationship between thimerosal and autism." Nonetheless, Kirby wrote, no one really knew for *sure* if thimerosal was safe:

> Meanwhile, the CDC has been unable to definitively prove or disprove the theory that thimerosal causes autism. . . . Several studies funded or conducted by the agency have been published in the past year, all of them suggesting that there is no connection between the preservative and the disease. The CDC insists that it has looked into the matter thoroughly and found "no evidence of any harm" from thimerosal in vaccines.

Then Kirby delivered his coup de grâce:

> But no "evidence of harm" is not the same as proof of safety. No evidence of harm is not a definitive answer; and this is a story that cries out for answers.

By daring officials to do the one thing they were incapable of—prove a negative—Kirby was making a mockery of science. And by successfully defining the terms of the debate, he rewrote a "story" that had, in fact, already provided the answers he said he so desperately hoped to find.

Convincing people there was still a legitimate debate was only one of the hurdles Kirby faced in reaching the public. He also had to come across as a reporter and not someone advocating for one side. In the book's opening pages, he confronted that issue head-on: He had, he acknowledged, focused on the "admittedly subjective point of view" of a community of "brave souls united in grief and hope." Still, that did not mean that he was favoring them over the "doctors, bureaucrats, and drug company reps." "[T]here are two sides to every good story, and this one is no exception," he wrote. "This is not an antivaccine book. . . . Neither should this book be viewed as partisan in nature. . . . This book looks at evidence presented on both sides of the thimerosal controversy."

In a jury trial, two of the most crucial jobs a lawyer faces are humanizing his clients and demonizing his opponents. Like an expert litigator, Kirby accomplished both jobs while making it seem as if he was merely describing the world as it was. Redwood, he wrote, was "an attractive woman" with "almost cat-like brown eyes." Compatriots like Sallie Bernard (a "tough businesswoman" with "upbeat emotions"), Liz Birt (a woman possessing a "fierce streak of determination and independence"), and Mark Blaxill (a man with a "remarkable aptitude for statistical analysis") sounded just as appealing. Mark Geier was "the kind of guy you would want to have dinner with"; in Kirby's hands, his basement full of old lab equipment was transformed into the headquarters for the "Genetic Centers of America, a private consulting firm in Silver Spring, Maryland." Then there was the "establishment": educators who "barked at" and "banished [children] to a small corner of the room," pediatricians who "poked and prodded" their patients "like some pet science project," and government scientists who "grew purple" until their "eyes bulged with annoyance."

Once he's succeeded in making his clients more appealing than their opponents, the lawyer's next task is to make his case sound plausible. The dozens of peer-reviewed studies, combined with the IOM committee's evisceration of the Geiers' work, had gone a long way toward undermining the mercury activists' claims. In order to square that circle, Kirby likened the dispute to a political campaign in which an "insurgent candidate" comes under "heavy fire from an entrenched opponent . . . the vitriol demonstrates that the challenge is being taken seriously, that it poses a realistic threat to the status quo."

In this political battle, Kirby employed a time-honored tactic of push pollers and ward politicians: He used an ominous-sounding claim—"Curiously, the first case of autism was not recorded until the early 1940s, a few years after thimerosal was introduced in vaccines"—to make his accusation sound as if it was idle speculation. In this case, Kirby both blurred the difference between correlation and causation and conflated the first time a disease is given a particular label with the first time it appears in a population. (It was a little like saying, "Curiously, schizophrenia was not identified as a disorder until the late 1880s, a few years after Alexander Graham Bell invented the telephone.") He also larded his writing with conditional statements and passive constructions: Eli Lilly "*reportedly* earn[ed] a profit" by licensing thimerosal to other drug companies; "the American health establishment . . . *understandably has an interest* in proving the unpleasant [thimerosal] theory wrong." When piled one atop another, they would have been enough to make any organization sound like a sinister cabal: The Girl Scouts of the USA *reportedly* uses winsome, underage children to peddle the high-fructose products it depends on for a significant portion of its income; the organization *understandably has an interest* in weakening child labor laws.

Kirby's most remarkable feat was his handling of criticism of a 2003 paper by the Geiers that had been published in the *Journal of American Physicians and Surgeons* and resulted in an American

Academy of Pediatrics statement titled "Study Fails to Show a Connection Between Thimerosal and Autism." The Geiers, Kirby wrote, "were made to sound like dimwits." Clearly, the AAP "felt the same disdain" for the researchers as it did for parents. Kirby went on to accuse "unnamed editors" of the AAP's journal, *Pediatrics*, of taking a "veiled swipe at their colleagues who publish the comparatively radical *Journal of American Physicians and Surgeons*," a publication at which the mainstream "looks down their noses" but which, Kirby said, was peer-reviewed, which presumably gave it a certain amount of credibility. According to Kirby, the reason for this haughtiness was the AAP's contempt for the Association of American Physicians and Surgeons, which published the journal. Many doctors, he wrote, "dismiss [the AAPS] as belonging to the bottom-feeding realm of homeopaths and chiropractors."

That summation made it sound as if the group was just one more organization testing the boundaries of modern medicine—but out of all the things the AAPS has been accused of, trafficking in nontraditional healing methods barely makes the list. It was founded in the 1940s as an extreme-right-wing organization, and over the years its leadership has overlapped with that of the ultraconservative John Birch Society. It has compared electronic medical records to the files kept by the German secret police, linked abortion to breast cancer, and claimed that illegal immigration leads to leprosy. For years, AAPS officials worked with Philip Morris on a junk science campaign attacking indoor smoking bans; as recently as the fall of 2009, it claimed cigarette taxes actually led to a "deterioration in public health."

And those are actually some of the group's more moderate stances. One month before the 2008 presidential election, it published an article on its Web site speculating that Barack Obama might be "deliberately using the techniques of neurolinguistic programming (NLP), a covert form of hypnosis." (The repetition of numbers in his speeches was one of Obama's "techniques of trance induction"; another was hand gestures that functioned as "hypnotic anchors.")

These tactics were most effective when employed on the weak-willed, non-Christian members of the cultural elite:

> Obama is clearly having a powerful effect on people, especially young people and highly educated people—both considered to be especially susceptible to hypnosis. It is also interesting that many Jews are supporting a candidate who is endorsed by Hamas, Farakhan, Khalidi, and Iran.

Indeed. In a four-hundred-plus-page book in which Kirby took pains to outline all the potentially compromising aspects of organizations like the CDC, the worst thing he could bring himself to say about the AAPS is that it is "considered by many experts to lie on the fringe."

While Kirby was careful to avoid looking like he had an allegiance to one side or the other in official statements and media appearances, he was not shy about trading on his ties with the anti-vaccine community. Nowhere was the symbiotic relationship between Kirby and the people about whom he had written more apparent than on a Yahoo!-hosted members-only forum called EOHarm. Soon after it was established, EOHarm took on the feel of a virtual autism advocacy conference. Like-minded parents talked about what avenues of research should be pursued ("moving the paradigm away from chasing the illusive [sic] autism gene . . . is of paramount importance") and rallied each other about the prospect of litigation ("once causation is established in vax court or state/federal court, then we will be able to place overwhelming political pressure to amend vica").* The forum was also a venue for Kirby to make thinly disguised requests

* The National Childhood Vaccine Injury Act is sometimes incorrectly referred to as the Vaccine Injury Compensation Act, or VICA. Some of the confusion stems from the fact that the official name of the program administered by the Vaccine Court is the National Vaccine Injury Compensation Program, or NVICP.

for support: When his publisher bought an ad in *The New York Times*, Kirby posted a downloadable copy with the message, "Please feel free to share, post, whatever. Some parents said they would buy space for it in their local papers, which is great, but I am certainly not asking anyone to lay out cash to promote the book, I just think it is a really cool ad!"

If there was ever any doubt that the anti-vaccine community saw Kirby as one of their own, and that he was more than happy for the embrace, it disappeared in the thousands of pages on the Yahoo! Message Board. (As of September 2010, the group's 2,320 members had posted a total of more than 106,000 messages.) In response to a message Kirby posted about future sales projections, one of the group's members wrote, "Two years ago this was the province of the loonie fringe. EOH has put us in the mainstream. Our main job is to destroy the credibility of the vaccine industry and that's just what EOH has done. . . . And don't forget the paperback is coming, and hopefully a movie too."

The impact of this grassroots support extended far beyond a community of die-hard activists: "Somehow," Kirby says, Deirdre Imus, the wife of radio and TV host Don Imus and a proponent of the thimerosal hypothesis, "got ahold of one of my galleys." The evening of March 9, 2005, Kirby spoke on the phone with the couple. The following day, he appeared as an in-studio guest on *Imus in the Morning*, which was nationally syndicated on WFAN–New York and also aired on MSNBC. At the time, Imus's show reached more than three million Americans, an audience that had earned the host a spot on *Time*'s list of the "25 Most Influential People in America."

The twenty-three-minute segment could hardly have gone better if Kirby had scripted it himself. Imus gave Kirby immediate credibility by introducing him as a "contributor to *The New York Times*, where he writes about science and health." (Since he'd started work on the book, the only health-related story Kirby had written for the *Times* was a June 21, 2004, piece about how "young men" were using Viagra "as an insurance policy against the effects of alcohol or for an increase in prowess.") After that introduction, the radio host prompted Kirby to hit all his main talking points. He asked about the book's name:

IMUS: Let me start with first—where did the title come from?

KIRBY: The title sort of emerged out of the text itself. . . . That actual term, it appears about seventeen times in the book, and the reason I chose it is "evidence" I think is the proper term. There is evidence of harm; there is a growing body of evidence of harm. Proof of harm is a loaded word and I didn't want to go quite that far.

IMUS: Evidence of harm linking thimerosal to autism.

KIRBY: To autism and other neurological disorders—ADD, ADHD, etc.

He asked Kirby why the CDC refused to acknowledge the error of its ways:

KIRBY: People at the CDC—the CDC recommended these vaccines very aggressively. . . . So the CDC has some blame to share here, and I don't think they're quite motivated to admit that they might have made a terrible blunder. . . . If what I write in the book is all true, we have just experienced one of the largest medical catastrophes of our time and put a generation of American children at terrible risk with possibly devastating results.

He even repeated Kirby's canard about needing to prove a negative:

IMUS: When you look at evidence on both sides, the studies that the CDC and IOM and others have done, where they suggest carefully that—I'll have to paraphrase what you wrote—where they found no causal link and no evidence of harm, what you point out I thought brilliantly is that that doesn't suggest it's completely safe or safe at all.

A quarter of the way through the interview, Imus, going along with Kirby's claim that he hadn't yet made up his mind on the subject, asked the author, "Which argument is most persuasive to you?" Kirby proceeded to summarize the theory that he focused the most attention on in his book. The typical person, he explained, has enough "sulfur-based proteins" to "sop up" the mercury that enters the body. In kids with autism, he said, "we now basically know" that those proteins are present in diminished amounts. The answer reflected an insight gained from the SafeMinds "Autism: A Novel Form of Mercury Poisoning" paper: Throw around some impressive-sounding scientific-sounding terms and very few people are going to feel comfortable calling you on the specifics. In this instance, Kirby's "we . . . basically know" would have been more accurately phrased as "there's no legitimate research that indicates this is the case."

After Imus announced Kirby's time on air was coming to a close, the author asked if he could "just mention a couple of other websites." He then gave a hat tip to the various organizations that had helped catapult him to one of the highest-profile appearances in the media world: He promoted the National Autism Association for "people interested" in mercury-related legislation, SafeMinds as a resource for "scientific research," and a group called Generation Rescue, which had been launched by a private equity manager and his wife in response to their son's diagnosis of autism, for anyone curious about "treatment and recovery."

Within minutes of arriving back home, Kirby discovered just how valuable a high-impact media appearance could be. Before he left his apartment, he says, he "happened to check Amazon." His book was not in the online retailer's top 30,000 sellers. After he returned from the interview he checked the site again: "It was at number eight."

There was no reason to expect Don Imus to do independent research about the vaccine debate—he made no bones about the fact that he was a radio talk show host and not a newsman. Tim Russert was another matter. As the doyen of the Washington press corps, he had developed a reputation as a relentless interrogator whose grasp of the issues allowed him to cut through the spin peddled by his high-profile guests.

In August, Kirby appeared on Russert's *Meet the Press* opposite Harvey Fineberg, the president of the IOM. While Fineberg, a medical doctor who had been the dean of Harvard's School of Public Health from 1984 to 1997, had the advantage of more than three decades of experience working in science and medicine, it was clear that Kirby was much better at making his case to a television audience. He spoke in easily digested sound bites that steered clear of eye-glazing details. He described the epidemiological studies that had been the foundation of the IOM committee's report as "rang[ing] from severely flawed to seriously questionable," without offering any further explanation. He claimed that in the fourteen months since the IOM report was published, the story had been "moving very, very fast." He repeated what had become a standard talking point among activists: There was "a small subset of children with a certain genetic predisposition, they are unable to properly process the mercury that they are exposed to," a theory whose appeal lies in the fact that it is both untestable and impossible to refute.

Fineberg, in contrast, sounded like he was participating in a graduate-level seminar in statistical analysis. He looked uncomfortable in his chair, his speech was filled with hesitations, and his language had enough qualifications to make his argument appear less strong than it was. The closest Fineberg came during the entire segment to making a definitive statement was, "We have now a growing body of evidence that while imperfect [is] altogether convincing and all reaching the same conclusion, even though they vary in their methods and in their approaches. And that conclusion was no association between the receipt of vaccines containing thimerosal and the development of autism."

Under these circumstances, it was Russert's responsibility to make sure that the truth was not a victim of his guests' respective skills in front of the camera. This was not, as Kirby had argued in his book, a political campaign—it was an important scientific debate in which one side had verifiable evidence and the other did not. Unfortunately, Russert didn't seem to have his bearings on the issue, which left the field open for Kirby to spout a series of baseless claims. "We know that certain children with autism, again, seem to have higher levels of mercury accumulating in their body," Kirby said. Later in the show, he built on that point: "Inorganic mercury basically gets trapped in the brain, and there's evidence to suggest that, in an infant brain, in the first six months to a year when the brain is still growing, when inorganic mercury gets trapped in that brain, you're going to have this hyper neuroinflammation, or the rapid brain growth that we see in autistic children." Kirby was confident this was true, he said, because of "a whole lot of new biology. This has all been published. None of the biology was published at the time of the IOM hearing. It has since been published, and I actually wonder if the IOM would consider reconvening a new committee or a new hearing to consider the evidence that's come out in the year and a half since the last report."*

Russert turned to Fineberg, who seemed simultaneously exasperated and astounded. It was as if he was being confronted by someone who argued that research done by people who believed the sun orbited around the earth had in fact proven that very thing. At the same time, he knew that simply stating that what Kirby was saying was ludicrous would make him seem arrogant and condescending—which would not have helped to sway the public to the side of science. Fine-

* Here, Kirby was referring to a series of studies done by the Geiers, the University of Kentucky's Boyd Haley, Northeastern University's Richard Deth, and the University of Arkansas's Jill James, all of which he also cited during his appearance on *Imus in the Morning*. In June 2010, the FDA accused Haley of multiple violations resulting from his marketing of an untested industrial compound used for heavy metal detoxification to parents of autistic children as a "supplement." The paper by Deth to which Kirby was referring had been rejected by three journals before it was published; one of Deth's claims was that massive B_{12} injections could help cure autism. In the text of her paper, James explicitly wrote, "[A]ttempts to interpret these findings are clearly speculative."

berg did his best to stammer out a response that was respectful and forceful at the same time: "Mr. Kirby's description about the certitude of this evidence, I think, exceeds the actual balance of evidence that is produced when you look at the totality." It wasn't a very convincing line, nor was his protestation that "other avenues of research looking at other possible causes today are much more promising ways to spend our precious resources."

In the months to come, Kirby would receive similarly credulous treatment from dozens of outlets. "The press was great, the reviews were great," he says. "I did not have a single hostile interview—not one." The narrative he laid out—proud, independent-minded mothers doing battle with greedy drug companies and corrupt government agencies—was, to quote an old journalistic cliché, a story that was too good to check. Reviewers treated the book more as if it were a John Grisham potboiler than a real-life story with enormous public health consequences. Again and again, they repeated Kirby's version of events as if it were the gospel truth. A full-page write-up in *The New York Times Book Review* was typical in this regard: Kirby was applauded for the "admirable job" he'd done "clarifying most of the scientific background" even as he resisted the temptation to "offer his own verdict on the debate." The book's "smoking gun," the *Times* wrote, was the *Journal of American Physicians and Surgeons* piece written by the Geiers. The IOM report, meanwhile, was relegated to a single sentence in the 1,200-word piece: "Despite their efforts, in May 2004 a committee from the Institute of Medicine found no 'causal relationship' between thimerosal-containing vaccines, or the MMR vaccine, and autism."

Over the years, Kirby's relationship with the movement he wrote about has become even more intimate. He was named a contributing editor at a SafeMinds-sponsored blog called *Age of Autism*, which bills itself as "The Daily Web Newspaper of the Autism Epidemic." (On Thanksgiving Day 2009, the site ran an illustration of health officials, science reporters, and vaccine advocates feasting on the corpses of dead babies.) He regularly covers autism and vaccines for the blog and news aggregator *The Huffington Post*. (Here are three headlines, chosen at random: "Autism, Vaccines, and the CDC: The Wrong Side of History," "The Autism Vaccine Debate—Anything But Over," and "Up to 1-in-50 Troops Seriously Injured . . . By Vaccines?")

In outlets such as those, as well as during media appearances and speeches, Kirby has helped to ensure that the story he says "cries out for answers" will never end. In the epilogue to his book, Kirby quoted a parent saying that if the thimerosal theory was correct, autism rates should start to go down by 2005, which was four years after thimerosal had been removed from childhood vaccines. Then in 2005, he told Don Imus, "It's going to take another two years before we know whether [thimerosal] has been causing the rise." In 2007, autism diagnoses were still going up; since then, Kirby has advanced an array of new theories to replace the ones that failed to materialize. His latest is that myelin, which forms a protective sheath around the part of a nerve cell that conducts electrical impulses, can be damaged by viruses, including the measles virus used in the MMR vaccine. Kirby has also displayed a recent preoccupation with aluminum, which is the latest bogeyman of vaccine denialists. "It is used in vaccines as an adjuvant to increase, to boost the immune response to the vaccine," he said as part of a recent presentation on "several pathways to autism." "Aluminum in and of itself is neurotoxic. Aluminum is very damaging to mitochondria. And aluminum is synergistic with mercury."

Kirby has also taken the lead in spreading misinformation about the Vaccine Court case of a girl named Hannah Poling. In 2000, when Hannah was nineteen months old, she received five vaccines during a wellness visit to her pediatrician. Not long afterward, she came down with a fever; by the time she was two years old, she'd begun a developmental decline that eventually left her unable to speak. In 2002, Hannah's parents, Jon Poling, a neurologist at Johns Hopkins, and his wife, Terry, filed a Vaccine Court claim.

Five years later, the Department of Health and Human Services conceded the Polings' assertion that as a result of an "underlying mitochondrial disorder," Hannah had a reaction to vaccines that "manifested as a regressive encephalopathy with features of [an autism spectrum disorder]." This was not seen as a particularly controversial decision: Since encephalopathies are recognized by the Vaccine Court as table injuries, in the absence of overwhelming evidence to the contrary, the court is mandated to grant an award for *any* encephalopathy that follows a vaccine within a set period of time.

Then, in February 2008, Kirby wrote about the case in a *Huffington Post* article headlined, "Government Concedes Vaccine-Autism Case in Federal Court." In his piece, Kirby described a study that had appeared in the *Journal of Child Neurology* in 2006 that seemed to support the notion that children with mitochondrial disorders could be especially susceptible to vaccine-induced encephalopathies. Kirby did *not* report that the piece's lead author was Hannah's father; that the paper's sole subject, an unidentified "19-month-old girl," was, in fact, Hannah Poling; or that the study was submitted for publication three years after the Polings had filed their initial claim and two years before the government's concession. When that news was eventually revealed, it came as a surprise to the *Journal of Child Neurology*'s editor, Roger Brumback, who called Jon Poling's behavior "appallingly troubling" in a published editorial. In the future, Brumback wrote, "statements from all authors concerning potential conflicts of interest" would appear with each article. "However, no written statement can substitute for honesty, good faith, and integrity on the part of authors."

None of that criticism or any of the detailed analysis of the case has tempered Kirby's rhetoric. At an autism conference in the spring of 2009, I spoke with Jon Poling at the back of the empty ballroom. "It's a no-fault system," he said. "So the ruling didn't say anything about the science. When they got up there and said, 'In no way have we said vaccines cause autism,' it's true. . . . It doesn't say anything about causation. The conclusion they came to is about compensation." A few hours later, David Kirby stood in front of hundreds of people at a lectern in that very room and gave his version of the case. "They conceded it twice," he said of the government's decision to award the Polings compensation. "And the short version of [the concession] is, in my own words, 'Hannah's autism was caused by a vaccine-induced exacerbation of her underlying mitochondrial dysfunction.' "

CHAPTER 18

A Conspiracy of Dunces

Almost exactly three months after *Evidence of Harm* hit bookstores, *Rolling Stone* and the online magazine Salon.com simultaneously published "Deadly Immunity," a 4,700-word story on mercury in vaccines written by Robert F. Kennedy, Jr. Kennedy, second-oldest son and namesake of the former attorney general and New York senator, described how he'd come to investigate the issue: "I was drawn into the controversy only reluctantly. As an attorney and environmentalist who has spent years working on issues of mercury toxicity, I frequently met mothers of autistic children who were absolutely convinced that their kids had been injured by vaccines. Privately, I was skeptical."

Then, he wrote, he began to look at the information these parents had collected. He pored over the transcript from the 2000 CDC-organized meeting at the Simpsonwood lodge outside Atlanta and spoke with members of SafeMinds and Generation Rescue. He also studied the work of the "only two scientists" who had managed to gain access to government data on the safety of vaccines: "Dr. Mark Geier, president of the Genetics Center of America, and his son, David." In the past three years alone, Kennedy wrote, "the Geiers have completed six studies that demonstrate a powerful correlation between thimerosal and neurological damage in children."

It wasn't long before Kennedy became convinced that he'd stumbled upon "a chilling case study of institutional arrogance, power and greed." If, as he believed to be the case, "our public-health authorities knowingly allowed the pharmaceutical industry to poison an entire generation of American children, their actions arguably constitute one of the biggest scandals in the annals of American medicine." Kennedy went on to quote SafeMinds' Mark Blaxill, whom he identified as the vice president of "a nonprofit organization concerned about the role of mercury in medicine," as Blaxill accused the CDC of "incompetence and gross negligence" and claimed that the damage done by vaccines was "bigger than asbestos, bigger than tobacco, bigger than anything you've ever seen."

In the article's final paragraph, Kennedy warned his readers of the scandal's likely effects on the future: "It's hard to calculate the damage to our country—and to the international efforts to eradicate epidemic diseases—if Third World nations come to believe that America's most heralded foreign-aid initiative is poisoning their children. It's not difficult to predict how this scenario will be interpreted by America's enemies abroad." In fact, he wrote, he was certain that the failure of a generation of "scientists and researchers . . . to come clean on thimerosal will come back horribly to haunt our country and the world's poorest populations."

Unlike David Kirby, Kennedy did not have the luxury of threading these indictments through hundreds of pages; as a result, the magnitude of the implied conspiracy was more immediately obvious. In order for what Kennedy was claiming to be true, scientists and officials in governmental agencies, nonprofit organizations, and publicly held companies around the world would need to be part of a coordinated multi-decade scheme to prop up "the vaccine industry's bottom line" by masking the dangers of thimerosal.

In Kennedy's telling, the plotting had been going on since the Great Depression, but it had begun in renewed earnest five years earlier "at the isolated Simpsonwood conference center," a location that Kennedy said was chosen because it was "nestled in wooded farmland next to the Chattahoochee River, to ensure complete secrecy." (In reality, the location was chosen because a series of previously scheduled conferences had booked up all the hotel rooms within fifty miles of Atlanta.) Kennedy relied on the 286-page transcript of the Simpsonwood conference to corroborate his allegations—and wherever the transcript diverged from the story he wanted to tell, he simply cut and pasted until things came out right. Again and again, he used participants' warnings about the reckless manipulation of scientific data by people with ulterior motives to do the very thing they were afraid would happen. The CDC's Robert Chen was one of the victims of Kennedy's approach. His actual quote is as follows:

> Before we all leave, someone raised a very good process question that all of us as a group needs to address, and that is this information of all the copies we have received and are taking back home to your institutions, to what extent should people feel free to make copies to distribute to others in their organization? We have been privileged so far that given the sensitivity of information, we have been able to manage to keep it out of, let's say, less responsible hands, yet the nature of kind of proliferation, and Xerox machines being what they are, the risk of that changes. So I guess as a group perhaps, and Roger [Bernier, the associate director of science at the National Immunization Program], you may have thought about that?

In Kennedy's hands, it became this:

> Dr. Bob Chen, head of vaccine safety for the CDC, expressed relief that "given the sensitivity of the information, we have been able to keep it out of the hands of, let's say, less responsible hands."

Even more egregious was Kennedy's slicing and dicing of a lengthy statement by the World Health Organization's John Clements. In this instance, Kennedy transposed sentences and left out words. Here is what actually appeared in the transcript, with italics added to indicate the sentences Kennedy used in his story:

> *And I really want to risk offending everyone in the room by saying that perhaps this study should not have been done at all*, because the outcome of it could have, to some extent, been predicted and we have all reached this point now where we are left hanging. . . .
>
> *There is now the point at which the research results have to be handled*, and even if this committee decides that there is no association and that information gets out, the work has been done and through Freedom of Information that *will be taken by others and will be used in other ways beyond the control of this group.* And I am very concerned about that as I suspect it is already too late to do anything regardless of any professional body and what they say. . . .
>
> My message would be that any other study—and I like the study that has just been described here very much, I think it makes a lot of sense—but it has to be thought through. What are the potential outcomes and how will you handle it? How will it be presented to a public and a media that is hungry for selecting the information they want to use for whatever means they have in store for them?

In "Deadly Immunity," that was changed to read:

> Dr. John Clements, vaccines advisor at the World Health Organization, declared flatly that the study "should not have been done at all" and warned that the results "will be taken by others and will be used in ways beyond the control of this group. The research results have to be handled."

To top it all off, Kennedy married together two separate comments made by the developmental biologist and pediatrician Robert Brent. In the first one, Brent said:

> Finally, the thing that concerns me the most, those who know me, I have been a pin stick in the litigation community because of the nonsense of our litigious society. This will be a resource to our very busy plaintiff attorneys in this country when this information becomes available. They don't want valid data. At least that is my biased opinion. They want business and this could potentially be a lot of business.

Thirty-eight pages later, Brent addressed the topic of "junk scientists":

> The medical/legal findings in this study, causal or not, are horrendous and therefore it is important that the suggested epidemiological, pharmacokinetic and animal studies be performed. If an allegation was made that a child's neurobehavioral findings were caused by thimerosal containing vaccines, you could readily find a junk scientist who would support the claim with "a reasonable degree of certainty." But you will not find a scientist with any integrity who would say the reverse with the data that is available. And that is true. So we are in a bad position from the standpoint of defending any lawsuits if they were initiated and I am concerned.

In a distortion that the editor of a high school newspaper would have balked at, Kennedy took these two statements, switched their order, and ran them together:*

> "We are in a bad position from the standpoint of defending any lawsuits," said Dr. Robert Brent, a pediatrician at the Alfred I.

* I attempted to contact Kennedy more than twenty times over an eighteen-month period. At various points, I was told that he was considering my interview request, that he was on vacation, that he was dealing with a family crisis, that he wasn't feeling well, that he was behind in his e-mails, and that he was on the verge of calling me back.

duPont Hospital for Children in Delaware. "This will be a resource to our very busy plaintiff attorneys in this country."

In the overall scheme of the piece, that type of quote massaging was considered so insignificant that it didn't warrant inclusion in the more than five hundred words' worth of "notes," "clarifications," and "corrections" that were eventually appended to the piece. (The misuse of Chen's quote wasn't acknowledged either.) Among the issues that were addressed were incorrect attributions, inaccuracies about which vaccines contained thimerosal at different points in time, a misrepresentation of the number of shots children had received in the 1980s, and a false claim about a scientist having a patent on the measles vaccine.

None of this put a dent in Kennedy's conviction that his allegations were valid, and in the weeks and months to come, he kept on repeating many of the errors *Rolling Stone* and Salon.com had already publicly acknowledged were wrong.* Just four days after a correction confirmed that the story had misstated the levels of ethylmercury infants had received—it was actually "40 percent, not 187 times, greater than the EPA's limit for daily exposure to methyl mercury"—Kennedy told MSNBC's Joe Scarborough, "We are injecting our children with *four hundred times* the amount of mercury that FDA or EPA considers safe." Kennedy also told Scarborough that children were being given twenty-four vaccines and that each one of them had "this thimerosal, this mercury in them." Those statements were not even remotely true: In 2005, the CDC recommended that children under twelve years old receive a total of eight vaccines that protected against a dozen different diseases. Only three of those vaccines had

* More dismaying than Kennedy's repetition of his discredited accusations was *Rolling Stone*'s insistence that the essence of the story remained correct. "It is important to note," the magazine's editors wrote in a statement that appeared in print and online, "that none of the mistakes weaken the primary point of the story." Five years later, the magazine appeared to have had a change of heart: In the spring of 2010, *Rolling Stone* removed the piece, along with all references to it and to the controversy it created, from its website.

ever contained thimerosal, and all had been manufactured without the preservative since 2001.

That Scarborough didn't ask Kennedy to produce evidence supporting his accusations is not surprising: Scarborough had long had a hunch that vaccines were to blame for his teenage son's "slight form of autism called Asperger's." Kennedy's research, it seemed, had confirmed his suspicions once and for all. "There's no doubt in my mind," Scarborough said, "and maybe it's two years from now, maybe it's five years from now, maybe it's ten years from now—we are going to find out thimerosal causes, in my opinion, autism."

CHAPTER 19

AUTISM SPEAKS

At the end of the twentieth century, there were only a handful of national autism advocacy organizations, the best known of which were Bernard Rimland's Autism Research Institute and its offshoot, Defeat Autism Now! Newer on the scene were the two groups that had begun in the mid-1990s, during a period of mounting frustration about the general state of autism research: The National Alliance for Autism Research, which Eric and Karen London had founded in 1994, and Cure Autism Now, which Hollywood producer Jonathan Shestack and his wife, television art director Portia Iversen, had started in 1995. Because of the environment in which they were launched, both groups had been relatively open-minded about the causes of autism—their leaders believed the best way to combat the neglect of the research community was to cast a wide net in search of answers.

The second-generation organizations that flowered in the early 2000s tended to follow a different model. They'd been started in large part in response to a specific set of circumstances—the fact that no one knew the effects of the amount of mercury being injected into children—and as a result, they tended to be more single-minded in their focus and more populist in their character. The best known of these organizations were SafeMinds and Generation Rescue. There was also Medical Interventions for Autism (MIA), which Liz Birt founded for the express purpose of raising money to support Andrew Wakefield; Talk About Curing Autism, which began when Lisa Ackerman posted a message on a Yahoo! Message Board inviting other parents in Southern California to join a "problem-solv[ing]" support group that emphasized alternative diets and treatments; Autism-One, which was started by parents named Teri and Ed Arranga and whose mission statement reads, "Parents are and must remain the driving force of our community. . . . AUTISM IS A PREVENTABLE/ TREATABLE BIOMEDICAL CONDITION. Autism is the result of environmental triggers"; and the National Autism Association, which preached, "Autism is a biologically based, treatable disorder."

The shared philosophical underpinnings of these groups could be seen in their repetition of a handful of key words and phrases: "Parents" and "community" were coded ways of signaling an opposition to an establishment that was beholden to the pharmaceutical industry; "the environment" and "environmental triggers" were euphemisms for vaccines; "alternative treatments" represented controversial methods for heavy-metal detoxification or severely restrictive diets; "biologically based" meant not genetic; and "treatable" and "curable" signified the crucial ingredient the organizations could offer that the mainstream could not: "hope."

By 2005, a preoccupation with vaccine safety and an opposition to traditional institutions were viewed by an ever-growing number of "autism advocates" as prerequisites for membership in their community. The organization most at risk of being a casualty of this ideological purge was the National Alliance for Autism Research, which, as Eric London puts it, remained steadfast in its opposition to the "DAN! model"—i.e., "decid[ing] what causes autism and fund[ing] those studies that proved it." Indications of this rift were evident as far back as 1998, when London wrote a critique of the methodology of Wakefield's *Lancet* paper in a NAAR newsletter. In his piece, London did his best to make it clear he was only addressing the way Wakefield had performed his research and was not writing about the legitimacy of the vaccine theory. "I said, Look, there's room to examine the hypothesis," London says. It didn't matter: "I received a death threat. Someone read my article and decided that, you know, vaccines were causing autism and people like me were defending vaccines."

The fissure between NAAR and those that disdained their approach had become a gulf by 2002, when *The New England Journal of Medicine* published an MMR study that NAAR had helped to fund. Because the results indicated a lack of causality between the vaccine and autism, NAAR was accused of shilling for the pharmaceutical industry; because the group hadn't insisted that mercury be included in the study, it was charged with betraying autistic children.* Eventually, the attacks became so fierce that the group felt compelled to release a statement justifying its work. "While [the thimerosal hypothesis] may merit additional research, it does not negate the validity of the Danish study's conclusions," it read. "Peer-reviewed research cannot and should not be refuted on the basis of hypothesis."

Even in the midst of all this intramural strife, NAAR remained one of the most successful autism-related organizations in the world. Its Walk for Autism Research had become the premier autism fundraising event in the country and its Autism Tissue Program—which collected donations from parents and made them available to qualified scientists for research—had assembled the largest collection of brain tissue from autistic children in the world.†

But by the middle of the decade, the autism advocacy movement had begun to resemble other burgeoning insurgencies throughout history whose rapid growth had put them at risk of breaking apart.

---

* Much was made of the fact that NAAR had received money from Merck, the manufacturer of the MMR vaccine: In 2000, the drug company made a one-time, unsolicited donation of $25,000, a sum that represented slightly more than one-tenth of one percent of NAAR's total revenue. That entire amount came after a former Cy Young Award–winning pitcher named Bret Saberhagen designated NAAR as his charity of choice in a Merck publicity campaign to promote its anti-baldness drug Propecia.

† The program operates in a manner similar to organ donation programs, with the majority of samples coming from people who died unexpectedly.

If a truce was going to be reached, it would have to be brokered by someone with political skills, celebrity cachet, and the ability to raise enough money to make all other concerns seem minor in comparison.

Katie Wright's son, Christian, was born on August 31, 2001, and his heartbreaking story resembles that of so many other parents around the world. "He developed typically until he was two and a half years old," Katie says. "He'd say, 'I love you, Mommy' and 'When do I get to drive the car'—all of that," she says. "He'd get excited when my mom came and he'd run down the driveway." Then, in what felt like an instant, the little boy Katie had known and loved simply disappeared. "Within a few months, he lost all of his skills, he became completely nonverbal," Katie says. "He didn't even recognize me anymore."

In February 2004, Christian was diagnosed with autism. If anything, the next several months were even worse than the previous ones had been, as Christian developed a host of physical problems that were every bit as pressing as his cognitive deficits: "He was having ten bowel movements a day," Katie says. "The skin was coming off of his backside. It burned through the carpet. It was just gross, and nobody could do anything." When his pain got especially bad, Christian would descend into violent, uncontrollable tantrums during which he'd throw himself against the floor. In those moments the only thing that would pacify Christian was milk—and before long he was drinking a dozen bottles a day. Less than a year earlier, Christian would use the phone to pretend to call his grandmother. Now he was so sick that his mother couldn't even get a good night's sleep.

In the year following Christian's diagnosis with autism, Katie and her husband, Andreas Hildebrand, traveled a disheartening path. They made appointments with specialists in "four or five different states," all to no avail. It was particularly frustrating, Katie says, when one high-paid doctor after another refused even to consider whether Christian's physical illness and his autism might be intertwined. "They didn't connect them," she says. "They were asking my husband and

I if we were related, if mental illness ran in my family, and I'm like, Can you talk about what's going on now? My husband and I aren't related. You're just barking up so many wrong trees." In desperation, Katie scheduled Christian for an endoscopy, but even that didn't provide any clues as to why he was in so much pain.

Katie and Andreas's inability to help their son did not mean that the Wright family was completely powerless: As the vice chairman and executive officer of General Electric and the chairman and CEO of NBC-Universal, Katie's father, Bob Wright, was one of the few people with the standing and influence to change the direction and the future of autism research. To that end, in February 2005, the Wrights announced they were starting a charity called Autism Speaks. "Too many parents go to bed each night praying that one day their child will look them in the eye, smile and say, 'Mommy,' " Suzanne Wright wrote in a *Newsweek* essay. "My daughter is one of them. My husband and I are launching Autism Speaks for her and for all the families stricken by the disorder."

From the outset, the Wrights made clear that one of their goals was to put together a "big tent" coalition as a way of stopping the internecine warfare that threatened to cripple the autism advocacy movement. Within a matter of months, they appeared to have made significant progress: That November, NAAR merged with the new charity, a move both sides said would allow them to continue their "joint commitment to accelerate and fund biomedical research into the causes, prevention, treatments and cure for autism spectrum disorders." Within a year, a similar "consolidation" took place between Autism Speaks and CAN, which had by that point become another group that supported the vaccine theory. Statements detailing the moves described them in virtually identical language: Each merger signified "the tipping point in the autism community, bringing together the best science, collaborative minds and impassioned advocates." In both arrangements the mergers could have more accurately been described as a swallowing up of the more established organization. The result was a mega-charity called Autism Speaks in which the president of Autism Speaks retained his position and Bob and Suzanne Wright continued in their roles as chairman and vice chair of the group's board.

To outsiders, those moves may have appeared to unite different factions for the sake of a greater good. To anyone fluent in the vocabulary of the autism world, however, it was apparent that the old divisions remained. In a press release announcing the CAN merger, Sallie Bernard, who was by then a CAN board member, made sure to signal that the new partnership would only improve CAN's ability to "creat[e] a network of parents, bonded by hope." Commenting on the newly expanded version of Autism Speaks, Eric London struck an equally independent note: "Our commitment to funding evidence-based science will always be our top priority."

It wasn't long before there were indications that London's promise had been made in vain: Former members of CAN outnumbered NAAR loyalists on the Autism Speaks board, and at a high-profile fund-raiser featuring Bill Cosby and Tom Brokaw, Bob Wright surprised the audience by announcing that the entire $1.4 million raised at the gala would go toward "environmental" research. But the biggest concern for those who wanted to move beyond the vaccine wars were the rumors that Katie Wright had come to believe vaccines had caused her son's autism—and that she was pressuring her parents to recalibrate the research priorities of Autism Speaks accordingly.

Katie Wright's introduction to the parent-led movement dedicated to nontraditional ways of thinking about autism occurred during a period in which she had all but given up hope for the future. "The child I knew [was] gone," she says. "So you have that, and then you have the lack of services, and you couple that with a feeling like there's no sense of urgency [to find effective treatments]." She was also suffering from a growing feeling of isolation. "I have a lot of friends in the city who have autistic kids," she says. "But not too many who have kids that are as severely impacted as mine." Whenever she would ask for advice, Katie says, "They would be like, 'I don't know—I don't know and that has nothing to do with my son.'"

The one place where Katie did find parents with experiences similar to her own was online. "So many of us, especially in the beginning, we can't leave our homes and the only time you have to seriously do research or discuss this is the middle of the night," she says. "You can't go to a doctor or leave your home—so yeah, it was so great for me, for sure . . . to find [other parents] in all parts of the country. It was so helpful."

Then, toward the end of 2005, Katie's mother met Jane Johnson, who had by that point become an active supporter of Andrew Wakefield's. "I was always complaining about [Christian's] terrible tummy issues," Katie says, "and Jane said, 'Why doesn't she come over and see me? I know a lot about that.'" Katie was skeptical. "I was thinking, I'm sure she's a nice lady but it's going to be another one of those dead ends." Nevertheless, Katie and her husband agreed to hear Johnson out, and, Katie says, "we were so happy we did." One of Johnson's first suggestions was that Christian see Arthur Krigsman, a New York gastroenterologist who subscribed to Wakefield's theory that a large majority of autistic children had underlying gut disorders that exacerbated or even caused their conditions.

In January 2006, Krigsman performed a second endoscopy on Christian, just to make sure the doctor who'd scoped him two years earlier had not missed anything. According to Krigsman, he had: Offering up the images from inside Christian's intestines as proof, Krigsman told Katie that her son had symptoms indicative of autistic enterocolitis, the condition Andrew Wakefield claimed to have discovered in 1998. "He was the first doctor who really understood what was going on with Christian's crazy gut," Katie says. Krigsman's proposed treatment was one that had grown popular among Wakefield's adherents: regular doses of anti-inflammatory medication, a gluten- and casein-free diet, and weekly chelation therapy.*

The bond between Krigsman and Wakefield went beyond their shared affinity for performing endoscopies on developmentally disabled children and their claims that they'd identified dozens of cases of intestinal trauma other doctors had seemingly overlooked: They also had in common a history of running into trouble with medical boards and employers. Krigsman's problems dated to 2001, when a review board at Manhattan's Lenox Hill Hospital rejected the first of several of his research proposals to scope autistic children because of concerns that the procedure's risks would outweigh its anticipated benefits. In November 2002, the hospital discovered that despite its review board's decisions, Krigsman had performed "invasive endoscopic procedures" on hundreds of children, many of whom were autistic. Krigsman later sued the hospital for restricting his privileges, and in 2004, he did not renew his application for employment.

Krigsman's difficulties with medical authorities did not end when his relationship with Lenox Hill did: In 2004, he was fined by the Florida Medical Board for failing to document continuing medical education required for his initial licensure, and in 2005, the Texas State Board of Medical Examiners found that he'd misrepresented himself to state officials and to the public. When finally he was permitted to practice medicine in Texas, that decision came only after a $5,000 fine and an order to provide a copy of the board's critical report to "all hospitals, nursing homes, treatment facilities, and other health care

---

* Chelation therapy is a process during which chemicals are introduced into the body to sever the bond between heavy metals and body tissue; its popularity as an alternative treatment for autism is based on the theory that autism is a result of mercury poisoning. Chelating agents can be administered through lotions, capsules, suppositories, or IV infusions. Chelation's risks include vomiting, convulsions, irregular heartbeats, and death. There has never been a clinical trial testing chelation for autism; one government-funded trial was halted after scientists at Cornell University and the University of California, Santa Cruz, found that rats without heavy metal poisoning who were chelated showed signs of cognitive impairment.

entities where Respondent has privileges, has applied for privileges, applies for privileges, or otherwise practices."

As significant as those charges sounded, they barely registered compared to the ones Wakefield was confronted with at the time. In February 2004, *The Times* (London) began running a series that was the result of months of dogged research by investigative reporter Brian Deer. In his first batch of stories, Deer exposed what the British press should have uncovered years earlier: Wakefield had not been a disinterested clinician while preparing his *Lancet* paper condemning the MMR vaccine; instead, he'd received multiple payments to examine children as part of a lawsuit that was being prepared against drug manufacturers. What's more, almost half of the twelve children in his study had been funneled to Wakefield by Richard Barr, the class action lawyer representing parents convinced that vaccines had injured their children. The most shocking revelation came later that year, when Deer reported that shortly before his piece in *The Lancet* was published, Wakefield had filed a patent for a measles vaccine that could be administered independently of those for mumps and rubella—which was just the product parents would be clamoring for if they became convinced that the MMR vaccine was more than their children's bodies could handle. The sum total of Deer's investigation raised concerns that the lumbar punctures, endoscopies, and heavy anesthesia that the children in Wakefield's study had undergone had exposed them, as one member of the British Parliament put it, "to unacceptable risks and unnecessary procedures."

Within days of the first of Deer's stories appearing in print, ten of Wakefield's twelve co-authors on the *Lancet* paper officially disassociated themselves from Wakefield's conclusions; Richard Horton, *The Lancet*'s editor, released a statement that said, "If we had known the conflict of interest Dr Wakefield had in his work, it would have been rejected"; and the General Medical Council (GMC), which is in charge of regulating and licensing doctors in the U.K., began an inquiry into Deer's allegations. (That December, the council formally announced that it would begin a "public hearing" regarding the possibility that

Wakefield and two other doctors who worked on the 1998 paper had committed "serious professional misconduct." In 2005, the GMC released a list of eleven preliminary charges, which included acting "in a manner likely to bring the medical profession into disrepute" and subjecting children to "unnecessary and invasive investigations.")

In an impartial setting, Deer's articles and the subsequent GMC inquiry would have raised concerns that Wakefield had used children as guinea pigs and fudged results in an effort to advance his career and fatten his wallet. Among Wakefield's supporters, the findings were viewed as just the latest effort on the part of cowed governments, powerful business interests, and mercenary journalists to suppress the truth. This reaction is not surprising: By that point, the anti-vaccine movement had come to embody the qualities that David Aaronovitch identifies in his book *Voodoo Histories* as characteristic of history's most enduring conspiracy theories, from the belief that the moon landing was a hoax to a conviction that HIV and AIDS were developed by drug companies in conjunction with the CIA:

> These include an appeal to precedent, self-heroization, contempt for the benighted masses, a claim to be only asking "disturbing questions," invariably exaggerating the status and expertise of supporters, the use of apparently scholarly ways of laying out arguments (or "death by footnote"), the appropriation of imagined Secret Service jargon, circularity in logic, hydra-headedness in growing new arguments as soon as old ones are chopped off, and, finally, the exciting suggestion of persecution.

Whenever Wakefield was involved, persecution wasn't just suggested, it was stipulated. True to form, he cast the litany of charges as yet another example of the personal sacrifices he made for the sickest and most helpless members of society. "I have [already] lost my job," he said. "But if you come in to me and say, 'This has happened to my child'—what's my job? What did I sign up to when I went into medicine? To look after your child. . . . I'm here to address the concerns of the patient. There's a high price to pay for that. But I'm prepared to pay it."

His supporters were as well: In 2004, with the help of several hundred thousand dollars raised by Liz Birt and a $1 million donation from Jane Johnson and her husband, Wakefield teamed up with Arthur Krigsman to set up a clinic in Austin, Texas. The treatment center was designed to be a one-stop shop for all things—research, treatment, parent education—relating to alternative ways of thinking about autism. Wakefield christened his new endeavor Thoughtful House, a choice that was typical of the savvy way he marketed himself and his ideas. It implied that his treatments, in contrast to those that autistic children received elsewhere, were caring and solicitous, and that he could deliver to parents an introspective, self-aware child.

One of the clinic's early patients was Christian Wright. "We tried the diet and it actually helped," Katie Wright says. "I got him healthy thanks to Thoughtful House. He no longer looks like he has AIDS or cancer and his T cell count is not so low. It's great—he's not in and out of the hospital and he's not covered in rashes." That did not mean, however, that Wakefield's methods were helping with all of Christian's problems: "Cognitively," Katie says, "I'm not so sure."

CHAPTER 20

Katie Wright's Accidental Manifesto

On 1:10 A.M. on March 26, 2007, a new thread was started on Yahoo!'s EOHarm forum by someone who wanted to know if Bob and Suzanne Wright were "pursuing [bio-medical interventions] with their grandson yet not publicizing this via Autism Speaks." Before the night was over, dozens of people had replied. No one seemed to know for sure what type of treatment Christian Wright was receiving, but that didn't stop increasingly caustic complaints about the ways in which the Wrights had failed the "autism community." One representative post read:*

> If they are utilizing bio-medical treatments, but choose not to disclose that to the pulic than to me that is lying by omission and very misleading. To think that they would deceive the autism community by actively setting forth one agenda for the rest of us, while pursuing cutting edge medial interventions for their own is bull shit and it's wrong. They obviously don't want

* The spelling and punctuation of quotations that appeared on the EOHarm forum have been left as they appeared so as to avoid the landscape of "sic"s that would result from the hurried spelling and disregard for grammar typical of Internet message boards.

> privacy for they wouldn't have put their story out into the public and started Autism Speaks. Can't have it both ways and get away with it.

As the day went on, the accusations became increasingly personal: The Wrights were just looking for attention, or they were so rich they didn't care what happened to "average" families with autistic children, or Katie Wright was being muzzled by her parents because they were cowards and refused to take an unpopular stance. "Remember when Katie backed out of the press conference last year?" read one message. "I think maybe her daddy told her too. I imagine Granny and Pappy pay lots of the bills for Christian's treatment." The author of a different message told the group's members, "What you all should be asking is, 'Where the hell is Katie Wright?' "

Barely twelve hours after the initial post in the thread had appeared, Wright herself provided an answer. "I can completely understand the community's frustration," she wrote, and went on to assure the group's members that she was on their side.

> Yes, my husband and I follow a true biomedical approach for Christian. He was a very sick little boy for years. Thanks to the work of wonderful DAN! doctors and parent mentors . . . Christian is much better. . . .
>
> It was not until we found Dr. Krigsman did these start to get better. I cannot pretend to imagine living through this horrible nightmare, only to have the situation complicated by not having the funds to get your child the treatment he/she needs. It is shameful.
>
> Our children are so sick and families are being destroyed in the struggle to find and pay for treatment. The parents, and some physicians, I have met through this horrible journey are the bravest people I know. There is no question that I believe my son regressed into autism due to environmental factors. No question.

Wright's willingness to reply to the group resulted in about an hour and a half's worth of appreciative posts—after which the criticism and accusations began once again. "Why must we mince words," wrote one parent. "I'm not being a wiseass (at least not here), but there seems to be some whacked-out religious conviction that no one can say vaccines." Another repeated the accusation that the care Christian was receiving betrayed a cowardly double standard on the part of Bob and Suzanne Wright:

> By treating their grandson with biomedical treatments they are privately acknoledging the vaccine played a role in Christian's regression. They chose to put their name and face with autism by starting Autism Speaks. They have been very successful in raising awareness for autism. . . . Yet, I have seen no grants awarded to vaccine or "environmental" research by Autism Speaks. . . .
>
> The fact that the Wrights have treated their son/grandson biomedically and then thru their organization whitewash that issue is not only hypocritical but shameful. . . .
>
> With power comes responsibility and I feel they have failed us thus far. Awareness could be a good thing but what good is it if more children are injured by vaccines while they are waiting for the "right time" to come forward?

Over the next several hours, the group's participants got increasingly worked up. When one member, echoing Wright's language, referred to "environmental factors," she was all but accused of collaborating with the enemy:

> Who's side of the fence are you on? The importance to using the word VACCINES is SO IMPORTANT, or dangerous, to the other side that they would rather get stung by a wasp then to use the word!!! Yet you berate us for wanting specific research, into the word they WON'T say!!! How do you know how many kids

> regressed or not? Have you done research? Please don't throw those comments out without facts.

At 12:30 the following morning, having faced a series of escalating challenges—"Do you believe your son was damaged by vaccines? Do you have vaccine records to prove it?"—Wright jumped back into the fray.

> I did not mean to be evasive. I believe that Christian's regression and subsequent autism was the result of receiving 6 vaccines during 1 office visit at 2 months of age. . . . His vaccines contained thirmesol. He received 31 vaccines total by 18 months. . . . It is devastating because so much of this is preventable.

Before signing off, she lauded a reporter at *Discover* magazine who had "really nailed" the story "regarding how autism was triggered." It was her fervent hope, she said, that more journalists would follow suit.

It didn't take long for Wright's plea to be answered: The following day, David Kirby wrote an article for *The Huffington Post* about the "fierce debate" within the autism community over the "potential role of environmental influences, particularly mercury, and very particularly, mercury in vaccines." Many of the people in the "upper echelons of Autism Speaks have rejected any environmental hypothesis and insisted that autism is purely a genetic disorder—though Bob and Suzanne Wright (and the organization itself) remain officially neutral on this crucial question." Then, invoking "the child who inspired Autism Speaks," he added his voice to those pressuring Christian's grandparents to take a stand.

> On Tuesday Christian's mom, Katie, posted an entry on Yahoo's EOH List (named after my book, "Evidence of Harm") that minced not a word. . . .
>
> So how will some Autism Speaks officials react to Katie's

> statements? They could fall back on two recent, but highly inconclusive studies that support the autism-is-genetic paradigm, and continue to reject the environmental hypothesis. But I wouldn't bet on it. . . .
>
> The Wrights' grandson is now, perhaps, the most famous toddler with autism in the world. And the whole world, including the world's largest autism charity founded upon his very diagnosis, should listen to his mom: "There is no question that I believe my son regressed into autism due to environmental factors," Katie wrote. "No question."

As always seemed to be the case, Kirby neglected to provide crucial pieces of context that would have made his story significantly less dramatic. He failed to mention that the last doses of childhood vaccines containing thimerosal had been used up in 2003, which had the effect of making a debate about the past appear to be a topic of urgent concern. His claim that two recently published genetic studies had been "inconclusive" was technically correct but missed the point: Though the studies had not identified the *precise* genes connected to autism, they had added to the collection of evidence suggesting the disorder had a genetic basis. His assertion that people in the "upper echelons of Autism Speaks" were insisting that autism was a "purely genetic disorder" was flat-out false: Virtually no one questioned whether environmental factors could play a role in the disease—what people *did* doubt was whether vaccines were that factor. (Kirby also employed the diversionary tactic of insinuating that any flaws in the genetic theory meant that those of its most vocal opponents were ipso facto correct.) Finally, there was his recasting of Katie Wright's late-night e-mails as the opening salvo in a with-me-or-against-me battle over the future of autism advocacy. As far as the readers of *The Huffington Post* knew, when Katie Wright "minced not a word" about the connection between her son's autism and the vaccines he'd received, it wasn't the product of two days of cyber-hazing—it was an impromptu manifesto.

• • •

On April 5, 2007, Katie Wright and Alison Singer, the executive vice president and spokesperson for Autism Speaks and the mother of an autistic daughter, were booked on *The Oprah Winfrey Show*, which is seen by seven million American viewers a day and is broadcast in 140 countries around the world. They were there to discuss *Autism Every Day*, a documentary that provided a raw and sometimes despairing look at life with an autistic child. Before they went on-air, Singer says, the show's producers told the two women to focus on the day-to-day experiences of families with autistic children—what it was like to go to the supermarket, how it affected relationships with friends—and not to get into the disputes about the disease's root causes. "They were pretty clear," Singer says. "We were not supposed to get into the science, because we were not scientists."

It was Wright's first public appearance since she'd been goaded, ten days earlier, to "be a big girl and step it up." Despite what her producers had said, Winfrey was going to give Wright a chance to do just that. According to Singer, several segments into the program, Winfrey turned to Wright during a commercial break. "I hear you wanted to say something about vaccines," Winfrey said. "I don't want you to go home feeling like you didn't get to say what you wanted to say." When they came back on the air, Wright was ready:

WRIGHT: We give thirty-seven vaccines to babies under the age of eighteen months. Nobody has shown that that's safe, a wise idea, and the multiple vaccines at once. My son is sick all the time, he has constant immune, immune reactions to everything, his allergies are exploding. I mean, if you look at food allergies, asthma, autism, it's all connected.

WINFREY: Well I'm going to let you—I said I was going to let you say it because I said when you got home, you would be crazed that you didn't say it. (Turns to audience.) She wanted to say it and I wanted you to get it out there. (Audience applauds.)

WRIGHT: Thank you, thank you so much.

WINFREY: Because you're a mother.

WRIGHT: Yeah.

WINFREY: You're a mother dealing with her child every day.

WRIGHT: Thank you. (Applause.)

WINFREY: We'll be back in a moment.

Wright's characterization of the vaccine program, which Winfrey did not challenge, was completely false. There *had* been studies showing vaccines were safe, just as there were reasons behind the structure of the recommended vaccine schedule. Her figure of thirty-seven vaccines—which was six more than what she'd claimed Christian had received in her post on the EOHarm forum—seemed to have been pulled from thin air: Even if you counted the MMR vaccine and the DPT vaccine as three separate vaccines each, there were only twelve vaccines administered to infants eighteen months or younger. Tallying them up by individual injections—meaning the polio vaccine, which requires four doses spread out over a period of a year, would be counted four times—still resulted in a total of only twenty-one shots.

In the coming weeks, Wright's statements about vaccines grew in intensity, but it wasn't until the annual AutismOne conference in Chicago in late May that the situation exploded into public view. In 2007, the marquee event of the conference's second-to-last night was the screening of an excerpt from a videotaped interview Wright had given to Kirby shortly after his article in *The Huffington Post* had appeared, which would be followed by a discussion session that both Wright and Kirby would participate in. (The full interview, along with "special commentary," was for sale at the conference as a DVD.) Shortly after the interview began, Kirby asked Wright to describe what had happened when Christian received his two-month shots. Her son, Katie said, had screamed "like someone was killing him." His temperature spiked to 104 degrees, where it remained for hours on end. Her doctor was unsympathetic: "She said just give him more Tylenol, every, whatever, four hours. I put him in an ice bath at her suggestion and that was horrible." Twelve hours later, Katie said, Christian finally passed out. "I went back to the top pediatrician and she assured me that was extreme but within the normal reactions," she said. "It was nothing to be concerned about."

For the next two years, she said, Christian thrived. Then, after he turned two, "he started losing language and he would give me blank looks." Initially, Katie said, she didn't believe Christian's autism had any connection to the vaccines he'd received—after all, the disease's first manifestations occurred more than twenty months after his extreme reaction. "I really fought it, because it's so painful," she said. "It's beyond painful to think that I, as his mother, took him to the pediatrician, held him down screaming and allowed this to happen to him. It's so hurtful for me to think about that. I fought it for probably a year." Then, she told Kirby, she began to learn more—"and *Evidence of Harm* was so pivotal to my education"—and it all clicked into place.

If that had been all that she'd said, Wright's statements would have posed a problem—the daughter of the founders of the largest autism charity in the world was taking an aggressive stance on an issue about which the organization had always insisted it was agnostic—but that was nothing compared to the statements she made during the post-screening Q&A session. First she claimed that the Autism Speaks science advisory board, of which Eric London was a member, was wasting money by awarding grants to people's "friends." Then she said that while parents like the Londons may have at one point been a valuable part of the community, "now that their children have grown it is time for them to step aside" so that Autism Speaks could focus exclusively on biomedical and environmental research.

"I'd just grown very frustrated," Katie says. "We'd been . . . traveling all over with Christian. I read, like, a thousand books. I was on the computer reading everything—I just felt such great frustration. His story wasn't getting out and I didn't even feel like anyone was studying him. . . . So, yeah, I just didn't care [about people's reactions]. I just really wanted someone to start helping Christian."

Wright's harsh characterization of some of Autism Speaks' most senior members forced resentments that had been hidden from public view out into the open. "We didn't realize that Katie was as anti-vaccine as she was, anti-NAAR," says Eric London. After Wright's interview with Kirby, London says, he began to feel that her attitude was basically, "We acquired NAAR and need to wait a little while to get rid of them." Bob and Suzanne Wright, however, were not as willing to see their coalition unravel. Within days, they'd posted a stiffly worded statement on the Autism Speaks Web site.

> Katie Wright is not a spokesperson for Autism Speaks. She is our daughter and we love her very much. Many of Katie's personal views differ from ours and do not represent or reflect the ongoing mission of Autism Speaks. Her appearance with David Kirby was done without the knowledge or consent of Autism Speaks. . . .
>
> Autism Speaks merged with NAAR because it believes in and supports its scientific mission, methods, and advisory board. We are proud of the accomplishments of NAAR and grateful to the families and volunteers who created it. They are a tremendously valued part of Autism Speaks. We welcome input from volunteers and parents/guardians of children with autism of all ages, including adults with autism. We apologize to our valued volunteers who were led to believe otherwise by our daughter's statement.

Katie Wright might not have been a spokesperson for Autism Speaks, but that point of detail was lost in the escalation that followed. What had started with a single question on an online forum had become a media free-for-all, with details of the spat appearing everywhere from a gossip column on foxnews.com ("Celebrity Autism Group in Civil War") to the front page of *The New York Times* ("Debate over Cause of Autism Strains a Family and Its Charity").

Today, the Wright family and Autism Speaks remain divided on the issue. "My daughter feels very strongly that vaccines played a serious role in her son's [autism]," Bob Wright told me in the fall of 2010, not long after Katie Wright posted an entry on the *Age of Autism* blog stating that Autism Speaks was "jeopardizing [its] relationship with the community of families" by continuing to fund research on the genetic causes of and behavioral treatments for autism. "She's my daughter. I love her to death and I understand her position." What's more, Wright says, he is not one of those people who feel it's time to turn the page. "The vaccine issue is one that's very important to me and very important to us at Autism Speaks because it could be resolved, one way or the other," he says. "[The government] ought to put together a series of very serious studies and publicly recognize that there are children that unfortunately are damaged by vaccines."

It's precisely that stance that has caused a growing number of the charity's former allies to break ranks. In January 2009, Alison Singer resigned after four years as one of the group's top executives. "It got to a point," she says, "where the science kept coming out . . . but there was always a group of parents that, when a vaccine study would come, would say, 'We don't care about the science because we know in our hearts that our child got autism from vaccines.' And that's just not how science is done. At some point, you have to say, 'This question has been asked and answered and it's time to move on.' " Five months later, Eric London resigned from Autism Speaks as well. "By preferentially investing and advocating for the use of limited financial resources on the 'biological plausibility' [of the causal relationship between vaccines and autism] argument, the organization is adversely impacting the advancement of autism research," he wrote in a statement announcing his departure. "The lowering of the vaccination rate has already led to deaths. If Autism Speaks' misguided stance continues, there will be more deaths and potentially the loss of herd immunity which would result in serious outbreaks of otherwise preventable disease. I further fear that if and when herd immunity is lost, there may be a societal backlash against the autism community."

CHAPTER 21

Jenny McCarthy's Mommy Instinct

Jenny McCarthy's career in the public eye began in October 1993, when, at the age of twenty-one, she was named *Playboy*'s Playmate of the Month. Not long thereafter, she was crowned the magazine's Playmate of the Year, and for much of 1994 she hosted *Hot Rocks*, an hour-long *Playboy* Channel TV show that aired "uncensored" music videos.

Her rise to mega-stardom began in earnest in 1995, when MTV hired her to co-host its new dating show, *Singled Out*. From the show's first episodes, it was clear the five-foot, seven-inch native Chicagoan's appeal had at least as much to do with her fearless sense of humor and the guileless way in which she broke taboos as it did with her bleached blond hair and her buxom physique: Every time she gleefully sniffed her armpits or bragged about the potency of her flatulence, she reminded the world that you didn't need to be an ethereal waif like Kate Moss or a supercilious beauty like Christy Turlington to be a sex symbol.

By 1996, McCarthy had become one of the most ubiquitous stars in the United States. That year she appeared on two more *Playboy* covers and bidding wars broke out whenever she announced her intention to work on a new project. In 1997, she had two eponymous shows running on TV at the same time—MTV's sketch comedy program *The Jenny McCarthy Show* and NBC's half-hour sitcom *Jenny*—and was paid $1.3 million by HarperCollins for her quasi-autobiography, *Jen-X*. Then, in an instant, her appeal seemed to evaporate. Despite a massive promotional campaign, *Jenny* tanked—its ratings were so bad that it was canceled after a single season—and her book flopped as well.

The stall in her career gave McCarthy a chance to focus on her personal life. In 1999, she married an actor and director named John Mallory Asher, and in 2002 she gave birth to the couple's first child, a boy named Evan. Shortly thereafter, she became an unlikely publishing phenomenon: A decade after the combination of her just-one-of-the-guys attitude and girl-next-door good looks made her the object of teenage boys' fantasies, she discovered that her willingness to openly and honestly tackle subjects other people were too timid (or uncomfortable) to address held a similar appeal for thirtysomething women trying to navigate their way through adulthood. In 2004, she released *Belly Laughs,* a book about pregnancy in which McCarthy addressed topics like butt-hole problems and pubic hair fiascos; it ultimately sold more than 500,000 copies. Her next book, 2005's *Baby Laughs*, didn't do quite as well, although its sales figure of 250,000 copies still made it an unqualified success.

Then things seemed to unravel once again. *Dirty Love*, a film McCarthy wrote and starred in and which Asher directed, was released to universally bad reviews: In his zero star write-up, Roger Ebert said the "hopelessly incompetent" film was "so pitiful, it doesn't rise to the level of badness." Instead of being refreshingly honest, McCarthy's attention-getting antics—like the scene in *Dirty Love* that featured her wallowing in a pool of her own menstrual blood—seemed increasingly desperate and contrived. Even *Life Laughs*, the third book in her trilogy on early motherhood, didn't do nearly as well as her previous two books. By the end of the year, her personal life had also hit a rough patch, and she and Asher filed for divorce. To top it all off, Evan, her perfect, blond-haired, blue-eyed little boy, was having problems of his own.

• • •

In the spring of 2006, McCarthy and her son were walking in downtown Los Angeles when a woman approached them. "You're an Indigo," the stranger said. "And your son is a Crystal." McCarthy barely had time to shout "Yes!" before the woman left as quickly as she'd come.

That chance encounter served as McCarthy's introduction to a New Age movement based on the belief that a group of spiritually advanced children known as Crystals are destined to lead humanity to its next evolutionary plateau. (Parents of Crystals recognize each other through the purplish aura they emit, hence their designation as Indigos.) It was only then, McCarthy would say later, "[that] things in my life started to make sense." Evan had always been a unique kid—he seemed less social and more intense than other children his age—and several doctors had already broached the topic of whether he had a behavioral or developmental disorder.

McCarthy's embrace of Crystal beliefs gave her the strength to reject doctors' efforts to squash Evan's spirit. "The reason why I was drawn to Indigo, and probably many other mothers [are] too, was the fact that my son was given a diagnosis for a behavior issue," she explained later. "I would not accept this negative label they were trying to put on my son and found out that he mirrored Indigo characteristics. . . . Once moms educate themselves, and find out what other mothers of Indigos do for behavior issues, we generally find the answers and solutions for everything."

That summer, McCarthy launched IndigoMoms.com, an online portal for Indigos looking to connect with one another. It included a social networking area called "Mommy+Me," a forum where McCarthy would answer readers' questions, and an e-commerce section that offered tank tops and baby doll tees for sale alongside one-year subscriptions to something called a "Prayer Wheel"; the services of McCarthy's sister, Jo Jo, a "celebrity makeup artist"; and Quantum Radiance Treatment by McCarthy's friend Nicole Pigeault.

IndigoMoms.com never really caught on, and by the end of 2006, McCarthy pulled the plug on the site. (It remains available only through a service that collects archives of Web pages.) As it turned out, by that point McCarthy was already deeply involved with another parent-led movement defined by its opposition to conventional medicine. In 2005, McCarthy had contacted Lisa Ackerman, the mother who'd founded Talk About Curing Autism five years earlier. "Jenny was looking for information to help her son Evan, who was recently diagnosed," Ackerman wrote in an essay titled "TACA and Jenny McCarthy." "Jenny is an extraordinary mom. She ran with every bit of information that she gleaned from TACA's website, individual mentoring and community outreach efforts and was back when she needed more. As Evan improved Jenny kept good on her promise to get involved." In fact, McCarthy got so involved that she donated a portion of the proceeds from *Life Laughs*, which was released several months before she launched her Indigo Web site, to the organization.

Shortly thereafter, McCarthy told Ackerman she'd decided to write her next book about autism—and, McCarthy vowed, when it came out she'd publicize it on *The Oprah Winfrey Show*. The narrative for this latest project would be in stark contrast to the Crystal Child one McCarthy had been promoting on *The Tonight Show* and in newspaper interviews: Now, McCarthy said, her mistreatment at the hands of Evan's doctors had begun when she'd tried to discuss with them her concerns about vaccines. The only two constants of McCarthy's competing story lines were her refusal to let the medical establishment victimize her and her promise of succor to anyone who followed her path. "I say, Okay, let's look at your choices," she says of the message she's currently pitching to the public. "You have a choice of listening to the medical community, which offers no hope, or you can listen to our community, which offers hope. . . . Our side at least gives you . . . somewhere to go."

True to her word, on September 18, 2007, one day after *Louder than Words: A Mother's Journey in Healing Autism* was released, McCarthy appeared as a guest on *Oprah*. That afternoon, with Ackerman looking on from the second row of the studio audience, McCarthy told Winfrey that her journey had begun with a flash of insight that sounded similarly dramatic to the one that had occurred in 2006 when a stranger told McCarthy that her son was a Crystal Child—except, McCarthy said, this epiphany had taken place two years earlier, when she awoke one morning with a terrifying premonition that something was wrong. Shortly thereafter, Evan, who was around two years old at the time, had the first in a series of what McCarthy described as life-threatening seizures. For months, McCarthy said, her requests for help for her child were dismissed by every doctor she approached. (At times, McCarthy said, her doctors' condescension would mutate into rage: She claimed that one pediatrician had become so incensed by her insistent questioning that he shouted at her to "leave the hospital—now!") It wasn't until Evan suffered a near-fatal heart attack that he was properly diagnosed as autistic—and even then, McCarthy said, she wasn't offered any help or support. "I got the, 'Sorry, your son has autism' [speech]," she told Winfrey. "I didn't get the here's-what-to-do-next pamphlet."

Winfrey, who praised *Louder than Words* as "beautiful" and "riveting," didn't ask McCarthy why she hadn't mentioned the seizures or the screaming doctors or the heart attack during her Indigo phase, when she'd claimed that treating Evan for a behavioral disorder would be akin to "taking away all the beautiful characteristics he came into this world with." In fact, neither Winfrey—nor, seemingly, anyone else—asked McCarthy about her involvement with the Indigo movement at all. Instead, Winfrey praised McCarthy's unwillingness to bow to authority, her faith in herself, and her use of the Internet as a tool for bypassing society's traditional gatekeepers:

McCARTHY: First thing I did—Google. I put in autism. And I started my research.

WINFREY: Thank God for Google.

McCARTHY: I'm telling you.

WINFREY: Thank God for Google.

McCARTHY: The University of Google is where I got my degree from. . . . And I put in autism and something came up that changed my life, that led me on this road to recovery, which said autism—it was in the corner of the screen—is reversible and treatable. And I said, What?! That has to be an ad for a hocus pocus thing, because if autism is reversible and treatable, well, then it would be on *Oprah*.

The ad McCarthy saw was for a wheat- and dairy-free diet. Within weeks of her putting Evan on this new regimen, McCarthy said, he'd doubled his language, his eye contact improved, he began smiling more, and he became more affectionate. "Once you detox them," McCarthy said, "your kids are going to get better. You're cleaning up their gut. You're cleaning up their brain. There is a connection."

Winfrey nodded in agreement—but how, she asked, did McCarthy know to try this specific diet as opposed to the "fifty other things" that showed up online?

McCARTHY: Mommy instinct.

WINFREY: Mommy instinct.

McCARTHY: Mommy instinct. . . . I went, okay—I know my kid. . . . I know what's going on in his body, so this is what makes sense to me. . . .

WINFREY: Okay—so this is what Jenny says really worked for her. It doesn't mean it will work for all children. . . . It worked for her. This is her book. She wrote the book. So she knows what she's talking about.

As it turned out, Mommy instinct had done more than just show McCarthy which of the many alternative "biomedical" treatments she should pursue—it had also given her insight into what had made Evan sick in the first place. Winfrey, in much the manner she'd done with Katie Wright five months earlier, prompted McCarthy to share that information with the audience:

WINFREY: So what do you think triggered the autism? I know you have a theory.

McCARTHY: I do have a theory.

WINFREY: Mom instinct.

McCARTHY: Mommy instinct. You know, everyone knows the stats, which being one in one hundred and fifty children have autism.

WINFREY: It used to be one in ten thousand.

McCARTHY: And, you know, what I have to say is this: What number does it have to be? What number will it take for people just to start listening to what the mothers of children who have autism have been saying for years? Which is that we vaccinated our baby and something happened. . . .

Right before his MMR shot, I said to the doctor, I have a very bad feeling about this shot. This is the autism shot, isn't it? And he said, "No, that is ridiculous. It is a mother's desperate attempt to blame something on autism." And he swore at me. . . . And not soon thereafter, I noticed that change in the pictures: Boom! Soul, gone from his eyes.

At that point, Winfrey picked up an index card. "Of course," she said, "we talked to the Centers for Disease Control and asked them whether or not there is a link between autism and childhood vaccinations. And here's what they said." As she started to read, the screen filled with text.

> We simply don't know what causes most cases of autism, but we're doing everything we can to find out. The vast majority of science to date does not support an association between thimerosal in vaccines and autism. . . . It is important to remember, vaccines protect and save lives.

When Winfrey appeared back on screen, she turned to McCarthy, who was ready with a response: "My science is named Evan, and he's at home. That's my science." There was little question that Winfrey's sympathies lay with the "mother warrior" who'd written a "beautiful new book" about how she'd cured her son of a supposedly incurable disease as opposed to the faceless bureaucracy that couldn't provide any answers.

Before the end of the show, Winfrey told viewers that McCarthy would be available to answer questions for anyone who logged on to a "special [online] message board just for this show so you can share your stories." One fan asked McCarthy what she would do if she could do it all over again. "The universe didn't mean for me to do anything else besides what I did," McCarthy answered, "but if I had another child, I would not vaccinate." A mother wrote in to say that she had decided not to give her child the MMR vaccine "due to the autism link." McCarthy was delighted. "I'm so proud you followed your mommy instinct," she wrote.

Within a week of her appearance on *Oprah*, during which time McCarthy had also broken the news about her relationship with the comedian Jim Carrey, McCarthy had repeated her story on *Larry King Live* and *Good Morning America*. On those three shows alone, she reached between fifteen and twenty million viewers—and that wasn't including people who watched repeats or saw the clips online. Print publications told her story as well: *People*, which is one of the largest general interest magazines in the country, ran an excerpt from *Louder than Words* under the headline "My Autistic Son: A Story of Hope." The media blitz's effects were felt immediately. Ackerman, who'd appointed McCarthy as TACA's celebrity spokesperson in June, said the group was so swamped with e-mails from parents pleading for information about how to cure their children that it scrapped its more cautious expansion plans and went national that fall. "Had to," she said. "Jenny forced us."

McCarthy's sudden ubiquity did more than give families affected by autism hope for a miracle cure—it also further legitimized a movement that still had not completely shed its reputation as being on the scientific fringe. Dan Olmsted, a former UPI reporter who is one of the editors of *Age of Autism*, gives McCarthy credit for single-handedly pushing vaccine skeptics into the mainstream: "To anybody who comes to this issue from the environmental and recovery side of this debate—the idea that something happened to these kids, and it's probably a toxic exposure—Jenny McCarthy is the biggest thing to happen since the word autism was coined."

The media's willingness to indulge McCarthy's campaign and its disinclination to provide an accurate representation of the issues at stake continued unabated in 2008. On World Autism Awareness Day that April, Larry King devoted his full hour-long broadcast to "Jenny McCarthy's Autism Fight." Also appearing on the show that evening were David Kirby and Jay Gordon, the celebrity pediatrician who'd been treating Evan ever since McCarthy became convinced he'd been harmed by the MMR vaccine. Together, the trio repeatedly shouted down David Tayloe, the president-elect of the American Academy of Pediatrics.

GORDON: David Kirby's book is entitled *Evidence of Harm*, okay? The evidence is there. We have to address the evidence. We do not have respect for the instincts of our parents. We don't have respect for the immune system. The immune system is a complicated, complicated system in the body, complex—

TAYLOE: But you need scientific evidence that—

GORDON: You need to prove it's safe!

McCARTHY: First!

GORDON: Yes!

KIRBY: There is a bill in Congress to study vaccinated versus unvaccinated populations in this country. Doctor, would you support that legislation? Would you?

TAYLOE: We support—

KIRBY: Do you?

TAYLOE: We are not afraid of the truth at the American Academy of Pediatrics—

McCARTHY: Well, will you support the unvaccinated/vaccinated study?

Two months later, McCarthy and Jim Carrey led a "Green Our Vaccines" rally in Washington that also featured Gordon and included a keynote address by Robert Kennedy. (In the TACA press release announcing the rally, Ackerman appeared to confuse Kennedy with his uncle, Massachusetts senator Edward Kennedy: "Having Senator Kennedy as part of the supporters for the Green Our Vaccines Rally is an honor.") McCarthy's rally-related appearances on *Good Morning America* and Fox News's *On the Record with Greta Van Susteren* didn't even feature anyone representing an opposing viewpoint.

By that time, McCarthy's autism activism had become a full-time job. She'd taken over Generation Rescue, which was rebranded as "Jenny McCarthy and Jim Carrey's Autism Organization." (After the couple's split in the spring of 2010, the Web site was listed either as "Jenny McCarthy's Generation Rescue" or "Jenny McCarthy's Autism Organization.") By the end of the year, she'd published *Mother Warriors: A Nation of Parents Healing Autism Against All Odds* and signed a deal with the licensing agency Brand Sense to create Too Good by Jenny, a line of products ranging from bedding to cleaning supplies that "will be positioned as providing safe, non-toxic surroundings for children." She'd also launched Teach2Talk Academy, a school for autistic children, and developed a series of Teach2Talk DVDs designed to improve autistic children's "imagination" and "empathy toward others" by having them mimic what they see on screen.

The following spring, McCarthy was booked as the keynote speaker at the annual Autism-One conference at the Westin O'Hare. A half-hour before she was scheduled to appear, most of the seats in the Westin's 7,400-square-foot Grand Ballroom had already been claimed. Twenty minutes later, people were sitting two and three deep along the walls and in the aisles. Before the start of the main event, the restless audience had to sit through a presentation by Sarah Clifford Scheflen, a thirty-one-year-old speech pathologist who was the co-founder of Teach2Talk. (In April 2010, Teach2Talk Academy was closed after McCarthy and Scheflen parted ways due to "different visions for the school.") Scheflen appreciated that putting autistic children in front of a TV might seem to some to be a counterintuitive way to teach children with developmental disorders how to interact appropriately with actual human beings.* "People always ask me, 'Why, Sarah? Why does it work with the TV and not one-on-one?' " she said. "Children tend to be strong visual learners. . . . Sometimes *I* get distracted when I see two pairs of shoes on the floor, and I think that's what happens when kids come into my office." A handful of people in the audience chuckled, although most seemed nonplussed by Scheflen's comparison. "So I would show the child the prerecorded model and then the child would watch the video and imitate that—because most children learn through imitation."

Within fifteen minutes of the start of Scheflen's presentation, murmured conversations had begun to break out throughout the hall. After a series of glitches with the hotel's AV system, a visibly flustered Scheflen began to peer offstage. "So, um, I think Miss Jenny—is she here?" Indeed she was. Scheflen quickly wrapped up, and when McCarthy stepped out from the wings, the crowd erupted. One woman started hopping up and down like a teenager at a Taylor Swift concert, shouting over and over, "We love you, Jenny!"

McCarthy was dressed casually—she had on a pink zip-up sweater and jeans, and her hair was pulled into a tight ponytail—and when she took the stage, she started clapping along with the audience. "How many people are back from last year?" Hundreds of hands shot up. "I love it! We have an overflow crowd in the other room, so I'll give a shout-out to them." Then, in lieu of a prepared speech, McCarthy told the audience she'd rather answer questions they had about her journey.

" 'Jenny,' " she said, reading off a note card, " 'will you please repeat

---

* The only widely acknowledged treatment for autism is applied behavioral analysis, or ABA, which involves hours of one-on-one therapy per day.

the five steps you said last year? That was a huge help to us.' " That, McCarthy explained, was a reference to the checklist she keeps on her refrigerator. "The first one I call cleaning the bucket. All these kids have a bucket of toxins, infections, funguses," she said. Before they can treat their children effectively, McCarthy explained, parents needed to properly identify the problem: "PLEASE go and get allergy testing. If you want to know what kind, it's in my book." Step two is cleaning out the fungus: "A lot of these kids are malnourished. . . . please, please don't forget about that." Step three is detoxification, either using chelation or any number of other therapies. "You know, I've been getting glutathione IVs every weekend because I'm starting to take care of myself," she said. (Glutathione is a naturally occurring antioxidant that is depleted in patients with wasting diseases such as sepsis, cancer, and AIDS. There has never been a clinical trial on the effects of glutathione infusions in healthy humans or developmentally disabled children.) "I was a cold sore, herpes monster. . . . I was a mess. I started glutathione and in the last three months I haven't had a cold sore."

After acknowledging the irony of her next recommendation, McCarthy explained that step four was drugs. "Even though some of us are so angry at the pharmaceutical companies," she said, "I'm grateful for the medicine that we do need to get our kids better." According to McCarthy, antifungals and antivirals were especially important; in fact, she said she knew one child who fully recovered from autism in less than three months as a result of antiviral therapy.

Finally, there was step five: positive thinking. "I love that one more than anything," McCarthy said, because it demonstrates the power parents have to change their lives. "If you think, 'My kid is going to get better,' he's going to get better. If you keep thinking, 'My kid is going to be sick,' he's going to be sick." (Presumably, a more detailed explanation of how that process worked was provided at one of the conference's earlier lectures, which covered a philosophy that involves wishing your way to better health.)

"I've come to trust other parents more than anyone during this journey," McCarthy said. "I salute you all for being here today, and for believing and trusting and working on your child, for spending the money and the time and the tears on autism." There are so many people, McCarthy said, who don't come to these conferences, a fact that left her dumbfounded. "It breaks my heart," she said, when she meets the parents of a child with autism and "they still refuse to do anything."

What McCarthy seemed to be saying was that it broke her heart when parents refused to do *everything:* buy books and DVDs; try chelation and hyperbaric chambers; take supplements, make gluten-free meals, and eat organic chickens; march on Washington, write elected officials, and sign petitions; use DAN! doctors, travel across the country for treatment at special clinics, and ignore anyone who suggests otherwise . . . and if none of that works, start all over again.

Besides being expensive and exhausting, one disadvantage of that approach is that it's impossible to know if any of it actually does any good. An indiscriminate attitude toward treatment also makes it hard to determine what changes are due to the natural rhythms of disease: Temporary ailments by definition get better and the symptoms of lifelong conditions almost always wax and wane, which means that even the most far-fetched cure is bound to look like a winner every now and again. In his book *Innumeracy*, the mathematician John Allen Paulos describes how proponents of pseudoscientific therapies rely on this reality to shade their products in the best light possible. "To take advantage of the natural ups and downs of any disease (as well as of any placebo effect)," Paulos writes, "it's best to begin your worthless treatment when the patient is getting worse. In this way, anything that happens can more easily be attributed to your wonderful and probably expensive intervention. If the patient improves, you take credit; if he remains stable, your treatment stopped his downward course. On the other hand, if the patient worsens, the dosage or intensity of the treatment was not great enough; if he dies, he delayed too long in coming to you."

James Laidler, a medical doctor who teaches in the biology

department at Portland State University, has firsthand experience with the lure of this approach. About a year after his oldest son was diagnosed with autism, Laidler's wife returned home from an autism conference flush with stories about how seemingly intractable cases of the disease had been "cured." While initially skeptical, Laidler agreed there was no harm in seeing if their son responded to some of the vitamins and supplements that had been recommended. Soon after that, the couple removed gluten and casein from their son's diet. The next thing they tried was hormone therapy. "Some of it worked—for a while—and that just spurred us to try the next therapy on the horizon," Laidler wrote in an essay about his experiences. When the Laidlers' second son started showing autistic-like symptoms, they decided to treat him as well. It was around this time that Laidler went with his wife to an autism conference and saw firsthand what had so impressed her. "[I] was dazzled and amazed," he wrote. "There were more treatments for autism than I could ever hope to try on my son, and every one of them had passionate promoters claiming that it had cured at least one autistic child."

This was how the family found themselves headed to Disneyland with forty pounds of preapproved food for their two boys, "lest a molecule of gluten or casein catapult them back to where we had begun." That's exactly what they were convinced would happen when, during an unobserved moment, their younger son ate a waffle he'd snatched off a table. "We watched with horror and awaited the dramatic deterioration of his condition that the 'experts' told us would inevitably occur," Laidler wrote. "The results were astounding—absolutely nothing happened." Over the next several months, the Laidlers stopped every treatment except for occupational and speech therapy. Not only did their sons not deteriorate, they "continued to improve at the same rate as before—or faster. Our bank balance improved, and the circles under our eyes started to fade." During those years in which he and his wife had been religious devotees of various biomedical treatments, Laidler wrote, they'd just been "chasing our tails, increasing this and decreasing that in response to every change in his behavior—and all the while his ups and downs had just been random fluctuation."

In some ways, the Laidlers were lucky: The cost of trying every new treatment that comes along can be more than time, money, and dashed hopes, a fact that is tragically illustrated by chelation, the favored cure for ridding the body of "environmental" toxins. A large part of chelation's appeal among parents lies with the way it tackles the putative problem head-on: It results in the literal expulsion—or "excretion," to use the phrase favored by its proponents—of the hypothesized poisons from autistic children's bodies. Unfortunately, as can be expected from a chemical cleansing process originally designed during World War I as a treatment for mustard gas exposure, chelation comes with a significant amount of risk. When Liz Birt's son, Matthew, was chelated, his condition seemed to worsen, and in one instance, chelation preceded a grand mal seizure. Colten Snyder, whose family's suit claiming the MMR vaccine had caused his autism was one of the Vaccine Court's initial Omnibus Autism Proceeding test cases, had an even worse experience: After his second round of chelation, a nurse wrote in his medical records that he went "berserk." He also became aggressive and noncompliant, became more prone to tantrums, and exhibited increased repetitive behaviors. After his third round of treatment, he was brought to a medical facility due to severe back pain, which is one of the procedure's known side effects.

Then there's the case of Abubakar Nadama, who moved with his mother to Pennsylvania from Batheaston, England, because chelation is not permitted for the treatment of neurodevelopmental disorders in the U.K. On August 23, 2005, Abubakar went into massive cardiac arrest and died while receiving intravenous chelation therapy from a sixty-eight-year-old ear, nose, and throat specialist named Roy Kerry. In a presentation at the 2009 AutismOne conference titled "Starting the Biomedical Treatment Journey," TACA's Lisa Ackerman told parents they couldn't let Abubakar's death dissuade them. "I'm going to not be politically correct again," she said. "There's a child that passed away from chelation and it was extraordinarily sad and tragic. . . . The guy that gave the chelator to that little boy gave the wrong dose and the wrong type, and the kid had a heart attack because the doctor erred." That didn't mean, Ackerman said, that parents should be "afraid"; after all, they were going to need to "step it up" if they wanted their kids to get better. Ackerman failed to mention that less than a year after his patient died, "the guy that gave the chelator" was recognized as a DAN!-approved clinician, a designation that is obtained by attending a thirteen-hour seminar conducted by the Autism Research Institute, signing a loyalty oath to the organization's principles, and paying an annual fee of $250. (In order to maintain certification, doctors must attend a continuing education seminar every two years.)

Another of Ackerman's recommendations that morning was to buy some of the "thousands" of supplements marketed to parents of autistic children. Her personal favorites were those produced by a company called Kirkman Labs. "Go get their handy-dandy resource guide called 'The Roadmap.' That will tell you what supplements do what," she said. "I'm a big fan of Kirkman's because they've been around forever and their products are tried and true." Less than nine months later, Kirkman did a voluntary recall of seven of their products because they contained high levels of antimony, a chemical element used in flameproofing, enamels, and electronics—and one that some anti-vaccine activists had recently been proposing as a potential cause of autism.

CHAPTER 22

Medical NIMBYism and Faith-Based Metaphysics

For decades, Jay Gordon—or "Dr. Jay" as he prefers to be called—has been more lenient than most of his peers about altering the vaccine schedule or skipping some shots altogether. He says that one reason for this stance was his exposure to "Vaccine Roulette" and "a relatively obscure European piece of literature" in the 1980s. Equally important is his stated philosophy, which is emblazoned across the top of every page of his Web site and which dovetails nicely with the "Mommy instinct" advocated by his most famous former client: "No one knows your child better than you do."

Over the past several years, Gordon has become a self-appointed spokesman for "traditional" doctors who support the anti-vaccine camp—and where vaccines are concerned, Gordon's belief is that it's more important to be loud than to be right. "I'm a member in good standing at Cedars-Sinai [Medical Center], at UCLA [Medical Center], the AAP, and so on," he told me when I asked him about Jenny McCarthy's influence on the national debate over immunization practices. "We need someone who's willing to yell about formaldehyde [being in vaccines]—even though it's wrong. 'There's formaldehyde in vaccines!'—Well, actually, I don't think there really is. . . . The role that [Jenny] plays is a higher visibility, she's one of the higher visibility actors, she has a very good story, very telegenic, well connected to the media. She also happens to be quite knowledgeable. She's not a ditz, she's a smart woman and the mother of a child who she says suffered a vaccine injury."

Perhaps this ends-justify-the-means approach to facts explains Gordon's habit of making statements that veer between inaccurate and highly speculative. In the Foreword to one of McCarthy's books, he wrote, "Vaccines can cause autism. . . . Period." Later, in response to a *Los Angeles Times* article about the dangers posed by unvaccinated children, Gordon wrote a story in *The Huffington Post* that included this line: "Unvaccinated children do not pose a threat to vaccinated children or their families."

An even better-known Southern California doctor who, despite having no specific training in immunology or public health, is an outspoken proponent of "working with" vaccine denialists is Bob Sears. Like Gordon, Sears prefers to be called by his first name only. He is, along with his father, William "Dr. Bill" Sears, the primary author of the more than a dozen books that make up the "Sears Parenting Library," including *The Baby Book* and *The Healthiest Kid in the Neighborhood.* (The elder Dr. Sears is best known for his attachment parenting philosophy, which claims that parents who are not sufficiently responsive to their infants' emotional needs put their children at risk for psychological and mental health problems later in life.)

"I became passionate about educating parents all over america [sic] when I discovered that there were no good, complete, unbiased sources of information out there for parents to read," Sears wrote in an e-mail in response to a question about his initial interest in the controversy. "Everything was either completely pro vaccine or antivaccine. I wanted to create something that would give parents both sides of the story." The result was 2007's *The Vaccine Book: Making the Right Decision for Your Child*, which includes an "alternative" vaccination schedule that's based on Sears's personal experiences as a pediatrician. (Sears's most recent best-seller about a topic in which he does not have specialized training is 2010's *The Autism Book: What*

Every Parent Needs to Know About Early Detection, Treatment, Recovery, and Prevention.)

Many of Sears's colleagues disagree with his claim that he presents "both sides of the story." "As a general pediatrician working in the community, I can say without hesitation that your book has done more to harm my efforts to educate families on vaccines and to give vaccines than to help them," Brian Bowman wrote in a letter titled " 'Front Line' Response to *The Vaccine Book*" that appeared in the December 30, 2008, issue of *Pediatrics*. According to Bowman, the time he spent talking to parents about vaccines "has shifted away from what has been discussed by Hollywood celebrities and more to parents wanting to follow your recommendations. You must understand that the timbre of your book, and your inability to offer the explicit truth regarding vaccines, their safety and the diseases they prevent supports the unfounded fears of parents. . . . [T]hey point at your book, and say that your book tells them it is not safe."

Sears's questionable assertions are by no means limited to his recommended schedule. In *The Vaccine Book*, he says that "natural" immunity is more effective than immunity gained through vaccination and implies that parents whose unvaccinated children come down with infections don't regret their decisions. The book's most startling passage, however, is included under the heading "The Way I See It." "Given the bad press for the MMR vaccine in recent years, I'm not surprised when a family . . . tells me they don't want the MMR," he writes. Because there's so little risk of getting infected, "I don't have much ammunition with which to try to change these parents' minds." He, does, however, advise them against talking to their friends about their concerns: "I also warn them not to share their fears with their neighbors, because if too many people avoid the MMR, we'll likely see the diseases increase significantly."

This brand of medical NIMBYism is consistent with an attitude toward medicine and personal health that's been bubbling up in America for more than fifty years. To the extent that one can generalize about a period that saw more medical advances than in all of previous history, the great thematic struggle of the twentieth century was the attempt to reconcile a mechanistic view of the body with emerging philosophies of mental well-being. As Anne Harrington writes in *The Cure Within*, her history of mind-body medicine, for decades the tacit assumption was that these two ways of thinking were incompatible—you either believed the body responded to intangible, unconscious forces (e.g., psychosomatic medicine's embrace of a Freudian approach to illness in the 1940s and 1950s) or that it operated according to immutable physiological laws (e.g., behavioral medicine's priority of "laboratory based rigor" in the 1970s).

Sometime in the 1980s, the mind-body connection became an object of fascination for the public at large—and, in doing so, what had been an either/or debate became an all-of-the-above smorgasbord. A century after Christian Science founder Mary Baker Eddy preached that spiritual fitness could cure bodily harm, a view emerged that a mix of holistic approaches chosen by whim could—and probably should—be employed alongside modern medicine, almost as a way of hedging one's bets. It's no accident that a proposition that did not require people to make a choice—between Eastern and Western approaches to medicine, between spiritual and physical philosophies of healing—was articulated just as the first baby boomers were reaching middle age: This was a generation whose cultural identity depended in part on a belief that you could turn on and tune in without actually needing to drop out. The premise of a slogan like "if it feels good, do it" is that by virtue of feeling good, "it" is by definition the *right* thing to do.

The result has been a buffet-style, nonjudgmental approach to wellness where any combination of ingredients—a serving of Prozac, a side portion of self-help, and a dessert of yoga and meditation—can supposedly be transformed into a coherent meal. The appeal of such an approach is obvious: Who wouldn't want to believe he could feel his way to the best course of treatment?

This type of faith-based metaphysics, indulged by doctors like Gordon and Sears and fetishized in the media, has no bigger cheerleader than Oprah Winfrey. The notion that physical health and personal fulfillment can result *only* from unifying the body with the mind, the physical with the psychological, the corporeal with the spiritual, is the thread that connects the assortment of New Age healers, "alternative" doctors, and past-their-prime celebrities Winfrey has promoted over the years. A recent example of the first category is Rhonda Byrne, the author of *The Secret*, who has been a guest on Winfrey's show and was featured on Oprah.com and in the pages of *O*, Winfrey's magazine. *The Secret* is based on something called the "law of attraction," which holds that "everything that's coming into your life you are attracting into your life."

The second category is embodied by Winfrey's "favorite ob-gyn," Christiane Northrup, who believes that illness is an example of being "in labor with yourself because everything that no longer serves your highest purpose and your optimal health starts to go away." (Northrup's prototypical example of this phenomenon is thyroid problems in women, which she says are often the result of energy blockages that stem from "the result of a lifetime of 'swallowing' words one is aching to say.") Finally there are those, like 1970s TV sitcom star Suzanne Somers, whose expertise seems to be based on nothing other than the fact that people recognize their names. In an appearance on *Oprah* to promote *Ageless: The Naked Truth About Bioidentical Hormones*, Somers detailed her health regimen, which, in addition to sixty daily supplements, includes injections of estrogen into her vagina that are meant to trick her body into thinking it's still in its thirties as opposed to its mid-sixties. "She's ready to take on anyone, including any doctor who questions her!" Winfrey told her audience after Somers bragged of her success. "We have the right to demand a better quality of life for ourselves—and that's what doctors have got to learn to start respecting."

While these empowering testimonials sound harmless enough, many of them are actually quite dangerous. On a psychological level, there's the implication that sickness—or any other kind of difficulty—is symptomatic of a personal failing. (It's probably easier for Winfrey, whose net worth is estimated at $2.7 billion, to meditate her way out of a funk than it is for a single parent who can't afford health care.) In terms of physical health, the stakes can literally be life and death: After one of Winfrey's viewers read about a breast cancer victim who claimed to have cured herself exclusively by using the law of attraction, the viewer decided to forgo surgery and chemotherapy and rely instead on the teachings of *The Secret* for a cure. In that instance, Winfrey, as *Newsweek* put it in a 2009 cover story on the talk show host, seemed "genuinely alarmed" when she discussed the incident on-air. "I don't think that you should ignore all of the advantages of medical science and try to, through your own mind now because you saw a *Secret* tape, heal yourself," she said.

Winfrey refused to talk to *Newsweek*, but she reiterated that point in a statement she gave to the magazine:

> The guests we feature often share their first-person stories in an effort to inform the audience and put a human face on topics relevant to them. I've been saying for years that people are responsible for their actions and their own well-being. I believe my viewers understand the medical information presented on the show is just that—information—not an endorsement or prescription. Rather, my intention is for our viewers to take the information and engage in a dialogue with their medical practitioners about what may be right for them.

Winfrey's suggestion that she is just a neutral disseminator of information is a dodge offered so frequently that it's easy to overlook how absurd it is. The gibberish that Northrop or Somers—or, for that matter, Jenny McCarthy—is peddling does not qualify as "information," and their appearances on Winfrey's show are nothing if not endorsements. A more frank reckoning with the message Winfrey promotes would have acknowledged that in her world, being responsible for your actions has less to do with determining how your behavior affects those around you than it does with making sure you're "living in harmony" with your true nature. This "I feel therefore it is" outlook may empower Winfrey's viewers to take charge of their lives, but it ignores completely the perils that face a society where everyone runs around intuiting their own versions of the truth.

The notion that people should base medical decisions on what is "right for them" is particularly problematic in a public health context, where individual choices cannot be cordoned off from each other. Consider the case of Julieanna Metcalf, a fifteen-month-old fully vaccinated girl who was taken to the hospital on January 23, 2008, with what her mother thought was a particularly bad case of the flu. It was only after extensive tests that doctors discovered that Julieanna had a compromised immune system that rendered the vaccine for Hib ineffective. By the time she got out of the hospital almost a month later, Julieanna had suffered multiple seizures and had had a buildup of fluid in the brain so dangerous it required emergency surgery. She'd also lost all her motor skills—including the ability to swallow—and will require multiple immune globulin injections each week for the rest of her life.

Even with her weakened immune system, Julieanna might not have caught Hib if everyone around her had had their shots, but the Minnesota community in which she lived was a place where the same ethos emanating from Gordon's and Sears's waiting rooms and Winfrey's couch had taken hold. The outbreak that ensnared Julieanna also resulted in the hospitalization of four other children. One was a baby who was too young to have been vaccinated. The parents of the three others had all chosen not to vaccinate their children; one of those, a seven-month-old girl, died of the disease.

Those realities are obviously hard to square with Bob Sears's downplaying of the dangers of vaccine-preventable diseases—but Sears didn't need to look to Minnesota for proof that those diseases actually *can* be "that bad." Ten days before Julieanna ended up in the hospital, a seven-year-old boy who was later revealed to be one of Sears's patients returned from a family vacation in Switzerland with the measles. While the boy's parents had made a choice not to vaccinate their child—as his mother explained in a *Time* magazine article, "We analyze the diseases and we analyze the risk of disease, and that's how my husband and I make our decision about what vaccines to give our children"—many of the people who paid the price for that decision had less say in the matter. Within days, the measles virus had spread to a swim school, a pediatrician's office, a Whole Foods, a Trader Joe's market, and a charter school; passengers on a plane headed to Honolulu were quarantined at a military base; and a ten-month-old child was hospitalized and required medical care for a full month. (According to Gordon, the hospitalization was an overly dramatic reaction: "My guess is that if this had happened in the 1960s, no one would have been hospitalized. They would have said, 'Oh well, an . . . outbreak of measles.' ") An additional forty-eight children who were too young to be vaccinated had to be quarantined in their homes, at an average cost of $775 per family. In total, the outbreak cost more than $175,000 to contain.

Eighteen months after one of Sears's patients had caused what turned out to be the largest measles outbreak in California since 1991, I wrote to the "media inquiries" e-mail address listed on Sears's Web site requesting "a time to speak with Dr. Sears for a book I'm working on for Simon & Schuster about vaccines."* A week later, I received a response with details about advertising on askdrsears.com, which, in addition to an online store selling the Sears Family Essentials line of "healthy snacks" and "supplements," at the time featured endorsements of products ranging from Vital Choice Wild Seafood ("My Favorite salmon!") to Meyenberg Goat Milk Products. The e-mail read:

> I'd like to put together a nice press kit for you. . . . We'd of course like to book as many [page view] impressions as you are willing to give us on a monthly basis, I believe you mentioned 8,000, would 10,000 be out of the question? We generally charge $15 cpm [cost

* That nationwide outbreak began in 1989 and lasted for two full years. In total, eleven thousand Americans were hospitalized and 123 died.

> per thousand page views] because of our specifically targeted audience, the vaccinebook.com is that in the price range you were expecting Please respond with any general ideas/questions. . . .
>
> Also, what sort of tracking will be used and how would be [*sic*] bill according to tracking?
>
> Thank you for your understanding and we look forward to the possibility of working with you.

Sears's willingness to work with advertisers and self-indulgent parents alike has proven to be very profitable. (*The Vaccine Book* has already sold more than 100,000 copies.) The philosophy it espouses has also directly contributed to the rising number of affluent enclaves in which vaccination rates have fallen so low that diseases such as measles and whooping cough are once again becoming endemic. In March 2008, before he had admitted that it was his unvaccinated patient that had brought measles back from Switzerland, Sears wrote on his Web site, "The recent measles outbreak (if you can call it that) . . . raises awareness of a growing trend among families to decline certain vaccines." According to Sears, this was a good thing: "I believe our nation can tolerate a certain percentage of unvaccinated children without risking the overall public health in any significant way. Since most children are vaccinated, our nation has enough 'herd immunity' to contain outbreaks like this one."

CHAPTER 23

Baby Brie

Ralph Romaguera first met Danielle Broussard when they were children: He was ten and she was six, or maybe he was twelve and she was eight—the specifics don't matter, Danielle says, because four years feels like an eternity when you're that young. The two kept running into each other over the years—their fathers were friends—but by the time they'd graduated from high school they'd lost touch. Then, one night when they were in their early twenties, they bumped into each other at a school dance. (Ralph was there working as a photographer; Danielle had agreed to chaperone as a favor to her mother, who was, Danielle says, "the school's disciplinarian.") They started dating soon after, and in 1998, after two years of courtship, the couple got married. Five years later, on January 13, 2003, Danielle gave birth to the Romagueras' first child. They named her Gabrielle, but from the day they brought her home from the hospital, they called her Brie.

Brie was not quite four weeks old when, at just after eleven A.M. on Saturday, February 8, her parents first noticed her cough. Initially, they didn't think much of it. Brie, like all infants, produced a steady stream of new and mysterious sounds, and, Danielle says, this "sounded like a regular cough." A couple of hours later, they'd grown more concerned. They called their pediatrician, who told the Romagueras that she suspected Brie had caught respiratory syncytial virus, or RSV, a common infection that's relatively harmless in older children and adults but that can make it hard for infants to breathe. Just to be on the safe side, she recommended that Brie get checked out at the emergency room. Worst-case scenario, she said, Brie would be sent home with some antibiotics and her cough would clear up within a few days.

But on Sunday Brie sounded worse than she had in the ER the night before—her coughing fits had become more violent, the lag time between them shorter and shorter—and by Monday night, it broke her parents' hearts just to hear her: Every fifteen minutes or so, Brie's attempt at exhalation would result in a series of high-pitched barks followed by an equally alarming strained intake of breath. The next day, Danielle took Brie to see her pediatrician. This time, a tiny mask was placed over the infant's face in order to administer medication via an aerosolized mist. That seemed to help, at least at first, but by Wednesday evening, Brie was once again gasping for air. By midnight, it seemed to Danielle and Ralph as if their daughter was on the verge of passing out each time she suffered through a new coughing attack. For the rest of the night, they took turns cradling Brie in one arm and gently pulling up on the back of her neck with the other. "We found," Danielle says, "that if we could kind of extend her head, it was helping open her airways. I guess she was getting more air that way." By the time the sun rose the following morning, Brie was turning blue around the lips. For the third time in less than a week, the Romagueras bundled their daughter into her car seat and took her to see a doctor. This time, there was no question that Brie needed immediate attention, and she was transported to River Parishes Hospital in LaPlace, Louisiana. By the time she arrived there, Brie was turning purple. Danielle was still filling out paperwork when she saw one of the emergency room's pediatricians sprinting down the hall: Brie's oxygen intake had fallen so precipitously that she needed to be airlifted to the Ochsner Foundation Hospital in downtown New Orleans.

After arriving at Ochsner, Brie began to vomit, which provided doctors with one clue as to a possible diagnosis: Because she'd been born three weeks early, they said, it was possible Brie had an underdeveloped epiglottis, which was causing food to become lodged in her windpipe. (The epiglottis, which is located at the back of the throat, is a thin, elastic piece of cartilage that covers the trachea during swallowing.) A little more than a week after Brie first started coughing, she underwent the corrective surgery her parents hoped would cure her.

"Here was this seven-pound baby girl," Danielle says. "She had two tubes going into her mouth—one to suction out the phlegm and one to feed her. It seemed so horrible at the time." Still, Danielle and Ralph felt lucky. For the eight or nine minutes in between Brie's coughing spells, she seemed like a normal baby. That was not the case for many of the other infants at Ochsner. "These poor kids are having open-heart surgery," Danielle thought to herself. "We're going to be out of here in no time."

Brie's initial surgery went smoothly. "That afternoon, when she was coming out of the anesthesia, she looked like she was doing great," Danielle says. But soon, the coughing started again. The Romagueras were back at square one. Over the next several hours, specialist after specialist came in for consultations. There was a pediatric cardiologist, a pediatric pathologist, a pediatric neurologist—the list seemed to stretch on forever. Finally, Dawn Sokol, a pediatric infectious disease specialist who'd treated everything from HIV/AIDS to osteomyelitis, came to see Brie. She immediately recognized what was wrong: "This baby has pertussis"—a disease that at the time was sufficiently rare that prior to Brie none of the other doctors had ever seen a case in an infant.

One of the reasons pertussis is so scary is that during its initial incubation period, its symptoms—low-grade fever, slight aches, mild cough—closely resemble those of other, relatively common illnesses, ranging from the common cold to the flu. The disease's trademark "whoop," which is the result of the body instinctively gasping for oxygen, doesn't come until the next phase. The whoops bring with them paroxysms and cyanosis, which are the clinical terms for coughing fits so severe they cause the skin to turn blue. By this point, the body is under such constant attack that a patient's white blood cell count can resemble that of someone with leukemia. Because children under six months of age oftentimes do not, as one research paper put it, "have the strength to have a whoop," they frequently stop breathing during the attacks.*

Up until the 1940s, whooping cough was one of the world's leading causes of infant mortality, but in the decades immediately following the widespread introduction of the pertussis vaccine, the total number of cases and the total number of deaths in the industrialized world declined by more than 90 percent. That trend reversed somewhat in the mid-1970s, when speculative or exaggerated reports about the dangers of the whole cell pertussis vaccine led to a sharp drop in its use. In Japan, pertussis vaccine uptake fell from 80 percent in 1974 to 10 percent in 1976; three years later, an epidemic there resulted in thirteen thousand infections and forty-one deaths. In the U.K. during that same period, there were between ten thousand and twelve thousand new cases and around three dozen deaths each year. In Sweden, after vaccination rates fell from 90 percent in 1974 to 12 percent in 1979, the Swedish Medical Society abandoned the whole cell vaccine—and in the years to come, the country's infection rates rivaled those in developing countries.

In the United States, the effects of media reports like "Vaccine Roulette" and various activist groups have been more diffuse; still, the combination of declining vaccination rates for children and waning immunity for adults led to a continuous rise in infections, from just over one thousand in 1976, to more than four thousand in 1986, eight

* The sound of a whooping cough fit is unmistakable. Throughout history, the disease has been given vividly descriptive names: "dog's bark" in Italian, "howling of wolves" and "braying of donkeys" in German, and "boisterous laughter" in Old English.

thousand in 1996, and approximately ten thousand in 2002. Brie's infection occurred at the start of a two-year period during which that steady increase became an explosion, with twelve thousand diagnoses in 2003 and 25,000 in 2004.

At the time, the Romagueras knew nothing about the growing number of vaccine skeptics or the reappearance of onetime childhood scourges. "The first thought that came into my head was, That's something that happened when my grandmother was a kid, not today," Danielle says. Still, the diagnosis left them feeling somewhat relieved: At least they knew what they were dealing with. "We didn't know at the time how severe pertussis can be for a child that size," Danielle says. "We were not thinking at all that we might lose our child." That realization would occur gradually, over the next several days, as a series of increasingly invasive treatments, ranging from an oxygen hood to a ventilator, proved incapable of providing Brie with sufficient air to survive. Finally, her doctors decided the only hope Brie had of making it through yet another surgery was to be hooked up to an ECMO machine, a contraption used by only a handful of hospitals in the country.

The best way to think of an ECMO—the initials stand for Extracorporeal Membrane Oxygenation—is as an external artificial heart and lung that's used when an infant is experiencing extreme organ failure so that the body can use all its energy to get better. It's less of a single device than a series of machines: One pumps blood into the body; another filters the blood; a third circulates oxygen into and carbon dioxide out of the lungs. Like most extreme life-saving measures, the use of an ECMO machine carries with it serious risks. As long as Brie was hooked up to the machine, she'd need two nurses and a respiratory therapist by her side at all times.

The Romagueras are intensely religious people, and the night of what they hoped would be the operation that would save their daughter's life, they stood in a hallway of Ochsner's pediatric ICU and prayed. For hours on end, they recited the Rosary: "Blessed is the fruit of thy womb, Jesus," they murmured. "Pray for us sinners, now and at the hour of our death." Finally, at around 5:30 in the morning, they received word: Brie had survived her surgery. Still, the doctors said, they strongly recommended that the Romagueras not visit their daughter. Because Brie's body was unable to expel waste, the interstitial space between her skin and the web of tissue that surrounds all the body's internal structures had filled with fluid, causing her to balloon to eight times her normal size. "They told us, 'This is not the same child,' " Danielle says. Two weeks after she'd checked into the hospital and six weeks after she was born, Brie was too big to fit into a diaper.

For two days, Danielle and Ralph heeded the doctors' advice, until waiting helplessly away from their child became too much to bear. Even after being warned in the starkest terms imaginable of what to expect, the Romagueras were shocked when they saw their daughter. It wasn't just that they could barely recognize her—it was that they had no point of comparison for the unconscious baby lying in front of them. "It's like, take a Cabbage Patch Kid and then double or triple it," Danielle says. "That's about how big she was."

The next days were an exhausting blur. That weekend, the doctors said they had some good news: Brie's kidneys had begun to do a little work on their own, raising the possibility that her other organs might begin to heal as well. Then, on Monday, doctors noticed a spot on the right side of her brain. This was not wholly unexpected—hemorrhaging is an unavoidable risk of the blood-thinner Brie had to be given so long as she was hooked up to the ECMO machine—and there was still a chance that Brie might escape any permanent damage. There was also a chance, the doctors said, that even if Brie survived, she might need to use a feeding tube for the rest of her life. For the time being, the only thing to do was monitor the situation: If the hemorrhaging could be halted, Brie could remain on the ECMO machine; if it did not, her brain would fill with blood and the machines would have to be turned off.

On Tuesday morning, Brie's EEG brought a sliver of hope: The bleeding had stopped. Wednesday's exam showed the same thing. On

Thursday, the Romagueras were picking at their lunch in the hospital cafeteria when Ralph's cell phone rang. The latest results were in. "They just told us to come back up," Danielle says.

By the time the Romagueras got back to Brie's unit, all four of the pediatric intensive care unit's doctors were waiting for them. The bleeding on the right side of Brie's brain was still clotted off, they said, but now there was extensive bleeding on the left side. It was time to turn off the machines that were keeping Brie alive. "They told us she could live for a little while once the machines were off," Danielle says.

> It took a doctor and three nurses to turn off the machines. Technically they weren't supposed to move her when she was on ECMO because of where the tubes were connected, but we hadn't held her in two weeks—since she'd been on the ventilator—so the doctor said, "We're going to get as many staff members as we need in here so mom and dad can hold their baby while she's still alive." . . . This doctor made sure that there were seven or eight people there to move her so that my husband and I could hold her before she passed away. And she survived probably for about a half an hour once the machines were turned off.

On March 6, seven weeks and three days after they became parents, Ralph and Danielle Romaguera held their daughter in their arms as she died. Had she survived, she would have been scheduled to receive the first dose of the pertussis vaccine four days later.

To this day, the Romagueras do not know how Brie contracted whooping cough. Because it often goes undiagnosed, it can be difficult, if not impossible, to track the disease through a population. What the Romagueras *do* know is that, prior to her hospitalization, the only place Brie had visited aside from her doctor's office was Danielle's parents' house. No one there was ill and nobody outside the family had been there at the time. According to CDC officials, the most likely scenario is that an infected child had coughed while in the waiting room during one of Brie's visits to her pediatrician, but, as Danielle says, "we have no idea where it started."

One result of the growing number of deaths like Brie's is a greater awareness about pertussis. Today, the AAP, the CDC, and many state health departments have audio recordings of whooping cough attacks on their Web sites, and it's unlikely there's a pediatrician in the country who isn't aware of the disease's warning signs. Even when properly diagnosed, 50 percent or more of infants with pertussis are hospitalized, but only around 2 percent develop permanent brain damage or die, making it all but certain that had Brie been diagnosed today, she would have survived.

Less than a year after Danielle and Ralph Romaguera buried their baby daughter, Danielle gave birth to a son named Trey. A couple of years after that, Michael was born. The pain of the Romagueras' loss remained strong, but for the most part they kept their anguish to themselves.

Then, not long after what would have been Brie's fifth birthday, Danielle noticed Jenny McCarthy appearing on TV to tell viewers about how vaccines had caused her son's autism. "It upset me," Danielle says. "I noticed a lot of TV shows that probably your everyday mom watches that allow celebrities to come on and state their side about how they don't vaccinate their kids because they believe it's connected to autism."

Danielle went online and found the contact information for some of the shows that had given McCarthy a platform to disseminate her message, including *The Oprah Winfrey Show*, *Good Morning America*, and *The View*. "I e-mailed them afterward and said, I can't believe you don't present the other side. There are people out there listening to you that are about to have a baby, and they're just listening to this side and thinking, 'I can't vaccinate my child.' And you're not telling these mothers what can happen if they don't." At times, Danielle's frustration grew to be so intense that she wished she'd let her husband document Brie's final days:

> One thing I remember was in the hospital, I kept telling him, Don't you dare take a picture of her. . . . I was like, This is not how I want to remember my child. . . . [But] sometimes I wish I had a picture to show people that this is what a baby looks like when they're hooked up to five machines and is ten times their size. This is what happens to a child who comes down with these diseases.
>
> It's gotten to this point, and I guess I shouldn't be, but it's gotten to be disgust. . . . [Jenny McCarthy] is giving medical advice to people who are listening. . . . It's been proven that vaccines don't lead to autism, but those studies don't get the kind of media coverage that someone like Jenny McCarthy does.

Danielle never heard back from any of the shows she wrote to. In the spring of 2009, *The Hollywood Reporter* broke the news that Winfrey's production company had signed McCarthy to a multiyear development deal. "Like other Winfrey protégés-turned TV moguls, among them Rachael Ray and Dr. Phil," it read, "McCarthy has been a frequent guest on *The Oprah Winfrey Show*, [where she] talked to the chat queen about her struggles with her son's autism in conjunction with the publication of her best-selling books." According to the story, the deal would start with a blog on Oprah.com and would eventually include a range of projects on different platforms—"including a syndicated talk show that the actress/author would host."

Just as McCarthy's latest venture was being announced, Danielle learned she was pregnant again. The couple's third son was born that December. They decided to name him Gabriel—in honor of the sister he'll never get a chance to meet.

CHAPTER 24

Casualties of a War Built on Lies

By the start of 2009, the debate over the connection between autism and vaccines had been raging for more than a decade. Already, the effects of this manufactured controversy—on the public's perception of vaccines and public health policy, on worldwide vaccination rates, and on the reemergence of infectious diseases—had arguably been greater than the combined effects of all the other immunization controversies of the previous one hundred years.

Another factor that differentiated this dispute from previous ones like Lora Little's campaign against smallpox inoculations and the debate over the DPT vaccine was the Omnibus Autism Proceeding, which had been set up after the United States' Vaccine Court was flooded with thousands of autism-related claims in the early 2000s. The implications of the Omnibus on the perceived legitimacy of worldwide vaccination efforts were obvious to everyone involved. The Special Masters presiding over the Vaccine Court had been trained to work as impartial arbiters. They had never been part of the public health sector and had no stake in vaccine development. Fairly or not, a ruling in favor of the families suing the government would be interpreted as conclusive proof that in certain circumstances vaccines could cause autism.

It was with these stakes in mind that in the summer of 2002, Gary Golkiewicz, the Chief Special Master, met with representatives from the Department of Health and Human Services, the Department of Justice, the Food and Drug Administration, and a Petitioners' Steering Committee (PSC) made up of five lawyers representing more than four thousand families to hash out a framework for the proceedings. Together, they decided that the Omnibus Proceeding would examine three separate "causation" theories: The SafeMinds hypothesis that thimerosal could cause autism, the Wakefield hypothesis that the MMR vaccine could cause autism, and a composite theory that thimerosal and the measles virus working together could cause autism. The petitioners would choose three test cases for each theory, and the trials for each one of the separate theories would occur simultaneously.*

With so much riding on the outcome, the PSC made clear from the outset that its clients would need significantly more time than would normally be allotted to assemble the necessary research to support their contentions. As a result, they were given a full year to submit a list of witnesses. The rest of the process was expected to drag out as well, and final rulings were not scheduled to be handed down until the summer of 2004.

It didn't take long before it became obvious that even that goal would not be met: It wasn't until 2005 that the discovery process, which produced hundreds of thousands of pages of government documents, was completed. That summer, the PSC received a further extension to assemble a list of experts willing to support the theories upon which it was basing its claims. When that deadline rolled

* The PSC ultimately decided not to present any test cases for the MMR theory (or "causation theory #2") under the assumption that the decisions in the joint theory cases would determine whether there was sufficient evidence to show that the MMR vaccine could cause autism in certain circumstances.

past, another extension was granted. Finally, on Valentine's Day 2006, more than three and a half years after the framework for the Omnibus had been laid out, the petitioners' lawyers handed in the names of the doctors and scientists who were prepared to back up their claims under oath.

Sixteen months later, on June 11, 2007, opening arguments in the first Omnibus Autism trial were made in a federal courthouse in Washington, D.C. The suit filed by the parents of Michelle Cedillo, who at twelve years old was largely confined to a wheelchair and continued to require around-the-clock medical care, was selected as the first test case for the dual-causation hypothesis. The Cedillos' claim was broken down into three interdependent parts: The thimerosal Michelle had received in the hepatitis B, DPT, and Hib vaccines led to the "dysregulation" of her immune system; her weakened immune system was unable to defend against the live measles virus she received as part of the MMR vaccine; and the resultant measles infection had overwhelmed her brain, thereby causing autism. Outside of the widely discredited research performed by a mutually reinforcing community of autism advocates, DAN! doctors, and fringe scientists, *none* of those claims had ever been shown to be true. Proving a single step in that proposed chain would have been a mammoth undertaking; providing sufficient evidence to support all three was a monumental challenge.

That Monday morning, when the proceedings were called to order, Michael and Theresa Cedillo wheeled Michelle into the courtroom. Large headphones, which one reporter compared to those worn by heavy machine operators, covered her ears. Throughout the brief period of time in which she remained present, Michelle grunted loudly and hit herself in the face. For close observers of the story, it seemed apparent that one tactic of the Cedillos' lawyers was to highlight the sheer magnitude of Michelle's suffering. *Vaccine* author Arthur Allen described both sides' opening statements in a blog post on his Web site:

> With their body language, the lawyers for the claimants and the defense in the case reflected what has always been true about the vaccines-cause-autism theory: one side appeals to the heart, the other to the brain. The claimants' lawyers, Tom Powers of Portland and Sylvia Chin-Caplan of Boston, spoke with inflection and warmth and turned to face the audience in the courtroom. The government's lawyer, the colorless-seeming Vince Matanoski, spoke to the special masters—the judges who will decide the case.

The logic behind that strategy became apparent not long after Chin-Caplan called H. Vasken Aposhian, her first witness, to the stand. Aposhian, a professor at the University of Arizona whom Allen described as "an elderly, hard-of-hearing toxicologist with a halo of white hair surrounding his large bald pate," had written an expert report detailing how thimerosal caused "injury to the developing brains of human embryos and young children" such as Michelle. Chin-Caplan kept her questions straightforward and direct, and for the most part Aposhian's replies were similarly matter-of-fact.

During the cross-examination, which was conducted by a Justice Department lawyer named Linda Renzi, Aposhian's responses weren't nearly as cogent: He seemed distracted and confused, and at times appeared to be unable to describe his own work. At one point, Renzi asked Aposhian about one of the two case studies upon which he'd relied to reach his conclusions:

RENZI: Would you like to see the paper by Davis that we're talking about?

APOSHIAN: If it's that one, yeah. Yes. Now I remember this one, yes. What is your question about it, please?

RENZI: There was a child in that study who actually had prenatal exposure to methylmercury. Is that correct?

APOSHIAN: Again, I haven't read this paper for years. If you say it's correct, I'll have to accept it.

RENZI: Well, you cited it in your report, which you wrote on February 16, 2007. So if you haven't read this article in years—

APOSHIAN: Which said what about a pregnant woman?

RENZI: Well, you cited the paper so I assumed you were aware of the study.

APOSHIAN: I don't think in my paper I say there was a pregnant woman involved. I could be wrong, but, again, I have to read so many papers.

. . .

RENZI: So you don't know whether, in the article that you cite, whether one of the family members was exposed to the methylmercury through consumption of the pig prenatally?

APOSHIAN: Since this is a court of law, I want to be absolutely truthful and I have the sneaking suspicion one may have been but I'm not positive.

A couple of minutes later, Aposhian seemed not to realize that he was holding a copy of the report he'd prepared specifically for the trial:

RENZI: In your report, on page six—

APOSHIAN: Page six of this article?

RENZI: Of your report. I'm sorry sir, of your report.

APOSHIAN: Could someone get me my report? If I had known that—

RENZI: I think we handed you your report sir. You should have it.

APOSHIAN: Thank you. Okay. Page six. All right. Now I have page six.

Even those farcical moments weren't as damning as when Renzi asked Aposhian if there was a *single* paper or peer-reviewed article that supported his thesis regarding thimerosal's effects on the human brain. "The hypothesis was made less than three or four weeks ago," he replied, "so the answer is no." It was a jaw-dropping acknowledgment that the mechanism through which Michelle Cedillo and thousands of other children's injuries were supposed to have occurred had been fleshed out for the first time less than a month before the Omnibus hearings had begun.

As unimpressive as Aposhian might have been, his testimony never quite crossed the line into the surreal. The same could not be said of Vera Byers, the doctor who claimed that Michelle's immune system had been damaged by vaccines. At one point while she was being cross-examined, Byers, whose primary employer was a Shanghai-based company that sought to get "classic Chinese herbs" approved for use in FDA clinical trials, began to giggle inexplicably. Later, when she was asked how she determined effective control readings for her experiments, she replied, "You're making faces at me. You are." After one of the three Special Masters observing the trial told Byers she could look at them if she preferred, she nodded at Special Master Patricia Campbell-Smith and said, "She's much more attractive. Thank you." Even when Byers managed to answer the question at hand, her testimony wasn't particularly helpful to Michelle's case: When asked how she knew that different types of mercury affected the body in different ways, she explained that she'd been "educated by Dr. Aposhian . . . not specifically for this trial, but for the investigation of mercury in the human body for litigation purposes."

On Friday, June 15, four days after the hearing began, the petitioners called Marcel Kinsbourne, their last witness, to the stand. Kinsbourne, a medical doctor who had no clinical practice and whose primary interactions with patients for the previous seventeen years had been in preparation for litigation, was well acquainted with the Vaccine Court: He'd already appeared as an expert witness in at least 185 separate cases—and that figure did not include the many cases he'd testified in that had been settled before a judgment was issued or any of the cases he was involved with that were still pending before the court.*

During the course of his testimony, Kinsbourne demonstrated the risks of relying on someone who, in the words of one of the Special Masters, "suffers from the stigma attached to a professional witness" and "derives considerable income from testifying in Vaccine Act cases." In the expert report he prepared for the Cedillo trial, Kinsbourne included a chart indicating all the various causes for autism. It was similar to a chart Kinsbourne incorporated into a chapter he wrote in a textbook titled *Child Neurology*—but as the Justice Department's Vince Matanoski pointed out, it was not identical.

MATANOSKI: You developed a chart on the concomitance of autism?

. . .

KINSBOURNE: There is this long, rather dreary list of names.

MATANOSKI: Yes, actually, I think we could show you what it looks like in your book chapter. . . . Under that you listed viral as one of the [causes]. You listed three different viral concomitances.

KINSBOURNE: Yes.

MATANOSKI: There they are: rubella, herpes, and cytomegalovirus.

KINSBOURNE: Okay.

MATANOSKI: Now, in your report you pretty much reproduce this chart.

KINSBOURNE: Yes.

MATANOSKI: There was one change. You added one. You added measles.

KINSBOURNE: Right.

---

* It wasn't just the American legal system that Kinsbourne was familiar with: Along with Andrew Wakefield, he had been one of the three highest-paid witnesses in a similar mass vaccine-related lawsuit in the U.K.

MATANOSKI: Anything happen in the last year to cause you to add measles to that chart?

KINSBOURNE: Nothing happened, but my mind being so much on this seemingly endless litigation I thought to myself, Hey, you didn't put in measles.

MATANOSKI: I see.

The tenor in the courtroom was dramatically different the following Monday, when the government began calling to the stand the ten witnesses who supported its position. Where Aposhian had seemed disoriented, Jeffrey Brent, a medical doctor who is one of around 250 board-certified medical toxicologists in the United States and is the former president of the American Academy of Clinical Toxicology, appeared measured and deliberate; where Byers's behavior had been occasionally bizarre, the London-based molecular biologist Stephen Bustin was devastatingly blunt. Finally, there was the testimony of Eric Fombonne, the director of the psychiatry department at the Montreal Children's Hospital, where he helped launch an autism clinic. Where Kinsbourne had relied on his interpretation of some of the Cedillos' home videos and on Theresa Cedillo's recollections of events to support his claim that Michelle had a dramatic onset of autism a week after her first MMR shot on December 20, 1998, Fombonne used contemporaneous notes from Michelle's doctor's visits to show that no such change had occurred. Unlike Kinsbourne, who had never before used videos to retroactively diagnose patients, Fombonne had years of experience in that area, and he pointed to numerous examples of Michelle's displaying behavior associated with autism before her first birthday, including repetitive "hand-flapping" and a failure to make eye contact, respond to her name, play with toys, use communicative gestures, use words, or even use nonverbal sounds to interact. Finally, Fombonne noted, Michelle's medical records indicated that her head circumference was abnormally large throughout the first year of her life, which is a well-known hallmark of autism.

As stark a contrast as the testimony of the two sides' expert witnesses provided, what stood out even more was the extent to which the Cedillos' case, and by extension those of thousands of other Omnibus families, were intertwined with the work of Andrew Wakefield. To start with, it was Theresa Cedillo's conversation with Wakefield at the 2001 ARI/DAN! conference that confirmed her and her husband's suspicion that Michelle's autism had been the result of a vaccine injury. Since 2003, Michelle had been under the care of Arthur Krigsman, Wakefield's partner at Thoughtful House, and hundreds of pages of Krigsman's notes were introduced into evidence to support the Cedillos' case. Even the test results the Cedillos relied on to demonstrate that Michelle had been infected with the measles virus—results that, as the Cedillos' lawyers wrote in a brief, were the "*sine qua non* in [Michelle's] quest for entitlement"—were tainted by their association with Wakefield: They came from the discredited (and defunct) lab in Dublin that Wakefield had colloborated with over a period of several years.

It wasn't just personal connections that yoked the Omnibus families to the infamous gastroenterologist. The entire dual-causation argument was dependent on Wakefield's theories being correct. Despite all the studies contradicting Wakefield's conclusions, despite indications that he'd had a financial stake in an alternative to the MMR vaccine, despite the disavowals of his co-authors and his forced departure from the Royal Free Hospital and the worldwide condemnation of his research methods, Andrew Wakefield remained the single most important figure in the vaccines-cause-autism movement. In the course of the Cedillo trial, the government called on a former graduate student of Wakefield's named Nicholas Chadwick to testify. Chadwick, who'd worked in Wakefield's lab in the late 1990s, said that he'd alerted Wakefield that he'd been unable to detect the presence of measles virus in the twelve children who were the basis of Wakefield's 1998 *Lancet* paper. What's more, Chadwick said, he'd discovered multiple instances in which patient samples had been contaminated by positive controls. In both cases, his warnings were ignored.*

In her closing argument on June 26, 2007, Sylvia Chin-Caplan tried to spin this latest humiliation as an indication of the legitimacy of her argument. "The final point I want to make is what this case is about," she said. "It is not about Andy Wakefield. It's not. It's about Michelle Cedillo. It's about the 4,800 families looking for justice. . . . And to hear a government's case that is based on a smear campaign, a character assassination, hearsay, innuendo, traveling around the world collecting information, using government resources against somebody who is not a party, who is not a witness, who is not offering evidence, is outrageous. It's not about Andy Wakefield. It's about the Cedillos."

With that, Chin-Caplan rested her case. It was a passionate plea that unintentionally exposed the moral and scientific bankruptcy of all those who'd encouraged the families of children with autism to blame vaccines. In the larger scheme of things, the Omnibus hearings were not about the Cedillos—they were about establishing whether there was any scientifically reliable evidence that connected autism to vaccines.

Six months later, on December 24, 2007, *Age of Autism* (sponsored by Lee Silsby Compounding Pharmacy—"The leader in quality compounded medicine for autism") named Michelle Cedillo its "Child of the Year." "We are pleased to announce our annual Age of Autism Awards during this holiday week of celebration," the announcement read. It went on to quote a nominating letter from a woman who "was herself the mother of an affected child":

---

* While testifying in a different Omnibus test case, Bert Rima, one of the world's leading measles experts and the head of the School of Biomedical Sciences at Queens University in Belfast, described a similar situation that had occurred in 1992, when Wakefield invited a number of prominent measles virologists to give him their opinion on his work. "I attended two of those meetings," Rima said, "and I came to the conclusion that whatever material was put in front of me was highly selective. When criticisms were made, they were not followed up."

Michelle braves each day waiting for relief. When relief doesn't come, she waits for the next. All the while, Michelle remains unimpressed by the glitter of fame, unmotivated by the promise of material wealth, and undeterred by the ignorance of those who refuse to understand. These Warrior Children who have engulfed our hearts and to whom we dedicate so much of ourselves are, truly, the hardest working children on this spinning orb. Through it all, they are able to somehow find the strength to persevere. And, like so many, Michelle Cedillo has proven herself to be the Warrior Child she never asked to be, but has become, nonetheless.

* * *

In a typical court proceeding, once closing arguments have concluded, the buzzer has, for all intents and purposes, sounded. That was not the case in *Cedillo v. Secretary of Health and Human Services*: Following the conclusion of the twelve-day hearing, it took more than a year of dueling corrections before both sides were satisfied with the accuracy of the 2,917-page trial transcript. By the time the Office of Special Masters released a statement on September 29, 2008, that the "evidentiary records are now *closed*," more than five hundred pages of post-hearing briefs and motions had been filed in the Cedillo trial alone.*

On February 12, 2009, seven years after the start of the proceedings, rulings were issued for the first three Omnibus Autism cases. In the preface to his 183-page decision, George Hastings, the Special Master who presided over the Cedillo trial, highlighted just how

* Including among those was an "interim attorney's fees" bill submitted to the Vaccine Court by Sylvia Chin-Caplan's law firm for a total of $2,180,855.29. Less than $20,000 of that was used to reimburse the Cedillo family for the out-of-pocket expenses they incurred during the trial. In addition to travel and lodging, those expenses included a personal check for $1,200 Marcel Kinsbourne demanded from the family as an initial retainer for his services.

much ground had been covered. "The evidentiary record," he wrote, "is massive. This record dwarfs, by far, any evidentiary record in any prior Program case." In addition to testimony from twenty-eight expert witnesses, there were 939 pieces of medical literature, many of which ran to hundreds of pages. Among the documents submitted as part of the Cedillo case were 7,700 pages of Michelle's medical records, which covered a range of subspecialties including epidemiology, gastroenterology, genetics, immunology, molecular biology, neurology, toxicology, and virology.

Despite having access to the same information, the three Masters who ruled on the dual-causation test cases had worked independently of each other to reach their verdicts—which made the unanimity of their conclusions all the more unassailable. In page after page of unexpectedly gripping legal writing, they laid out in a way that no journal article or literature review ever had just how one-sided this dispute actually was. Their blunt assessment of each side's appointed experts articulated what the trial's observers had been saying to themselves all along: At every point of comparison, the government's witnesses were "far better qualified, far more experienced, and far more persuasive than the petitioners' experts." One of the Masters characterized Vera Byers's testimony as "disjointed," "often unclear," and "stray[ing] into matters beyond her expertise," and wrote that her "insistence that it was acceptable" to use figures for adult immune systems to determine appropriate readings for children and infants was "frankly, incredible." The best that could be said about H. Vasken Aposhian's testimony was that it was "reasonably coherent," although he was "at times unfocused and sometimes non-responsive." And a "fair assessment" of the differences between the information in Marcel Kinsbourne's expert report and his published work was that he "was unwilling to say measles was a cause of autism in a publication for his peers, but was willing to do so in a Vaccine Act proceeding."

After making clear that the Cedillo case had been a rout, Hastings emphasized that the petitioners had not been asked to clear a particularly high bar. "I have not required a level of proof greater than 'more probable than not,' which has also been described as '50 percent plus a feather,' " he wrote. "I have looked beyond the epidemiologic evidence to determine whether the *overall evidence—i.e.*, medical opinion and circumstantial evidence and other evidence considered *as a whole*—tips the balance even slightly in favor of a causation showing as to any of Michelle's conditions." It did not: "This is a case in which the evidence is so one-sided that any nuances in the interpretation of the causation case law would make no difference to the outcome of the case."

Hastings devoted the final three paragraphs of his ruling to the Cedillos themselves. It would be an abrogation of his duty, he wrote, to find in the family's favor—and unfortunately, he was unable to hold those most at fault responsible.

> The record of this case demonstrates plainly that Michelle Cedillo and her family have been though a tragic and painful ordeal. I had the opportunity, in the courtroom during the evidentiary hearing, to meet and to observe both of Michelle's parents, and a number of other family members as well. I have also studied the records describing Michelle's medical history, and the efforts of her family in caring for her. Based upon those experiences, I am deeply impressed by the very loving, caring, and courageous nature of the Cedillo family. Those family members clearly have done a wonderful job of coping with Michelle's conditions, and in caring for her with great love. I admire them greatly for their dedication to Michelle's welfare.
>
> Nor do I doubt that Michelle's parents and relatives are sincere in their belief that the MMR vaccine played a role in causing Michelle's devastating disorders. Certainly, the mere fact that Michelle's autistic symptoms first became evident to her family during the months after her MMR vaccination might make them wonder about a possible causal connection. Further, the Cedillos have read about physicians who profess to believe in a causal connection between the MMR vaccine and both autism and chronic

gastrointestinal problems. They have visited at least one physician, Dr. Krigsman, who has explicitly opined that Michelle's own chronic gastrointestinal symptoms are MMR-caused. And they have even been told that a medical laboratory has positively identified the presence of the persisting vaccine-strain measles virus in Michelle's body, years after her vaccination. After studying the extensive evidence in this case for many months, I am convinced that the reports and advice given to the Cedillos by Dr. Krigsman and some other physicians, advising the Cedillos that there is a causal connection between Michelle's MMR vaccination and her chronic conditions, have been *very wrong*. Unfortunately, the Cedillos have been misled by physicians who are guilty, in my view, of gross medical misjudgment. Nevertheless, I can understand why the Cedillos found such reports and advice to be believable under the circumstances. I conclude that the Cedillos filed this Program claim in good faith.

Thus, I feel deep sympathy and admiration for the Cedillo family. And I have no doubt that the families of countless *other* autistic children, families that cope every day with the tremendous challenges of caring for autistic children, are similarly deserving of sympathy and admiration. However, I must decide this case not on sentiment, but by analyzing the evidence. In this case the evidence advanced by the petitioners has fallen far short of demonstrating such a link. Accordingly, I conclude that the petitioners in this case are *not* entitled to a Program award on Michelle's behalf.

It was hard not to interpret Hastings's harsh language as an indication of what might have occurred had Theresa Cedillo never attended an ARI/DAN! conference, never met Andrew Wakefield or Arthur Krigsman, never gotten caught up in a theory that, when all the cards were on the table, proved to be a bit of chicanery barely worthy of a boardwalk three-card monte dealer.

As Sylvia Chin-Caplan said in her summation, there were many people who behaved outrageously during the seven-year run-up to the Cedillo trial, but it wasn't, as she tried to argue, the government's lawyers or their expert witnesses. It also wasn't the Special Masters, whom Chin-Caplan accused of being "arbitrary and capricious" and "simply unfair" at an AutismOne presentation she gave with Krigsman titled "Autism and Vaccines in the US Omnibus Hearings." The people who behaved outrageously were all those who stoked those 4,800 families' sense of injustice, who dangled hopes of sudden recoveries, who sold bogus "supplements" and $70,000 miracle cures, who told parents that if their children weren't getting better it was because they weren't trying hard enough, who built their medical practices on the backs of parental anxiety, who wrote books and newspaper articles and broadcast TV shows and radio programs based on what made a good story and not on the truth, and who convinced the Cedillos to act as front-line troops in a war built on lies.

Epilogue

Almost exactly one year after the Cedillo verdict was issued, a Fitness to Practise Panel convened by the U.K.'s General Medical Council concluded its own hearings on the autism-vaccine debate—and unlike the Omnibus trials, these did focus explicitly on Andrew Wakefield. The panel's concern was, as its name implied, fairly narrow: Its job was to determine whether Wakefield had acted ethically and responsibly while conducting research on children. (Two of Wakefield's collaborators on the 1998 *Lancet* paper were also under investigation.) The hearings, which were the longest and most expensive in the council's century-and-a-half-long history, included 148 days of testimony spread out over two and a half years.

On January 28, 2010, the morning the panel's ruling was to be issued, Wakefield appeared outside the GMC's headquarters wearing a dark gray suit, crisp white shirt, and bright red tie. He was accompanied by his wife, whose long blond hair was accented by a knee-length black leather coat and tall black leather boots. While a small coterie of his supporters disrupted the council's actual meeting by shouting "Bastards!" and "Bullshit!" at the panel's members, Wakefield remained on the street outside, where he kissed despondent mothers and hugged autistic children. At one point, the crowd, which was speckled with signs reading, "Guilty of helping damaged kids" and "We're WITH Wakefield—Crucified for helping sick kids with autism," broke into spontaneous chants of "For he's a jolly good fellow."

Ever since 2004, when the fruits of Brian Deer's investigative reporting began appearing in *The Times* (London), Wakefield and his allies had steadfastly insisted that the actual issue at hand was not anything as pedestrian as medical ethics or conflicts of interest—it was the suppression of a doctor trying to help families who were being ignored by a pitiless establishment. "My only concern was for the clinical well-being of this child," Wakefield told the press one morning before testifying about his treatment of a research subject. "It was my duty as a physician and a human being to respond to the plight of this mother."

Despite his lofty rhetoric, Wakefield failed to call on even a single parent to give evidence in his defense. When Rochelle Poulter, a fervent supporter of Wakefield's whose son, Matthew, had been identified as "Child 12" in the pages of *The Lancet*, did take the stand, it was the prosecution, not Wakefield's lawyers, who had requested her presence. Matthew's health problems had begun in 1994, when he began soiling himself multiple times every day. Two years later, when Matthew was five, he was diagnosed as being on the autism spectrum. Poulter and her husband were just coming to grips with this new reality when Poulter ran into an acquaintance at a mother-and-toddler play group. "[She] asked me if he had received his MMR jab, because she'd heard there might be a link with autism and bowel disorders," Poulter said. Despite the fact that Matthew's first MMR shot had been administered more than a year and a half before his gastrointestinal problems had started and almost four years before he was diagnosed with a developmental disorder, when Poulter heard about Wakefield's theory, "The pieces fell into place. I cried."

Soon thereafter, Poulter contacted Wakefield's research team at the Royal Free Hospital. Even though a slew of preliminary tests showed that Matthew had little or no signs of an ongoing GI disorder—Wakefield himself acknowledged that while Matthew had some "features of autism," he had "rather minimal gastrointestinal symptoms"—Wakefield included the boy in his study. On the afternoon Matthew was admitted to the hospital, a physician making rounds wrote on Matthew's chart that he was "not to have MRI or LP [lumbar puncture]." Nevertheless, over a four-day period, Matthew was subjected to "a colonoscopy, a barium meal and follow-through, an MRI scan of his brain, a lumbar puncture . . . an EEG and other neurophysiological tests, and a variety of blood and urine tests." Despite an employment contract that explicitly prohibited his involvement in the clinical management of patients' care, Wakefield signed the hospital request forms for a number of these procedures.

When the GMC ruled that Wakefield's actions had been "contrary to the clinical interests of Child 12," Rochelle Poulter was aghast. "I insisted that the hearing be informed that I was completely happy with the treatment my son had received and that I did not have any complaint against any of the doctors," she wrote in a testimonial of her support. "To this day I do not really know why I was asked to attend and give evidence." Her confusion was typical of the willful incomprehension of Wakefield's supporters: As long as they had no grievance with Wakefield's methods, they felt that any ethical violations he had committed in the course of treating their children should be ignored.

The final GMC report had none of the narrative drama or barely sublimated moral outrage of the Special Masters' decisions in the Omnibus trials. (A typical passage read, "At all material times you were, a) A UK registered medical practitioner, **Admitted and found proved** b) Employed by the Royal Free Hospital School of Medicine. . . . **Admitted and found proved** c) An Honorary Consultant in Experimental Gastroenterology at the Royal Free Hospital; **Admitted and found proved**.") The formal language did not diminish the record of callous opportunism chronicled in its pages. It wasn't only that Wakefield had been working with Richard Barr, the lawyer representing families who believed their children had been injured by vaccines—he'd also been "dishonest," "misleading," and "in breach of your duty" when he accepted £50,000 from the U.K.'s Legal Aid Board to do work that had already been financed by the National Health Service. It wasn't only that his claim that the children he had written about in *The Lancet* had been "consecutively [i.e. randomly] referred" was "dishonest," "irresponsible," and "resulted in a misleading description of the patient population in the Lancet paper"—he'd also subjected children who "did not meet the criteria for either autism or disintegrative disorder" to invasive and dangerous medical procedures. It wasn't only that he had precipitated a worldwide vaccine scare whose repercussions were still being felt—he'd also positioned himself to profit on the panic by developing an oral measles vaccine that would be produced and distributed by a company he and the father of one of his test subjects had been in the process of founding.

Finally, there were the events that took place at Wakefield's son's birthday party, where he'd not only "caused blood to be taken from a group of children for research purposes" and "paid those children who gave blood £5 for doing so"—he'd also "described the incident referred to above in humorous terms," and "expressed an intention to obtain research samples in similar circumstances in the future." "You showed a callous disregard for the distress and pain that you knew or ought to have known the children involved might suffer," the panel wrote in its conclusion. "In the circumstances you abused your position of trust as a medical practitioner [and] your conduct . . . was such as to bring the medical profession into disrepute."

The GMC report didn't contain any surprises—Brian Deer had already chronicled most of the particulars in even greater detail—but for the first time it appeared as though some of Wakefield's stalwart defenders had had enough: On February 17, Jane Johnson, who'd gone from being Thoughtful House's main financial backer to the co–managing director of its board, released a terse statement announcing that Wakefield had resigned. (The following day, Arthur Krigsman said he was leaving Thoughtful House as well.) When I asked Johnson why after all these years and all the red flags Wakefield was only now making his exit, she said the GMC decision had put him at greater risk of hurting the clinic "in terms of the local medical community with whom we're trying to build bridges, with the Texas

Medical Board. . . . Being associated with Andy is a lightning rod for negative attention."

A couple of months after his departure from the treatment center he'd helped to launch, I called Wakefield at his home in Austin. It was our fifth conversation over the course of the previous year, and in that time, his countenance had become noticeably less cocksure. Our first interview had taken place the previous April, during an ARI/DAN! conference at the Renaissance Waverly Hotel in Atlanta. Wakefield was wearing sun-bleached cowboy boots and weathered jeans, and his blue-and-white-striped Oxford shirt was unbuttoned far enough to show off his deeply tanned chest. Throughout the hour I spent with him that morning, the only time his right leg stopped jackknifing was when admirers came over to introduce themselves. When I visited him at Thoughtful House four months later, he was noticeably more self-conscious; after I sat in on a brief staff meeting, he asked if we could conduct the remainder of our interview in a sandwich shop down the road as opposed to his windowless basement office. Now he just seemed dazed. He maintained his defiance—"Their case was pathetic. . . . It was impossible to work out quite how they had the gall to get away with it"—but I got the impression that he was talking more to himself than to me.

When the conversation turned to his plans for the future, Wakefield said he was putting the finishing touches on a book titled *Callous Disregard*, which, in a manner reminiscent of David Kirby's *Evidence of Harm*, took a quote from the ruling against him and flipped it on its head. "Basically it's about why this [research] was closed down and why we have been treated the way we have—because the government is trying to cover up the fact that it introduced an unsafe MMR vaccine," he said. "I haven't been able to speak about it because of the legal proceedings, and now I've decided, To hell with that, and I've written a book on it."

Before we hung up I asked Wakefield if I could read a prepublication copy of his manuscript, which was scheduled to go on sale that

May. Of course, he replied—in fact, he said, he was keen on sending me a copy. "You might want to read it and decide whether it has a market in Israel, to see if anyone wants to find out what went wrong and why it went wrong," he said. "You may want to decide if it's worth translating into Hebrew." It was one of the few times in my life I have been stunned into silence. The only context in which Wakefield had ever known me was as a reporter writing about a controversy that he'd help to start; in fact, on several occasions he'd stressed that he was limited in what he could share with me because he needed to marshal material for his memoir-cum-exposé. We'd never once had a conversation about our personal lives, never mind our religious backgrounds. (As it happens, I am Jewish. I do not, however, speak Hebrew.) I awkwardly tried to change the subject. Then, just as we were getting off the phone, he reminded me to stay in touch regarding "getting involved with the translation." "Certainly your English is very good, so that's not a problem," he said. "But you may be too busy to do it—or you might have some recommendations."

On Monday, May 24, 2010, Andrew Wakefield's name was officially struck off of the U.K.'s medical register, which left him without a job or the ability to practice his chosen profession. Later that week, he received a standing ovation at the AutismOne conference in Chicago, where he also headlined a rally, gave two presentations, took part in an *Age of Autism* panel, posed for pictures with Bob Sears, and held a book signing. Wakefield might have been a lightning rod for negative attention from state medical boards and public health agencies, but it appeared as if his support among his core followers was as strong as ever. As Jay Gordon wrote in a blog post on his Web site, "I spent Saturday at an incredible conference in Chicago. Any thoughts I ever had about wavering in my support of Andrew Wakefield have dissolved. Jay."

In the years since the conversation with my friend that launched this project, there's been a dramatic rise in the number of communities where vaccination rates have fallen below the 90 to 95 percent threshold needed to maintain herd immunity. An overwhelming percentage of those are left-leaning, well-educated enclaves demographically similar to the neighborhood in which I live. The city that's gotten the most attention as of late is Ashland, Oregon, which is home to a nationally renowned Shakespeare festival and the Ashland Independent Film Festival and has a vaccine exemption rate of around 30 percent, which is the highest in the country. Just north of San Francisco, Marin County, which has the fifth-highest average-per-capita income in the United States, has an exemption rate more than three times that of the rest of California. A recent *Los Angeles Times* investigation identified two hundred Southern California schools where outbreaks are more likely "in large part because of parents choosing not to immunize. . . . Most are schools in affluent areas." One of those schools is the Ocean Charter School in Del Rey, California, where an entire century's worth of medical advances have effectively been reversed: Since the 2007–2008 school year, between 40 and 60 percent of incoming kindergarteners have been exempted from vaccines. Administrators told the *Times* those figures were no surprise, because the school's "nontraditional curriculum" attracted "well-educated parents who tend to be skeptical of mainstream beliefs." "They question traditional knowledge," the school's assistant director said, "and feel empowered to make their own decisions for their families, not deferring to traditional wisdom."

The situation is much the same throughout the rest of the country. Between 2005 and 2010, the rates of unvaccinated children doubled in New York and Connecticut and rose by 800 percent in New Jersey. Meg Fisher, the head of the AAP's section on infectious diseases, said she almost never came across parents who asked for exemptions when her practice was located in inner-city Philadelphia. Now that she works in the suburbs of New Jersey, she encounters them all the time.

The consequences of these trends are as tragic as they are predictable. In 2009, six unvaccinated children in southeastern Pennsylvania were infected with Hib, a disease that was assumed to have been eliminated in the United States twenty years ago. Two of them died. In May 2010, a mumps outbreak that began the previous summer, when an eleven-year-old unvaccinated boy from Brooklyn was infected during a trip to England, was tracked all the way to Los Angeles. In October, the California Department of Public Health announced that a statewide whooping cough epidemic had already caused more than 5,500 infections, which put the state on pace to record the highest number of cases since 1950, when the pertussis vaccine was just entering widespread use. By that point, nine children had already died. Eight of them were infected when they were less than two months old, which is the age at which infants are scheduled to receive their first dose of the DPT vaccine.

It's tempting to place the blame for this state of affairs squarely on the shoulders of people like Andrew Wakefield; after all, it would be hard to think up a character more sinister than someone who pays children for their blood. But that's the easy way out: Wakefield might have provided the spark, and any number of other charlatans and hucksters might have fanned the flames, but it's the media that provided—and continues to provide—the fuel for this particular fire. In February 2010, a month after the GMC issued its ruling and a year after the dual-causation Omnibus decisions were handed down, a columnist for *The Boston Globe* wrote about how she was worried that vaccinating her son could lead to "the moment he'd slip away. . . . For every scientific study that rejects a link [between vaccines and autism], there's a heartbreaking, unprovable, irrefutable anecdotal story that says otherwise." Four months later, the *Globe* used the occasion of Wakefield's loss of his medical license to run another column on the issue of autism and vaccines. The main subject of that piece was Wakefield ally Richard Deth, an "undeterred" Northeastern University professor of pharmacology who "believes in the possibilities of outside-the-mainstream therapies and research" and is "intrigued by the use of special diets and supplements" to treat autism. Deth's conclusions, the columnist wrote, "didn't seem very controversial."

She was right. Deth's conclusions aren't controversial: virtually everyone else in his field agrees that they're wrong. This is not information that requires a lot of research to uncover. Deth's testimony as an expert witness in one of the Omnibus cases prompted George Hastings to write, in his ruling in the Cedillo trial, a nine-page, point-by-point summary of the various deficiencies of Deth's theories. Hastings concluded his analysis by writing, "There were also a number of other specific points concerning which Dr. Deth's presentation was again shown to be erroneous, too numerous to detail here."

The type of journalism that relies on the reporter's notion of what does or doesn't "seem" correct or controversial is self-indulgent and irresponsible. It gives credence to the belief that we can intuit our way through all the various decisions we need to make in our lives and it validates the notion that our feelings are a more reliable barometer of reality than the facts.

Make no mistake: the repercussions of this outlook extend far beyond this specific issue. According to NASA's Goddard Institute for Space Studies, 2005 was the hottest year ever recorded; 1998, 2002, 2003, 2006, 2007, and 2009 are tied for second. (As of October 15, 2010 is on pace to take over the top spot.) During that time, the percentage of the population that said global warming is *not* a problem has doubled. Over the past several years, a number of states have either introduced or passed laws *mandating* that students be tutored in misinformation: In February 2010, both houses of the Kentucky legislature began considering a "science education and intellectual freedom" bill that "encourages" teachers to promote "critical thinking skills" about the "advantages and disadvantages of scientific theories . . . including but not limited to the study of evolution, the origins of life, global warming, and human cloning." Louisiana has already passed a nearly identical law, and in 2009, the Texas Board of Education passed a curriculum that requires schools to teach "all sides" of evolution and the "strengths and weaknesses" of global warming.

Just as I was finishing my research for this book, my wife gave birth to our first child. Like hundreds of thousands of new parents around the world, vaccines scare me—but when I sneak into my son's room at night to watch him sleep, I don't worry that the day he gets his MMR shot will be the day he "slips away." Instead, I worry that he might be one of those children for whom a given vaccine isn't effective, or that he'll come into contact with someone infected with Hib or measles or whooping cough before he's old enough to have gotten all his shots. I worry that he'll end up in a pediatric ICU because some parent decided the Internet was more trustworthy than the AMA and the AAP.

As Max grows older, I hope that, similar to the parents of children at the Ocean Charter School, he will feel empowered to make his own decisions and will have the self-confidence to challenge traditional wisdom. I also hope that he learns the difference between critical thinking and getting swept up in a wave of self-righteous hysteria, and I hope he considers the effects of his actions on those around him. Finally, for his sake and for that of everyone else alive, I hope he grows up in a world where science is acknowledged not as an ideology but as the best tool we have for understanding the universe, and where striving for the truth is recognized as the most noble quest humankind will ever undertake.

Afterword

On January 5, 2011, a week before the hardcover edition of *The Panic Virus* was published, the *British Medical Journal* ran the first in a series of articles in which Brian Deer claimed to have uncovered evidence that Andrew Wakefield had altered the patient histories of some of the children in his 1998 *Lancet* study. In other words, according to Deer's latest reporting, Wakefield had done more than lie about whether the children in his study had been "consecutively referred" to his clinic—he'd committed outright fraud.

Three months later, an investigation by the Maryland State Board of Physicians found that Mark and David Geier's Lupron protocol "endangers autistic children and exploits their parents by administering to the children a treatment protocol that has a known substantial risk of serious harm and which is neither consistent with evidence-based medicine nor generally accepted in the relevant scientific community." In late April, Maryland suspended Mark Geier's license to practice medicine. By the end of the summer, Geier's license had been suspended in Washington State, Virginia, and California as well.

These charges and accusations had little impact on Wakefield's or the Geiers' most ardent supporters. On April 24, not long after Wakefield announced he would headline a rally titled "The Masterplan: The Hidden Agenda for a Global Scientific Dictatorship" with a cohort of 9/11 Truthers, One World Government conspiracists, and anti-fluoridationists, one of the co-founders of Generation Rescue was quoted in *The New York Times Magazine* as saying, "To our community, Andrew Wakefield is Nelson Mandela and Jesus Christ rolled up into one. He's a symbol of how all of us feel." The Geiers, for their parts, were treated as heroes at the 2011 AutismOne conference in Chicago. Their research on "the role and treatment of elevated male hormones in autism spectrum disorders" was one of the focal points of the debut issue of a new AutismOne journal unironically called *Autism Science Digest*.

Unfortunately, there is no restart button when it comes to public consciousness: The fact that the researchers whose work forms the foundation of the anti-vaccine movement have been revealed as frauds (or worse) does not mean the anxiety they've helped plant in people's minds will suddenly melt away. The first half of 2011 provided still more evidence that when fear and misinformation are injected into a population, the effects are almost impossible to eradicate. The nationwide pertussis outbreaks that had begun in 2010 continued to spread. On April 4, the Blue Mountain School, a small private school about forty miles southwest of Roanoke, Virginia, was shut down for a week after roughly half of its students were infected with pertussis. According to local health officials, many of the parents of the infected students had chosen not to have their children vaccinated.

Pertussis was not the only vaccine-preventable disease that experienced a wide-spread resurgence in 2011: By the early spring, there had been measles outbreaks across the country, in areas ranging from the Northeast to the Pacific Northwest and from the Gulf Coast to Southern California. The largest outbreak was in Minnesota, where anti-vaccine activists had targeted a community of Somali immigrants that appeared to be experiencing higher-than-expected rates of autism. That outbreak began when a deliberately unvaccinated child returned from Africa infected with the disease; by the time it had run its course, more than a dozen children had been hospitalized.

On June 22, the CDC issued an official health alert about the surge in measles infections. From 2001 to 2008, the U.S. had had an average of around 50 measles cases a year. Over the first twenty-four weeks of 2011, the CDC had verified 156 measles infections, which put the country on track to record over 330 cases by the end of the year. As recent history has shown, these figures can explode in an incredibly short time. In 2006 and 2007, France had an average of 40 measles cases per year. In the first five months of 2011, the country recorded more than 10,000 infections, including 360 cases of severe measles pneumonia, 12 cases of encephalitis and six deaths.

In addition to the obvious health implications, each new measles case poses an enormous financial burden on the country's public health infrastructure. A pair of independent studies conducted after 2008 measles outbreaks in California and Arizona found that containing each new case of infection cost an average of more than $38,000. When there was rumor of a single infected visitor in Maryland, the state's Department of Health and Mental Hygiene announced that it was investigating potential exposures at a grocery store, a liquor store, a high school graduation ceremony, an Applebee's, and a Baltimore Orioles game.

If there's one thing I've learned over the past three years, it's that virtually everyone involved in this issue is frustrated. According to a May 2011 report in *American Journal of Preventative Medicine,* "pediatricians were more likely to report that their job was less satisfying because of the need to discuss parents' questions or concerns about vaccines and to perceive that when parents disagreed with their recommendations it showed a lack of respect for their medical judgment and experience." A CDC-sponsored survey found that the ease with which "immunization-related concerns and misperceptions" can be disseminated means that "most parents—even those whose children receive all of the recommended vaccines—have questions, concerns, or misperceptions about [vaccines]." And, of course, families who believe that their children were injured by vaccines feel continue to feel forsaken by their doctors and their government.

But it's expectant parents and parents of very young children who are the most important players in this drama—and in many ways, we're the ones who have the most legitimate reason to complain about the way we're being treated. A significant majority of new parents actively seek out information about vaccine safety before their children are vaccinated, but typically the first time the topic comes up with a medical professional is when our pediatricians have a needle in their hands. The first several months of a baby's life are overwhelming, exhausting, and nerve-racking—not exactly the best time to process a lot of new information.

It wasn't until my wife and I were expecting our second child that I realized how few chances new parents get to discuss vaccines with medical professionals. Throughout both of her pregnancies, we kept meticulous notes about every conversation we had with our health-care providers. We knew the difference between nuchal translucency screening and amniocentesis, and we knew the relative risks of various procedures, both nationwide and at the hospital we were using.

No such notes exist for the period after our son was born. By the time we showed up for his two-month checkup, I barely had enough energy to brush my teeth. That was when, in the midst of a fifteen-minute "wellness" appointment, he was scheduled to receive vaccines that would protect him against rotavirus, pneumococcal disease, diphtheria, tetanus, pertussis, Hib, polio and hepatitis B. It's true that there were signs in our pediatricians' office stressing the importance of vaccines—but at no point did anyone broach the topic or offer to answer any concerns we might have had.

I hope my book can be a resource for new parents—but for public attitudes to truly change, we need to find a way for health professionals to play a more active role in the discussion. One way to do this is to initiate conversations during standard prenatal care: If my wife and I had had a scheduled appointment before our son was born to discuss issues of infant health—such as the dangers of infectious diseases and the importance of vaccines—we would have written that information down and reviewed it when we got home. If we'd had questions, we could have discussed them at a time when we were able to actually process the answers.

At a prenatal appointment, with no baby to distract or soothe, parents could ask how vaccines work and could discuss the rumors they'd come across. They could discuss early warning signs for developmental disabilities, review the studies showing that there is no connection between vaccines and autism, and learn about herd immunity. Finally, they could hear about the dozens of infants who have recently been hospitalized with measles or have died of whooping cough.

After I proposed this type of program in a *Washington Post* editorial last June, I heard all about the logistical hurdles to setting up this type of system, including the fact that for the most part, the obstetricians who treat pregnant women are not trained in pediatric care. These do not strike me as convincing. Squabbling over treatment turf instead of looking for new ways to tackle the problem is shortsighted—and the costs of getting it wrong are potentially life and death.

Acknowledgments

In January 2009, a twenty-seven-year-old former Obama staffer named Kevin Hartnett got in touch with me for advice about building a career as a writer. I told him in no uncertain terms that he was making the worst decision of his professional life. For the sake of his marriage and his sanity, I urged him to seek work in an industry that wasn't in the midst of an epic collapse.

Two weeks later, I asked Kevin if he was interested in helping me with some research for a book I was working on. Since that day, there is no one I have relied on more. To say that *The Panic Virus* would not have been completed without him would be downplaying his contributions: He transcribed hour upon hour of interviews, waded through hundreds of pages of dense scholarship, prepared reports on topics ranging from memory manipulation to tort reform, and helped wrestle an unwieldy manuscript into submission. Over the past six months, he has been as close to a writing partner as I have ever had. Oftentimes, I would write until two or three in the morning. Kevin would start reading my copy at seven A.M, and by the time I was back at my computer, my frenzied scribblings would be well on their way to coherence. I have no doubt that his name will be on the cover of many excellent books of his own in years to come.

I also owe a debt of gratitude to Kevin's wife, Caroline, and their son, James, for their indulgence and understanding. Kristin Hartnett Sheppe, Kevin's younger sister, performed the herculean task of organizing a bibliography that at one point stretched to well over one hundred pages long.

For the first ten months of my son's life, I was working between ten and twelve hours a day. That would not have been possible without the help of my wife's and my families. For three weeks, my parents, Jim and Wendy, fed and housed us as we struggled to keep our heads above water. My mother-in-law, Cathy, cheerfully pretended there was nothing she'd rather do than drive fifteen hours in order to share a foldout couch with a mildly senile cat and help care for an infant with a raging case of colic. My sister, Abby, her partner, Laura, and my brother, Jake, all logged an impressive number of hours changing diapers, giving baths, and taking our dog on six A.M. trips to the park.

When this project was in its infancy, Kurt Andersen helped me shape a series of barely connected ideas into a coherent structure. Without his input and advice, this book would never have seen the light of day. In addition to coming up with a title, crafting an introduction, and providing daily insights, my agent Scott Moyers supplied me with a steady diet of encouragement and advice. Caroline and Helmut Wymar loaned my wife and me the use of "78" on Nantucket. The only downside to the weeks we spent there was that eventually we had to leave. Hanya Yanagihara gave me invaluable advice on the book's structure.

For two books in a row, I've had the good fortune of working with the people at Simon & Schuster. David Rosenthal had faith that I could write a book about a subject that was dramatically different from any I had written about before. My editor Bob Bender performed the impossible task of keeping my neuroses in check while improving my manuscript and overseeing a ridiculously truncated publication schedule. His assistant, the unflappable Johanna Li, somehow kept straight the dozens of overlapping files I sent in on an hourly basis. (She also kept me well stocked with mysteries.) Mara Lurie translated my scribbling and incorporated my many last-minute edits, re-edits, and un-edits. And Jonathan Karp championed the project during some of its most difficult moments.

In the course of my research, there were dozens of people who shared their time and expertise. I owe a special debt of gratitude to Bob Chen at the Centers for Disease Control and Prevention, Jane Johnson at the Autism Research Institute, and Lisa Randall at the Immunization Action Coalition. Arthur Allen, whose book *Vaccine* is a remarkably engaging work of scholarship, gave much-appreciated feedback on parts of my manuscript.

Most generous of all were the family members who agreed to speak with me about subjects that were difficult and painful, including Vicky Debold, Peter and Emily Hotez, Kelly Lacek, Kevin Leitch, Toni and Dana McCaffery, Lyn Redwood, Danielle Romaguera, Anissa Ryland, Alison Singer, and Bob and Katie Wright. I am grateful to every one of them for their candor and trust.

The Panic Virus is the most difficult project I have ever embarked on. The research was intellectually strenuous and the writing was physically and emotionally draining. I would not have had the strength to undertake such a challenge without the love, nourishment, and support of my wife, Sara. The sacrifices she made and the commitment she displayed were humbling. I owe this book, and everything else I have done over the last three years, to her.

Notes

INTRODUCTION

PAGE

1 *On April 22, 2006, Kelly Lacek:* Kelly Lacek, interview with author, May 7, 2009.

3 *Oftentimes, a Hib infection:* "*Haemophilus influenza* type b (Hib) Vaccine—What You Need to Know," Centers for Disease Control and Prevention, n.d., http://www.cdc.gov/vaccines/pubs/vis/downloads/vis-hib.pdf.

3 *due to a condition called epiglottitis:* K. Tanner et al., "*Haemophilus influenzae* type b Epiglottitis as a Cause of Acute Upper Airways Obstruction in Children," *British Medical Journal* 2002;325(7732): 1099.

3 *As recently as the 1970s:* "What Would Happen if We Stopped Vaccinations?," Centers for Disease Control and Prevention, n.d., http://www.cdc.gov/vaccines/vac-gen/whatifstop.htm#hib.

3 *the disease all but disappeared in the United States:* "Disease Listing—*Haemophilus influenzae* Serotype b (Hib) Disease," Centers for Disease Control and Prevention, October 10, 2009, http://www.cdc.gov/ncidod/dbmd/diseaseinfo/haeminfluserob_t.htm.

3 *"I must have read somewhere":* Kelly Lacek, interview with author, May 7, 2009.

4 *a tracheotomy, which involves cutting into the windpipe:* P. Oliver et al., "Tracheotomy in Children," *Survey of Anesthesiology*, 1964;7(2): 9–11.

4 *The physicist Stephen Hawking:* Stephen Hawking, "Prof. Stephen Hawking's Disability Advice," *Professor Stephen W. Hawking*, n.d., http://www.hawking.org.uk/index.php?option=com_content&view=article&id=51&Itemid=55.

5 *"They said something about not catching it":* Kelly Lacek, interview with author, May 7, 2009.

5 *The roots of this latest alarm dated back to 1998:* See citations for Chapters 8 and 9.

5 *The medical establishment was so determined to discredit him:* Andrew Wakefield, "Correspondence: Author's Reply: Autism, Inflammatory Bowel Disease, and MMR Vaccine," *The Lancet* 1998;351(9106): 908.

5 *Within months, vaccination rates across Western Europe:* "Q&A: Measles," graph titled "MMR Immunisation Levels—Children Immunised by 2nd Birthday," *BBC News*, November 28, 2008, http://news.bbc.co.uk/2/hi/health/7754052.stm.

6 *Then, a year later, the Centers for Disease Control and Prevention:* Centers for Disease Control and Prevention, "Notice to Readers: Thimerosal in Vaccines: A Joint Statement of the American Academy of Pediatrics and the Public Health Service," *Morbidity and Mortality Weekly Report* July 9, 1999;48(26): 563–65.

6 *The move had been hotly debated:* Gary Freed et al., "The Process of Public Policy Formulation: The Case of Thimerosal in Vaccines," *Pediatrics* 2002;109(6): 1153–59.

6 *In the year following the CDC/AAP recommendations:* See citations for Chapter 11.

7 *Soon, vaccination rates began to fall in the United States as well:* "Vaccines and Immunizations—Statistics and Surveillance: Immunization Coverage in the U.S.," Centers for Disease Control and Prevention, n.d., http://www.cdc.gov/vaccines/stats-surv/imz-coverage.htm.

8 *Taitz, who believes that the Federal Emergency Management Agency:* Benjamin L. Hartman, "Orly Taitz: Obama Policies Are 'Clear and Present Danger to Israel,'" *Haaretz*, August 18, 2009, http://www.haaretz.com/news/_orly-taitz-obama-policies-are-clear-and-present-danger-to-israel-1.282161.

8 *Hugo Chávez controls the software:* Gabriel Winant, "What Orly Taitz Believes: The Head Birther Talks About Obama's Boyfriends, the Long Arm of Hugo Chávez and How the Web Is Rigged Against Her," Salon.com, August 13, 2009, http://www.salon.com/news/feature/2009/08/13/orly_taitz/print.html.

9 *Lou Dobbs were pimping the story:* James Rainey, "A Natural-Born Canard About Obama," *Los Angeles Times*, July 22, 2009, D1.

9 *"those who think with their head":* Stephen Colbert, "The Word—Truthiness," *The Colbert Report*, October 17, 2005, video, http://www.colbertnation.com/the-colbert-report-videos/24039/october-17-2005/the-word---truthiness.

9 *"in what we call the reality-based community":* Ron Suskind, "Faith, Certainty, and the Presidency of George W. Bush," *The New York Times Magazine*, October 17, 2004, 52.

11 *"No amount of experimentation can ever prove me right":* Alice Calaprice, ed., *The New Quotable Einstein* (Princeton: Princeton University Press and He-

brew University of Jerusalem, 2005), 291. (This quotation appears in Chapter 19, "Attributed to Einstein." According to Calaprice, "This may be a paraphrase of sentiments expressed in 'Induction and Deduction,' December 25, 1919, *CPAE* [*Collected Papers of Albert Einstein*], Vol. 7, Doc. 28.")

12 *This was "not a close case": Cedillo v. Sec'y of Health and Human Services*, No. 98-916V (Ct. Fed. Cl., February 12, 2009), 172.

13 *has barred journalists:* Ken Riebel, "Listening to Parents at AutismOne," *Autism News Beat,* May 29, 2010, http://www.autism-news-beat.com/archives/1030.

13 *kicked out parents:* Ken Riebel, "Expelled!," *Autism News Beat,* May 26, 2008, http://www.autism-news-beat.com/archives/62.

13 *asked security to remove a public health official:* Sullivan, "AutismOne–Generation Rescue Conference Expells Registered Attendees," *Left Brain/Right Brain,* June 2, 2010, http://leftbrainrightbrain.co.uk/2010/06/autismone-generation-rescue-conference-expells-registered-attendees/.

14 *"The great majority of children suffering":* "About Us," AutismOne, n.d., http://conference.autismone/?goto=aboutus.

14 *"A picture caption on Tuesday":* "Corrections," *The New York Times*, April 30, 2010, http://www.nytimes.com/2010/04/30/pageoneplus/corrections.html?pagewanted=all. The correction referred to the following article: Neil Genzlinger, "Vaccinations: A Hot Debate Still Burning," *The New York Times*, April 28, 2010.

14 *a four-hour-long "vaccine education" seminar:* Vicky Debold, Barbara Loe Fisher, and Louise Kuo Habakus, "Vaccine Education Seminar," presentation, AutismOne conference, Westin O'Hare, Chicago, May 22, 2009.

14 *"autism and vaccines in the US [legal system]":* Arthur Krigsman and Sylvia Chin-Caplan, "Autism and Vaccines in the US Omnibus Hearings: Legal and Gastrointestinal Perspectives of the Michelle Cedillo Case," presentation, AutismOne conference, Westin O'Hare, Chicago, May 24, 2009.

14 *"the toxic assault on our children":* Alice Shabecoff, "The Toxic Assault on Our Children," presentation, AutismOne conference, Westin O'Hare, Chicago, May 22, 2009.

14 *"Down syndrome, vaccinations, and genetic susceptibility to injury":* Laurette Janak, "Down Syndrome, Vaccinations, and Genetic Susceptibility to Injury: What Does the Research Show?," presentation, AutismOne conference, Westin O'Hare, Chicago, May 22, 2009.

14 *explained how vaccines are a "de facto selection":* Fisher, "Vaccine Education Seminar."

14 *premiere of a documentary called* Shots in the Dark: "New Documentary Film Release: A Shot [Shots] in the Dark," AutismOne, September 2, 2009, http://www.autismone.org/content/new-documentary-film-release-shot-dark.

15 *Suppose an individual:* Leon Festinger, *When Prophecy Fails* (London: Pinter & Martin, 2008), 1. Originally published 1956 by University of Minnesota Press.

15 *"The little group":* Ibid., 171.

16 *In a speech delivered at eight in the morning:* Lisa Ackerman, "Starting the Biomedical Treatment Journey," presentation, AutismOne conference, Westin O'Hare, Chicago, May 22, 2009.

16 *the father-son team of Mark and David Geier stood on stage:* Mark Geier and David Geier, "New Insights into the Underlying Biochemistry of Autism: The Mercury Connection," presentation, AutismOne conference, Westin O'Hare, Chicago, May 23, 2009.

16 *a cost of more than $12,000:* Trine Tsouderos, "Autism 'Miracle' Called Junk Science," *Chicago Tribune*, May 21, 2009, 1.

17 *"His dad is a big guy like myself":* Ibid.

17 *"below the ethical standards": Aldridge v. Sec'y of Health and Human Services*, 1992 WL 153770 (Ct. Fed. Cl., June 11, 1992), footnote, 9.

17 *"intellectually dishonest": Marascalco v. Sec'y of Health and Human Services*, WL 277095 (Ct. Fed. Cl., July 9, 1993), 5.

17 *"not reliable": Haim v. Sec'y of Health and Human Services*, WL 346392 (Ct. Fed. Cl., August 27, 1993), footnote, 15.

17 *"fills me with horror":* Tsouderos, "Autism 'Miracle' Called Junk Science."

17 *"If someone like Mark Geier comes up":* Kevin Leitch, interview with author, May 5, 2009.

18 *in a series of groundbreaking papers in the 1970s:* Daniel Kahneman and Amos Tversky, "Subjective Probability: A Judgment of Representativeness," *Cognitive Psychology* 1973;3: 430–54; Daniel Kahneman and Amos Tversky, "Judgment Under Uncertainty: Heuristics and Biases," *Science* 1974;185(4157): 1124–31; Daniel Kahneman and Amos Tversky, "Prospect Theory: An Analysis of Decisions Under Risk," *Econometrica* March 1979;47(2): 313–27. See also: *Preference, Belief, and Similarity—Selected Writings, Amos Tversky*, edited by Eldar Shafir (Cambridge: Massachusetts Institute of Technology Press, 2004), Chapter 7, "Belief in the Law of Small Numbers," 193–202; Chapter 9, "Extensional Versus Intuitive Reasoning: The Conjunction Fallacy in Probability Judgment," 221–56.

18 *A recent Hib outbreak in Minnesota:* Centers for Disease Control and Prevention, "Invasive *Haemophilus influenzae* Type B Disease in Five Young Children—Minnesota, 2008," *Morbidity and Mortality Weekly Report*, January 23, 2009;58(Early Release): 1–3.

19 *Among those infected was Dana McCaffery:* Toni and David McCaffery, interview and e-mails with author, 2009–2010.

19 *A decade after the World Health Organization:* World Health Organization,

"Measles Eradication Still a Long Way Off," *Bulletin of the World Health Organization* 2001;79(6), http://www.who.int/mediacentre/factsheets/fs288/en/index.html.

19 *In Great Britain, there's been more than a thousandfold increase:* "Agency Publishes Annual Measles Figures for 2008," Health Protection Agency (U.K.), February 9, 2009.

19 *outbreaks in many of the country's most populous states:* Centers for Disease Control and Prevention, "Update: Measles—United States, January–July 2008," *Morbidity and Mortality Weekly Report*, August 22, 2008;57(33): 893–96.

19 *"felt safe in making the choice":* "How My Son Spread the Measles," *Time*, May 25, 2008.

20 *Before the MMR vaccine was introduced:* Nancy Shute, "Parents' Vaccine Safety Fears Mean Big Trouble for Children's Health," usnews.com, March 1, 2010.

20 *During the 1964–1965 rubella epidemic:* Pan American Health Organization, "Public Health Burden of Bubella and CRS," *EPI Newsletter,* August 2008; xx(4).

20 *On the fourth morning of Matthew Lacek's coma:* Kelly Lacek, interview with author, May 7, 2009.

20 *"We just celebrated* [*Matthew's*] *7th birthday":* Kelly Lacek, e-mail to author, "Subject: Re: from Seth Mnookin/via Trish at PKids," April 12, 2010.

CHAPTER 1: THE SPOTTED PIMPLE OF DEATH

PAGE

23 *In the three thousand years:* Jessica Reaves, "The New Worry: Smallpox," *Time*, October 18, 2001.

23 *The term "smallpox" was coined:* A. Geddes, "The History of Smallpox," *Clinics in Dermatology* 2006;24(3): 152–57.

23 *The Plague of Antonine:* Nicolau Barquet and Pere Domingo, "Smallpox: The Triumph over the Most Terrible of the Ministers of Death," *Annals of Internal Medicine* 1997;127(8): 635–42.

23 *Between 1694 and 1774, eight reigning sovereigns:* Donald Henderson and Bernard Moss, "Smallpox and Vaccine," *Vaccines*, ed., Stanley Plotkin and Walter Orenstein (Philadelphia: W. B. Saunders, 1999).

24 *The virulence of smallpox:* Ibid., 2.

24 *In 1717, Lady Mary Wortley Montagu:* Edgar M. Crookshank, *History and Pathology of Vaccination: Vol. 1, A Critical Inquiry* (Philadelphia: P. Blakiston, Son, & Co, 1889), 32.

24 *clearly considered those risks worth taking:* Paul Strathern, *A Brief History of Medicine* (London: Robinson, 2005), 179.

24 *In London, her doctor:* Stephanie True Peters, *Epidemic! Smallpox in the New World* (New York: Benchmark, 2005), 37.

25 *In 1706, a "Coromantee" slave:* Ibid., 34.

25 *Mather, whose wife:* Barbara Rogoff, *The Cultural Nature of Human Development* (New York: Oxford University Press, 2003), 106.

25 *still, it wasn't until 1721:* Margot Minardi, "The Boston Inoculation Controversy of 1721–1722," *The William and Mary Quarterly* 2004;61(1): 48.

25 *Not long after he and Boylston:* Mitchell Breitwieser, "Cotton Mather's Pharmacy," *Early American Literature* 1981;16(1): 42–49.

25 *After a dormant period:* Elizabeth Fenn, *Pox Americana* (New York: Hill & Wang, 2001), 16–19.

26 *One eighteenth-century account:* Noble Cook, *Born to Die* (Cambridge: Cambridge University Press, 1999), 116.

26 *Throughout the 1700s:* Barquet and Domingo, "Smallpox," 635–42.

26 *In a 2001 paper:* Valerie Curtis and Adam Biran, "Dirt, Disgust and Disease," *Perspectives in Biology and Medicine* 2001;44(1): 17–31.

27 *On September 28, 1751:* Fenn, *Pox Americana*, 1–2.

27 *by the time it was at thirty:* Ibid., 47, 265.

28 *Throughout December:* Ibid., 46, 62–72.

28 *"entire ruin of the Army":* Bruce Chadwick, *The First American Army* (Naperville, Illinois: Sourcebooks, 2005), 100.

28 *By Christmas Day:* Fenn, *Pox Americana*, 64.

28 *In the end, repeated rumors:* Ibid., 91.

29 *In 1763, the commander:* Barquet and Domingo, "Smallpox," citing J. Duffy, "Smallpox and the Indians in the American Colonies," *Bulletin of the History of Medicine* 1951;25: 324–41.

CHAPTER 2: MILKMAID ENVY AND A FEAR OF MODERNITY

PAGE

30 *When naturally occurring:* Collette Flight, "Smallpox: Eradicating the Scourge," *BBC: British History In-Depth*, n.d., http://www.bbc.co.uk/history/british/empire_seapower/smallpox_01.shtml.

30 *As soon as the immune system:* William T. Keeton and James L. Gould, *Biological Science*, 4th ed. (New York: W. W. Norton, 1986). See Chapter 4, "Structure of the Cell Membrane," 100–4; Chapter 13, "Internal Transport in Animals," 344–50; Chapter 27, "Transcription and Translation," 709–13.

31 *One popular rhyme:* R. S. Bray, *Armies of Pestilence: The Effects of Pandemics on History* (Cambridge, U.K.: James Clark, 2004), 114.

31 *An English scientist and naturalist named Edward Jenner:* "Jenner and Smallpox," The Jenner Museum, n.d., http://www.jennermuseum.com/Jenner/cowpox.html.

31 *Jenner tried inoculating Phipps:* Stefan Riedel, "Edward Jenner and the History of Smallpox and Vaccination," *Proceedings of the Baylor University Medical Center* 2005;18(1): 21–25.

31 *Nelmes was infected:* "Campus Curiosities (10): Edward Jenner's Cow, St. George's, University of London," *The Times Higher Education* (London), July 22, 2005.

32 *The relative safety of the cowpox vaccine:* Catherine Mark and Jose Rigau-Perez, "The World's First Immunization Campaign," *Bulletin of the History of Medicine* 2009;83(1): 63–94.

33 *Anne Harrington calls the feeling:* Anne Harrington, *The Cure Within* (New York: W. W. Norton, 2008), 139.

33 *the American Revolution's promise:* William Goetzmann, *Beyond the Revolution* (New York: Basic Books, 2009), 95.

34 *When the American Medical Association was founded:* "Our History," American Medical Association, n.d., http://www.ama-assn.org/ama/pub/about-ama/our-history.shtml.

34 *gravitated toward the Eclectic Medicine movement:* John S. Haller, Jr., *A Profile in Alternative Medicine* (Kent, Ohio: Kent State University Press, 1999), 16–20.

34 *A recent DAN! conference:* Autism Research Institute/Defeat Autism Now! conference, Waverly Renaissance, Atlanta, April 16–19, 2009.

34 *Lora Little, whom the journalist and author:* Arthur Allen, *Vaccine* (New York: W. W. Norton, 2007), 104–5.

35 *In her 1906 tract:* James Colgrove, *State of Immunity* (Los Angeles: University of California Press, 2006), 61.

35 *Here's Barbara Loe Fisher:* Fisher, "Vaccine Education Seminar."

36 *For Little, that tragedy:* Allen, *Vaccine*, 105.

36 *For Fisher, a television program:* Barbara Loe Fisher, "In the Wake of Vaccines," *Mothering*, September/October 2004.

36 *In 1910, Paul Ehrlich:* Paul Ehrlich, *Studies in Immunity* (New York: John Wiley & Sons, 1910).

36 *in the 1930s:* "Sulfa Drugs," *Encyclopedia*, 2003, n.d., http://www.encyclopedia.com/doc/1G2-3409800539.html.

36 *and in 1941:* Lennard Bickel, *Howard Florey: The Man Who Made Penicillin* (Melbourne: Melbourne University Press, 1996).

36 *In the Spanish-American War:* Allen, *Vaccine*, 159.

36 *Over that same span:* Ibid., 162.

37 *the American government for the first time assumed a central role:* "A Short History of the National Institutes of Health: WWI and the Ransdell Act of 1930," National Institutes of Health, n.d., http://history.nih.gov/exhibits/history/docs/page_04.html; Elizabeth Etheridge, *Sentinel: A History of*

the Centers for Disease Control (Los Angeles: University of California Press, 1992), xv.

37 *"in a futile attempt"*: "Editorial," *The New York Times*, April 1, 1894.

37 *what physician Benjamin Gruenberg described:* Colgrove, *State of Immunity*, 62.

37 *By the time the country joined the Allied cause:* Allen, *Vaccine*, 159.

38 *as Arthur Allen wrote:* Ibid., 158.

CHAPTER 3: THE POLIO VACCINE: FROM MEDICAL MIRACLE TO PUBLIC HEALTH CATASTROPHE

PAGE

39 *On June 6, 1916:* Paul Offit, *The Cutter Incident* (New Haven: Yale University Press, 2005), 4.

39 *It had been a quietly persistent presence:* David Oshinsky, *Polio: An American Story* (New York: Oxford University Press, 2006), 10–11.

39 *The virus quickly made up for lost ground:* Ibid., 8–43.

40 *polio victims in the first decade and a half:* Ibid., 10–20.

40 *Even more terrifying:* "Day Shows 12 Dead by Infant Paralysis," *The New York Times*, July 2, 1916, 6.

40 *Since 1911, when New York's Department of Health:* City of New York Department of Health, "Some Increase in Poliomyelitis," *City of New York Department of Health Weekly Bulletin*, July 18, 1931, 199.

41 *One local expert:* "Day Shows 12 Dead by Infant Paralysis."

41 *In Staten Island:* "Paralysis Figures Rise in Manhattan," *The New York Times*, July 26, 1916, 5.

41 *a local newspaper reported:* "Oyster Bay Revolts over Poliomyelitis," *The New York Times*, August 29, 1916, 1.

41 *In Hoboken, New Jersey:* "31 Die of Paralysis; 162 More Ill in City," *The New York Times*, July 15, 1916, 1.

41 *By early August:* "Defense League of 21,000 Citizens Fights Paralysis," *The New York Times*, July 9, 1916, 1.

41 *city counselors accused:* "Oyster Bay Revolts over Poliomyelitis."

41 *26,212 people had been infected:* Barry Trevelyan et al., "The Spatial Dynamics of Poliomyelitis in the United States," *Annals of the Association of American Geographers* 2005;95(2): 276.

42 *In New York City alone:* City of New York Department of Health, "Some Increase in Poliomyelitis," 199.

42 *For the next thirty years:* Trevelyan, "The Spatial Dynamics of Poliomyelitis in the United States," 276.

42 *polio ranked second:* Offit, *The Cutter Incident*, 32.

42 *a leading medical writer:* Leonard Engel, "The Salk Vaccine: What Caused the Mess?," *Harper's*, August 1955, 30.

42 *which had been founded in 1937:* David Rose, *March of Dimes* (Charleston, South Carolina: Arcadia, 2003), 9.

42 *In 2003, an academic paper:* Armond Goldman et al., "What Was the Cause of Franklin Delano Roosevelt's Paralytic Illness?," *Journal of Medical Biology* 2003;11: 232.

42 *Year after year, its celebrity-studded:* Oshinsky, *Polio*, 55.

43 *When, in November 1953:* Ibid., 174.

43 *The Salk trials began:* Ibid., 187.

43 *At the end of the year:* Ibid., 188.

43 *That summer and fall, another 25,000:* Trevelyan et al., "Spatial Dynamics of Poliomyelitis," 276.

43 *At one point, the* New York World-Telegram and Sun: Engel, "The Salk Vaccine: What Caused the Mess?," 28.

44 *On April 12, 1955:* Oshinsky, *Polio*, 201–3.

44 *one newspaper described the morning:* "Salk Polio Vaccine Proves Success," *The New York Times*, April 13, 1955, 1.

44 *At the back of the auditorium:* Oshinsky, *Polio*, 203.

44 *Reporters on site:* Val Adams, "Release Broken by N.B.C. on Polio," *The New York Times*, April 13, 1955, 40.

44 *In ballrooms and conference halls:* Offit, *The Cutter Incident*, 55.

44 *Six thousand crammed into:* "Fanfare Ushers in Verdict on Tests," *The New York Times*, April 13, 1955, 1.

44 *At 10:20 A.M. as spotlight clicked:* Offit, *The Cutter Incident*, 54.

44 *Salk's vaccine, he announced:* Joe Palca, "Salk Polio Vaccine Conquered Terrifying Disease," National Public Radio, April 12, 2005.

45 *Air raid sirens were set off:* Offit, *The Cutter Incident*, 56; Engel, "The Salk Vaccine: What Caused the Mess?," 28.

45 *"one of the greatest events":* Harold M. Schmeck, Jr., "Dr. Jonas Salk, Whose Vaccine Turned Tide on Polio, Dies at 80," *The New York Times*, June 24, 1995.

45 *"Gone are the old helplessness":* "Dawn of a New Medical Day," *The New York Times*, April 13, 1955, 28.

45 *"It's a wonderful day":* Offit, *The Cutter Incident*, 58.

45 *At five that afternoon:* Oshinsky, *Polio*, 208.

46 *failed to generate immunity:* Ibid., 204.

46 *Republicans spoke sotto voce:* Engel, "The Salk Vaccine: What Caused The Mess?," 29.

46 *Ten days after Francis's:* Morris Kaplan, "Tighter Controls over Vaccine Due," *The New York Times*, April 27, 1955, 1.

46 *In the midst of this frantic activity:* Engel, "The Salk Vaccine: What Caused the Mess?," 29.

46 *By April 26, exactly two weeks:* Offit, *The Cutter Incident*, 68–78.

46 *"This action does not indicate":* "Govt. Experts Push Polio Vaccine Probe," *The Spartanburg Herald* (South Carolina), April 29, 1955, 1.

47 *The foundation said it had "no control":* "Text of the Statement by Basil O'Connor," *The New York Times*, May 9, 1955, 15.

47 *they had not been shown an advance copy:* Oshinsky, *Polio*, 240.

47 *President Eisenhower, at a loss:* "Transcript of President's Press Conference on Domestic and Foreign Affairs," *The New York Times*, May 5, 1955.

47 *Eisenhower's emphasis on damage control:* Engel, "The Salk Vaccine: What Caused the Mess?," 32.

48 *On May 7, Scheele announced:* Offit, *The Cutter Incident*, 98.

48 *At one point Eisenhower attributed:* William Blair, "Eisenhower Sees Polio's Early End with Salk Shots," *The New York Times*, May 12, 1955, 20.

49 *Hobby told a Senate committee:* Oshinksy, *Polio*, 218.

49 *it took only a week for the government:* Offit, *The Cutter Incident*, 120.

49 *"The nation is now badly scared":* "Confusion over Polio," *The New York Times*, May 22, 1955, E1.

49 *The government had not required the pharmaceutical companies:* Offit, *The Cutter Incident*, 61.

49 *Cutter had discarded a full third:* Ibid., 67.

49 *The children of several friends:* Sabin Russell, "When Polio Vaccine Backfired," *San Francisco Chronicle*, April 25, 2005, A1.

50 *With a vacation planned:* Offit, *The Cutter Incident*, 3.

50 *On November 22, 1957:* Ibid., 134.

50 *If anyone was capable of convincing a jury:* Jim Herron Zamora, " 'King of Torts' Belli Dead at 88," *San Francisco Chronicle*, July 10, 1996.

51 *a waitress named Gladys Escola: Escola v. Coca Cola Bottling Co.*, 24 C2d 453 (Cal. Sup. Ct. 1944).

51 *In order to win that case:* "The Cutter Polio Vaccine Incident," *Yale Law Journal* 1955;65(2): 262.

52 *"There is no doubt in my mind":* Offit, *The Cutter Incident*, 142.

52 *"If you find that the vaccine":* Lawrence Davies, "2 Polio Victims Win Vaccine Suit but Cutter Is Held Not Negligent," *The New York Times*, January 18, 1958, 1.

52 *The jurors made their frustration known:* Offit, *The Cutter Incident*, 150.

52 *The $147,300 the Gottsdankers:* Russell, "When Polio Vaccine Backfired."

52 *For the remainder of their lives:* Ibid.

52 *"He was a scientist":* Ibid.

53 *a contemporaneous account that ran in . . .* Harper's*:* Engel, "The Salk Vaccine: What Caused the Mess?," 27.

54 *The Epidemic Intelligence Service:* Alexander D. Langmuir, "The Epidemic In-

telligence Service of the Center for Disease Control," *Public Health Reports* 1980;95(5): 470–71.

54 *It had been the EIS that had first traced:* "The Salk Verdict," *Time*, November 28, 1955.

54 *But in the same investigation:* Lawrence K. Altman, M.D., "An Elite Team of Sleuths, Saving Lives in Obscurity," *The New York Times*, April 6, 2010, D5.

54 *Seven years later:* Donald A. Henderson et al., "Public Health and Medical Responses to the 1957–58 Influenza Pandemic," *Biosecurity and Bioterrorism*. 2009;7(3): 265.

54 *Given this situation, Langmuir chose to bury:* Altman, "An Elite Team of Sleuths, Saving Lives in Obscurity."

CHAPTER 4: FLUORIDE SCARES AND SWINE FLU SCANDALS

PAGE

55 *In its 1962 annual report:* Colgrove, *State of Immunity*, 155.

55 *One of the first indications of this shift:* D. T. Karzon, "Immunization and Practice in the United States and Great Britain: A Comparative Study," *Postgraduate Medical Journal* 1969;45: 148.

55 *Soon, state legislatures:* Colgrove, *State of Immunity*, 176–77.

56 *Before the measles vaccine:* Philip J. Landrigan and J. Lyle Conrad, "Current Status of Measles in the United States," *The Journal of Infectious Diseases* 1971;124(6): 620–22.

56 *A 2006 study by researchers at the Harvard School of Public Health:* Kimberly Thompson and Radbound Tebbens, "Retrospective Cost-Effectiveness Analyses for Polio Vaccination in the United States," *Risk Analysis* 2006;26(6): 1423.

56 *Just as the benefits:* Colgrove, *State of Immunity*, 150–60.

57 *The first hint that fluoride:* Christopher Toumey, *Conjuring Science* (New Brunswick, New Jersey: Rutgers University Press, 1994), 64.

57 *The next piece of evidence came from Minonk:* Ibid.

58 *By 1950, the children in the communities:* Ibid.

58 *In the vast majority of places:* Ibid.

58 *Fifty-nine percent of the time:* Ibid, 65.

58 *One of New Jersey's most prominent:* "Prevent Mandated Water Fluoridation in NJ, Please Make Three Calls Today," New Jersey Coalition for Vaccine Choice, December 10, 2009.

58 *What we know as "fluoride":* Toumey, *Conjuring Science*, 66.

59 *that amount was fifty bathtubs' worth:* "Fluoride Facts and Myths," Kern County Children's Dental Health Network, n.d., http://www.kccdhn.org/stories/storyReader$97.

59 *Consuming as little as a tenth:* T. D. Noakes et al., "Peak Rates of Diuresis in Healthy Humans During Oral Fluid Overload," *South African Medical Journal* 2001;91(10).

60 *in the words of the social anthropologist Arnold Green:* Toumey, *Conjuring Science*, 67.

60 *In 1942, an American scientist:* Colgrove, *State of Immunity*, 111.

60 *a decade in which more than 60 percent:* Jeffrey Baker, "The Pertussis Vaccine Controversy in Great Britain, 1974–1986," *Vaccine* 2003;21: 4003–10.

60 *After mass DPT inoculations began:* Ibid.

60 *While whole cell vaccines can be both effective and safe:* Paul Offit, interview with author, February 3, 2009.

61 *In Britain, this veneer of nonchalance:* Baker, "The Pertussis Vaccine Controversy in Great Britain, 1974–1986," 4004.

61 *After decades of close to one hundred percent employment:* Mark Tran, "Unemployment," *The Guardian* (London), May 15, 2002.

61 *One particularly flimsy example:* Baker, "The Pertussis Vaccine Controversy in Great Britain, 1974–1986," 4004.

61 *Within months, vaccination rates began to fall:* Ibid.

62 *On February 4, 1976:* Joel Gaydos et al., "Swine Influenza A Outbreak, Fort Dix, New Jersey, 1976," *Emerging Infectious Diseases* 2006;12(1): 24.

62 *the 1918 flu pandemic:* C. W. Potter, "A History of Influenza," *Journal of Applied Microbiology* 2006;91(4): 575.

62 *"the greatest medical holocaust":* Ibid.

63 *On March 11, CDC director David Sencer:* Richard Neustadt and Harvey Fineberg, *The Swine Flu Affair: Decision-Making on a Slippery Slope* (Washington, D.C.: Department of Health, Education, and Welfare, 1978), 11–14.

63 *"I certainly thought of it":* Ibid., 24.

63 *"each and every American":* Harold M. Schmeck, Jr., "Ford Urges Flu Campaign to Inoculate Entire U.S.," *The New York Times*, March 25, 1976, 1.

64 *did nothing to mollify:* Neustadt and Fineberg, *The Swine Flu Affair*, 45.

64 *when a jury ordered Wyeth Pharmaceuticals:* Colgrove, *State of Immunity*, 189–90.

64 *"particularly if you are talking about":* Neustadt and Fineberg, *The Swine Flu Affair*, 45.

64 *an article had appeared in* Pediatrics: Richard Krugman, "Immunization 'Dyspractice': The Need for 'No Fault' Insurance," *Pediatrics* 1975;56(2): 159–60.

65 *On August 12, Ford signed:* Neustadt and Fineberg, *The Swine Flu Affair*, 53.

65 *On October 1, the government's massive:* Harold Schmeck, "Swine Flu Shots Will Start Today for the Elderly and Ill in 2 Cities," *The New York Times*, October 1, 1976, 14.

65 *By the end of December:* Arthur Silverstein, *Pure Politics, Impure Science: The Swine Flu Affair* (Baltimore: Johns Hopkins University Press, 1982).

65 *among the forty million:* Stephanie Beck, "When Politics, and Swine Flu, Infect Health," *San Francisco Chronicle*, April 30, 2009.

65 *it occurs at an incidence:* A. H. Ropper, "The Guillain-Barré Syndrome," *New England Journal of Medicine* 1992;326(17): 1130–36.

66 *On December 16, it was called off:* Neustadt and Fineberg, *The Swine Flu Affair*, 1.

66 *"Any program conceived by politicians":* Richard Cohen, "Science as Fiction Makes Skeptical Fan," *The Washington Post*, November 14, 1976.

66 *By 1977, uptake of the pertussis vaccine:* Baker, "The Pertussis Vaccine Controversy in Great Britain, 1974–1986," 4004.

CHAPTER 5: "VACCINE ROULETTE"

PAGE

67 *On April 19, 1982:* Vincent Fulginiti, "'Red Book' Update," *Pediatrics* 1982;70: 819–22.

67 *"For more than a year":* Lea Thompson, "DPT: Vaccine Roulette," WRC-TV, Washington, D.C., April 19, 1982, unofficial transcript.

69 *"[DPT] was without a doubt":* Paul Offit, interview with author, February 3, 2009.

69 *A UPI wire dispatch:* Michael Conlon, "Report Says Vaccines May Cause Brain Damage," United Press International, April 20, 1982.

69 The Washington Post *gave Thompson:* Sandy Rovner, "Risks, Benefits & DPT," *The Washington Post*, April 30, 1982, C5.

70 *At the awards ceremony:* Ann Trebbe, "Local TV Honors Its Own; Channels 4 and 7 Sweep the Emmys," *The Washington Post*, June 27, 1983, B1.

70 *What it found was a dispatch rife:* Elizabeth Gonzalez, "TV Report on DPT Galvanizes US Pediatricians," *Journal of the American Medical Association* 1982;248(1): 12–22.

71 *the extent to which Thompson misstated:* Ibid.

71 *in 1977, he published a paper:* Gordon Stewart, "Vaccination Against Whooping Cough: Efficacy Versus Risks," *The Lancet* 1977;309(8005): 234–37.

71 *he'd been for years the go-to guy:* Baker, "The Pertussis Vaccine Controversy in Great Britain, 1974–1986," 4004, 4006.

71 *When she highlighted the conclusions:* Gonzalez, "TV Report on DPT Galvanizes US Pediatricians," 21.

72 *Stewart questioned whether these outbreaks:* Thompson, "DPT: Vaccine Roulette."

72 *"I think much of the potshotting":* Gonzalez, "TV Report on DPT Galvanizes US Pediatricians," 22.

72 *her station didn't have the budget:* Ibid., 20.

72 *Thompson and her employer's role:* Allen, *Vaccine*, 254.

73 *One of these parents was Barbara Loe Fisher:* Fisher, "In the Wake of Vaccines."

73 *Many years later, Fisher would describe:* Barbara Loe Fisher, "Statement to the IOM Immunization Safety Committee," January 11, 2001.

73 *"When we got home":* Ibid.

74 *"She lies":* Amy Wallace, "Epidemic of Fear," *Wired*, November 2009, 166.

74 *In her e-mail to me Fisher wrote:* E-mail to author, "Subject Re: checking in," April 19, 2010.

74 *In December 2009:* Complaint with Demand for Jury Trial, *Barbara Loe Arthur v. Paul A. Offit et al.*, No. 09-CV-1398 (U.S. Dist. Ct., E.D. VA), March 10, 2010.

74 *The case was summarily dismissed:* Memorandum Opinion, *Barbara Loe Arthur v. Paul A. Offit et al.*

74 *Fisher, as she told a government vaccine safety committee:* Fisher, "Statement to the IOM Immunization Safety Committee."

74 *Fisher founded the National Vaccine Information Center:* "About National Vaccine Information Center," National Vaccine Information Center, n.d., http://www.nvic.org/about.aspx.

75 *Today, every state save for Mississippi:* "Vaccine Laws," National Vaccine Information Center, n.d., http://www.nvic.org/vaccine-laws.aspx.

75 *In 2010,* Pediatrics *released a study:* Gary L. Freed et al., "Parental Vaccine Safety Concerns in 2009," *Pediatrics*, March 1, 2010, http://pediatrics.aappublications.org/cgi/content/abstract/peds.2009-1962v1.

75 *"The genie is not going back":* J. B. Handley, "Tinderbox: U.S. Vaccine Fears up 700% in 7 years," *Age of Autism*, March 17, 2010, http://www.ageofautism.com/2010/03/tinderbox-us-vaccine-fears-up-700-in-7-years.html.

CHAPTER 6: AUTISM'S EVOLVING IDENTITIES

PAGE

76 *In 1943, Leo Kanner:* Leo Kanner, "Autistic Disturbances of Affective Contact," *The Nervous Child* 1943: 217–50.

77 *Six years later, in Kanner's second major paper:* Leo Kanner, "Problems of Nosology and Psychodynamics in Early Infantile Autism," *American Journal of Orthopsychiatry* 1949;19(3): 416–26.

77 *In 1959, Bettelheim published:* Bruno Bettelheim, "Joey: 'A Mechanical Boy,' " *Scientific American*, March 1959, 117–26.

78 *to compare the households in which autistic children:* Bruno Bettelheim, *The Informed Heart* (Chicago: University of Chicago Press, 1961).

78 *By the time his book* The Empty Fortress: Bruno Bettelheim, *The Empty Fortress* (New York: Free Press, 1967).

78 *claimed they'd been physically and mentally* abused: Richard Bernstein, "Accusations of Abuse Haunt the Legacy of Dr. Bruno Bettelheim," *The New York Times*, November 4, 1990, 4–6.

80 *In the United States, this shift was illustrated:* G. N. Grob, "Origins of DSM-I: A Study in Appearance and Reality," *The American Journal of Psychiatry* 1991;148(4): 421–31.

80 *when the gay rights movement:* R. L. Spitzer, "The Diagnostic Status of Homosexuality in the DSM-III: A Reformulation of the Issues," *The American Journal of Psychiatry* 1981;138: 210–15.

80 *Jonathan Metzl writes about one:* Jonathan Metzl, *The Protest Psychosis: How Schizophrenia Became a Black Disease* (Boston: Beacon Press, 2010).

80 *The evolution of the DSM's handling of autism:* Richard Roy Grinker, *Unstrange Minds* (Philadelphia: Basic Books, 2007), 111–35.

80 *The notion that scientific progress:* Thomas Kuhn, *The Structure of Scientific Revolutions*, 3rd ed. (Chicago: University of Chicago Press, 1996). Originally published 1962. See also: Thomas Kuhn, *The Copernican Revolution* (Cambridge: Harvard University Press, 1985). Originally published 1957.

81 *the heading of "schizophrenic reaction, childhood type," which listed:* American Psychiatric Association, *Diagnostic and Statistical Manual, Mental Disorders* (Washington, D.C.: American Psychiatric Association Mental Hospital Service), 28.

81 *In the DSM-II:* American Psychiatric Association, *DSM-II: Diagnostic and Statistical Manual of Mental Disorders*, 2nd ed. (Washington, D.C.: American Psychiatric Association), 35.

81 *It wasn't until 1980, with the publication of the DSM-III:* "Diagnostic Criteria for Autism Through the Years," *Unstrange Minds*, n.d., http://www.unstrange.com/dsm1.html.

81 *it was folded into the newly expanded class:* American Psychiatric Association, *Diagnostic and Statistical Manual of Mental Disorders*, 4th ed. (Washington, D.C.: American Psychiatric Association, 2000), 59–60.

81 *the number of members of the American Academy of Child and Adolescent Psychiatry:* Grinker, *Unstrange Minds*, 108

82 *In 1964, at the precise time the medical community:* Bernard Rimland, *Infantile Autism* (New Jersey: Prentice Hall, 1964).

82 *In 1965, Rimland founded:* "About Us," Autism Society of America, n.d., http://www.autism-society.org/site/PageServer?pagename=asa_home.

82 *The Autism Research Institute:* "ARI Mission Statement," Autism Research Institute, n.d., http://www.autism.com/gen_mission.asp.

83 *Lobbying efforts in states:* Saul Spigel, "Medicaid Autism Waivers and State Agencies Serving People with Autism," Connecticut General Assembly, April 10, 2007, http://www.cga.ct.gov/2007/rpt/2007-R-0319.htm.

83 *As the anthropologist Roy Richard Grinker explains:* Grinker, *Unstrange Minds*, 130–31.

83 *there are huge variations:* Lynn Waterhouse et al., "Diagnosis and Classification in Autism," *Journal of Autism and Developmental Disorders* 1996;26(1): 59–86.

83 *The most remarkable example of the arbitrary nature:* Grinker, *Unstrange Minds*, 141.

83 *the rise in the number of children:* Craig J. Newschaffer et al., "The Epidemiology of Autism Spectrum Disorders," *Annual Review of Public Health* April 2007;28:240.

84 *including advanced maternal age:* Janie F. Shelton, Daniel J. Tancredi, and Irva Hertz-Picciotto, "Independent and Dependent Contributions of Advanced Maternal and Paternal Ages to Autism Risk," *Autism Research* February 2010:3(2): 30–39.

84 *paternal age as related to maternal age:* Maureen Durkin et al., "Advanced Paternal Age and the Risk of Autism Spectrum Disorders," *American Journal of Epidemiology* December 2008;168(11):1268–76; Shelton, Tancredi, and Hertz-Picciotto, "Independent and Dependent Contributions of Advanced Maternal and Paternal Ages to Autism Risk."

84 *proximity to families with autistic children:* Ka-Yuet Liu, Marissa King, and Peter S. Bearman, "Social Influence and the Autism Epidemic," *American Journal of Sociology* March 2010;115(5): 1387–434.

84 *the growing use of a type of muscle relaxant:* F.R. Witter et al., "In Utero Beta 2 Adrenergic Agonist Exposure and Adverse Neurophysiologic and Behavioral Outcomes," *American Journal of Obstetrics and Gynecology* December 2009;201(6): 553–59.

84 *From 1989 through 2005:* Christine Russell, "Covering Controversial Science," Joan Shorenstein Center Working Paper Series, Spring 2006.

84 *In 2008, CNN got rid of:* Curtis Brainard, "CNN Cuts Entire Science, Tech Team," *Columbia Journalism Review*, December 4, 2008.

84 *in a 2009 survey:* Phil Galewitz, "Survey Shows 'Battered' Health Journalists Press On," Association of Health Care Journalists, March 11, 2009, http://www.healthjournalism.org/resources-articles-details.php?id=94.

84 *Take a 2009 piece:* Michael Johnson, "Watch What You Think," *International Herald Tribune*, March 3, 2009.

85 *early research has shown promising results:* Tom M. Mitchell et al., "Predicting Human Brain Activity Associated with the Meanings of Nouns," *Science*, May 30, 2008;320(5880): 1191–95.

85 *As Vaughan Bell:* Vaughan Bell, "Cigarette Smoking Lady Cops to Read Minds," *Mind Hacks*, March 15, 2009, http://mindhacksblog.wordpress.com/2009/03/15/cigarette-smoking-lady-cops-to-read-minds.

CHAPTER 7: HELP! THERE ARE FIBERS GROWING OUT OF MY EYEBALLS!

PAGE

87 *In 2007, a sixty-year-old woman:* Robert Accordino et al., "Morgellons Disease?," *Dermatologic Therapy* 2008;21: 8.

88 *a disease that was invented in 2002:* Brigid Schulte, "Figments of the Imagination?," *The Washington Post*, January 20, 2008, W10.

88 *Leitao could be suffering:* Elizabeth Devita-Raeburn, "The Morgellons Mystery," *Psychology Today*, March 1, 2007.

88 *After using a RadioShack microscope:* Chico Harlan, "Mom Fights for Answers on What's Wrong with Her Son," *Pittsburgh Post-Gazette*, July 23, 2006, A1.

88 *The name "Morgellons" is taken from a seventeenth-century monograph:* "Frequently Asked Questions," Morgellons Research Foundation, n.d., http://www.morgellons.org/faq-home.htm#item2.

88 *For the first four years of its existence:* "Media and Public Relations," Morgellons Research Foundation, n.d., http://www.morgellons.org/media.htm.

88 *That all changed in May 2006:* Martin Savidge, "Controversial Morgellons Disease and Its Sufferers," *Today*, NBC, July 28, 2006; Elizabeth Cohen, "Morgellons," *Paula Zahn Now*, CNN, June 23, 2006.

88 *On ABC's* Good Morning America: Cynthia McFadden, "Mysterious Skin Disease Causes Itching, Loose Fibers; Morgellons Has Plenty of Skeptics," *Good Morning America*, ABC, July 28, 2006.

88 *Several weeks later:* Page Bowers, "Itching for Answers to a Mystery Condition," *Time*, July 28, 2006.

89 *there were rumors that Morgellons was the product:* "Horror Illness Is Viral . . . Marketing?," *Sploid*, May 23, 2006.

89 *In the coming year, the number of hits:* Accordino, "Morgellons Disease?," 10.

90 *Schwartz began treating Morgellons:* Wendy Brown, "Doctor Now Focuses on Disputed Skin Disease," *The Santa Fe New Mexican*, December 14, 2005, A2.

90 *In 2006, Schwartz's license:* Polly Summar, "Doctor Agrees to Retire License," *Albuquerque Journal*, July 15, 2008, 1.

90 *That March, Savely had left Texas:* Howard Witt, "A Mystery Ailment Gets Under Skin," *Chicago Tribune*, July 25, 2006, C1.

90 *Wymore, whose last published study:* Randy Wymore, "Tissue and Species Distribution of mRNA for the I_{Kr}-like K^+ Channel," *Circulation Research* 1997;80: 261–68.

91 *He claimed to have "physical evidence":* Randy Wymore. "A Position Statement from Randy S. Wymore on the Topic of Morgellons Disease and Other Morgellons-Related Issues," Oklahoma State University's Center for Health Sciences Center for the Investigation of Morgellons Disease, June 19, 2007,

http://www.healthsciences.okstate.edu/morgellons/docs/Wymore-position-statement-2-19-07.pdf.

91 *On her official OSU staff page:* "Faculty & Staff: Rhonda L. Case," Oklahoma State University, n.d., http://www.healthsciences.okstate.edu/college/clinical/pediatrics/casey.cfm.

91 *advertised itself as having six specialties:* Union Square Medical Associates, n.d., http://www.usmamed.com/03viagra/index.htm; http://www.usmamed.com/04weightloss/index.htm; http://www.usmamed.com/02fertility/index.htm; http://www.usmamed.com/07hyperbaric/index.html; http://www.usmamed.com/05hiv/index.htm; http://www.usmamed.com/06lyme/lyme.htm.

91 *In October 2006, Stricker was the sole:* Michael Mason, "Is It Disease or Delusion? U.S. Takes on a Dilemma," *The New York Times*, October 24, 2006, F5.

92 *In a 2007 study, researchers at the University of Michigan:* Norbert Schwarz et al., "Metacognitive Experiences and the Intricacies of Setting People Straight," *Advances in Experimental Social Psychology* 2007;39: 127–61.

92 *On January 20, 2008, the* Washington Post*'s:* Schulte, "Figments of the Imagination?"

93 *one of four recognized caffeine-induced psychiatric illnesses:* R. Gregory Lande, "Caffeine-Related Psychiatric Disorders," *eMedicine from WebMD,* n.d., http://emedicine.medscape.com/article/290113-overview.

95 *Two days after her article ran:* Brigid Schulte, "Morgellons Disease. Live Discussion with Brigid Schulte," washingtonpost.com, January 22, 2008.

95 *More than three dozen members of Congress:* Schulte, "Figments of the Imagination?"

95 *In a statement announcing a study:* "Morgellons: Statement of Work," Centers for Disease Control and Prevention, 2007, http://www.morgellons.org/docs/CDC_Morgellons_RFQ_2007.pdf.

96 *The CDC's "Unexplained Dermopathy Project":* "Unexplained Dermopathy (also called 'Morgellons')," Centers for Disease Control and Prevention, November 5, 2009, http://www.cdc.gov/unexplaineddermopathy/investigation.html.

CHAPTER 8: ENTER ANDREW WAKEFIELD

PAGE

99 *In September 1992, British officials:* Liz Hunt, "New Vaccine for Children Aims to Curb Meningitis," *The Independent* (London), October 1, 1992, 9.

99 *estimates ranged from one in six thousand:* Carol Midgley and Vikki Orvice, "Why Did They Wait?," *The Daily Mail* (London), September 21, 1992, 18.

99 *as nothing tempered the panic:* Angella Johnson, "Doctors Angered by Jabs Ban Hitch," *The Guardian* (London), September 16, 1992, 2.

99 *or that one in four hundred:* Celia Hall, "Children Received Vaccine Despite Meningitis Link," *The Independent* (London), September 16, 1992, 2.

99 *in 1988, the year the single-dose MMR vaccine:* Carol Midgley, "Vaccination: The Dilemma Now Facing Every Parent," *The Daily Mail* (London), September 16, 1992, 14.

99 *Within twenty-four hours of the announcement:* Ibid.

100 *Britain found itself facing a potential measles epidemic:* Liz Hunt, "Measles Campaign to Avert Epidemic," *The Independent* (London), July 29, 1994, 7.

100 *With epidemics in Great Britain historically occurring:* Jenny Hope, "The Flaw That Let Measles Back in Force, *The Daily Mail* (London), August 2, 1994, 38.

100 *Kenneth Calman, the government's chief medical officer:* Hunt, "Measles Campaign to Avert Epidemic."

100 *Well aware of the need for positive publicity:* Bryan Christie, "Measles Jab for 800,000 Scots Children," *The Scotsman*, September 30, 1994.

100 *The media, while quick to cover any unease:* Annabel Ferriman, "Why Another Needle, Mummy?," *The Independent* (London), October 11, 1994, 23.

100 *Ampleforth and Stonyhurst:* Liz Hunt, "Vaccine Ban Blow to Fight Against Epidemic," *The Independent* (London), October 27, 1994, 6.

100 *days later, Muslim leaders:* Liz Hunt and Andrew Brown, "Muslims Urged to Boycott Rubella Vaccine," *The Independent* (London), October 29, 1994, 2.

100 *In London, a group of homeopaths:* Deborah Jackson, "Please Be Sick After the Party," *The Independent* (London), November 15, 1994, 27.

101 *Jackie Fletcher, a suburban housewife:* Jan Roberts, "Vaccination: Do You Know the Risks?," *The Independent* (London), April 12, 1994, 20.

101 *Fletcher embarked on a strategy:* Ibid.

101 *From the outset, Barr claimed:* "Measles: A Spot of Bother," *The Economist*, October 29, 1994, 98.

102 *In late 1995, Fletcher told reporters:* "Measles Jab Row over 85 Ill Children," *The Independent* (London), October 1, 1995, 1.

102 *The coverage was at times unintentionally humorous:* Jill Palmer, "My Son Went Bald After His Measles Jab," *The Daily Mirror* (London), January 29, 1996, 25.

102 *In late 1996, Barr let the press:* Grania Langdon-Down, "A Shot in the Dark," *The Independent* (London), November 27, 1996, 25.

102 *As Fletcher's husband, John, told* The Guardian: John Illman, "Painful Choice of Risks: Measles Kills. But Preventative Vaccination Causes Problems, Too," *The Guardian* (London), November 2, 1994, T12.

103 *a Canadian-trained surgeon:* Jeremy Laurance, "Not Immune to How Research Can Hurt," *The Independent* (London), March 3, 1998, 14.

103 *He was a former amateur rugby player:* Brian Deer, "Truth of the MMR Vaccine Scandal," *The Times* (London), January 24, 2010.

103 *an autoimmune disorder that causes the body:* M. Comalada and M. P. Peppelenbosch, "Impaired Innate Immunity in Crohn's Disease," *Trends in Molecular Medicine* 2006;12(9): 397–99.

103 *The disease's trademark inflammations:* Laurance, "Not Immune to How Research Can Hurt."

103 *"[It was] an interesting idea":* Transcript of record, *Hazlehurst v. Sec'y of Health and Human Services*, No. 03-654V (Ct. Fed. Cl., February 12, 2009), 630.

104 *His "eureka" moment:* Laurance, "Not Immune to How Research Can Hurt."

104 *According to an investigative journalist:* Brian Deer, "The MMR-Autism Fraud—Our Story So Far," n.d., Briandeer.com, http://briandeer.com/solved/story-highlights.htm.

104 *That spring, he and several co-authors:* Andrew Wakefield et al., "Evidence of Persistent Measles Infection in Crohn's Disease," *Journal of Medical Virology* 1993;39(4): 345–53.

104 *That study created such a furor:* Deer, "The MMR-Autism Fraud—Our Story So Far."

104 *Wakefield's next attention-getting paper:* N. P. Thompson and A. J. Wakefield, "Is Measles Vaccination a Risk Factor for Inflammatory Bowel Disease?," *The Lancet* 1995;345(8957): 1071–74. See also: A. J. Wakefield, "Crohn's Disease: Pathogenesis and Persistent Measles Virus Infection," *Gastroenterology* 1995;108(3): 911–16.

105 *In the years to come, teams in Japan: Snyder v. Sec'y of Health and Human Services*, No 01-162V (Ct. Fed. Cl., February 12, 2009), 117.

105 *the chemical solutions he'd used:* Ibid., 122.

105 *actually stemmed from contamination:* Ibid., 120.

105 *as Thomas MacDonald explained later:* Transcript of record, *Hazelhurst v. Sec'y Health and Human Services*, 630a.

CHAPTER 9: THE *LANCET* PAPER

PAGE

106 *On February 26, 1998:* Andrew Wakefield, press conference, Royal Free Hospital, London, February 26, 1998.

106 *a dense academic paper in* The Lancet: Andrew Wakefield et al., "Ileal-Lymphoid-Nodular Hyperplasia, Non-Specific Colitis, and Pervasive Developmental Disorder in Children," *The Lancet* 1998;351: 637–41.

106 *the Royal Free's PR team gave hints:* "New Research Links Autism and Bowel Disease," Royal Free Hospital School of Medicine, February 26, 1998.

107 *The next step in Wakefield's hypothesis:* Wakefield et al., "Ileal-Lymphoid-

Nodular Hyperplasia, Non-Specific Colitis, and Pervasive Developmental Disorder," 640.

107 *The opioid excess theory:* Jaak Panskepp, "A Neurochemical Theory of Autism," *Trends in Neurosciences* 1979;2: 174–77.

107 *the five experts who addressed the media:* Jeremy Laurance, "I Was There When He Dropped His Bombshell," *The Independent* (London), January 29, 2010.

107 *"With the debate over MMR that has started":* Jo Revill, "Scientists Warning Prompts Fears over Measles Vaccine," *The Evening Standard* (London), February 26, 1998, 1.

108 *the press conference descended:* Laurance, "I Was There When He Dropped His Bombshell."

108 *"If this were to precipitate a scare":* Ian Murray, "Measles Vaccine's Link with Autism Studied," *The Times* (London), February 27, 1998.

108 *As scientists around the world already knew:* Michael Fitzpatrick, *MMR and Autism: What Parents Need to Know* (London: Routledge, 2004), 119–20.

108 *After an initial peer review:* Richard Horton, *Second Opinion: Doctors, Diseases, and Decisions in Modern Medicine* (London: Granta, 2003), 207–8.

108 *"Usually, when they publish a commentary":* Frank DeStefano, interview with author, July 2, 2009.

108 *Only two years had passed:* Jeremy Laurance, "Second Opinion: Doctors, Diseases and Decisions in Modern Medicine by Richard Horton," *The Independent*, July 3, 2003, 15.

109 The Lancet *conspicuously declined:* Laurance, "Not Immune to How Research Can Hurt."

109 *"Reported adverse events":* Richard Horton, "Correspondence: Autism, Inflammatory Bowel Disease, and MMR Vaccine," *The Lancet* 1998;351(9106): 908–09.

110 *As Chen and DeStefano demonstrated:* Robert T. Chen and Frank DeStefano, "Vaccine Adverse Events: Causal or Coincidental?," *The Lancet* 1998;351(9103): 611–12.

111 *Several years later, when the dean of research:* Transcript of record, *Hazelhurst v. Sec'y Health and Human Services*, 634.

111 *While his detractors were explaining:* Peter Richmond and David Goldblatt, "Correspondence: Autism, Inflammatory Bowel Disease, and MMR Vaccine," *The Lancet* 1998;351(9112): 1355–56.

112 *He condemned his critics:* Laurance, "Not Immune to How Research Can Hurt."

112 *And Wakefield had answers:* "Helpline Puts GPs in the Firing Line," *Pulse*, March 14, 1998, 3.

112 *Even when Wakefield was called on to defend:* Andrew Wakefield, "Correspon-

dence: Author's reply: Autism, Inflammatory Bowel Disease, and MMR Vaccine," *The Lancet* 1998; 351(9106): 908.

113 *a* Guardian *feature on Karen Prosser:* Sarah Boseley, "Undetected Bowel Illness Led to Baby's Misery," *The Guardian* (London), February 27, 1998, 5.

114 *It, too, found his research lacking:* "Group Concluded No Reason for Change in MMR Vaccine Policy," Medical Research Council, March 24, 1998.

114 *One of the few voices of support:* Barbara Loe Fisher, "Correspondence: Autism, Inflammatory Bowel Disease, and MMR Vaccine," *The Lancet* 1998;351(9112): 1357–58.

114 *Ten co-signers:* Helen Bedford et al., "Correspondence: Autism, Inflammatory Bowel Disease, and MMR Vaccine," *The Lancet* 1998;351(9106): 907.

114 *three more from the Barnsley Health Authority's:* David Black et al., "Correspondence: Autism, Inflammatory Bowel Disease, and MMR Vaccine," *The Lancet* 1998;351(9106): 905–06.

114 *"Ban Three-In-One jab":* Jenny Hope, "Ban Three-in-One Jab Urge Doctors After New Fears," *The Daily Mail* (London), February 27, 1998, 13.

114 *"Doctors Link Autism to MMR Vaccination":* Jeremy Laurance, "Doctors Link Autism to MMR Vaccination," *The Independent* (London), February 27, 1998, 1.

114 *"Research Rejects Autism Link with Vaccine":* Jennifer Trueland, "Research Rejects Autism Link with Vaccine," *The Scotsman*, May 1, 1998, 10.

114 *"Triple Jab Is Safe":* "Triple Jab Is Safe, Says Medical Chief," *The Daily Mail* (London), March 13, 1998, 25.

115 *a letter that was sent to* The Lancet: Andrew Rouse, "Correspondence: Autism, Inflammatory Bowel Disease, and MMR Vaccine," *The Lancet* 1998;351(9112): 1356.

115 *The first public hints of the answer:* Andrew Wakefield, "Correspondence: Author's reply: Autism, Inflammatory Bowel Disease, and MMR Vaccine," *The Lancet* 1998;351(9112): 1357.

115 *the Dawbarns pamphlet, which recommended:* "JABS Briefing Note 9th April 2008," Justice, Awareness and Basic Support, April 9, 2008.

116 *On November 27, 1996,* The Independent *ran:* Langdon-Down, "Law: A Shot in the Dark."

116 *planning on suing the manufacturers:* "Fitness to Practise Panel Hearing," General Medical Council (U.K.), January 20, 2010, 6.

116 *Wakefield's lab had been paid £50,000:* Ibid., 4.

116 *the initial research proposal:* Andrew Wakefield, "Submission to the UK Press Complaints Commission," n.d., http://www.cryshame.co.uk//index.php?option=com_content&task=view&id=135.

116 *Wakefield received £435,643:* Brian Deer, "MMR Scare Doctor Got Legal Aid Fortune," *The Sunday Times* (London), December 31, 2006, 12.

116 *Wakefield had filed a patent:* Brian Deer, "MMR Scare Doctor Planning Rival Vaccine," *The Sunday Times* (London), November 14, 2004, 8; "Patent Application Number 9711663.6," The Patent Office, the Department of Trade and Industry (U.K.), June 6, 1997, http://briandeer.com/wakefield/vaccine-patent.htm.

117 *"Measles Turned My Son into an Autistic Child":* Ken Oxley, "Measles Turned My Son into an Autistic Child," *The Daily Record* (Glasgow), March 4, 1998, 2.

117 *"I Want Justice for My James":* James Wildman, "I Want Justice for My James—Study Boosts Mum's Battle," *Gloucestershire Echo*, February 28, 1998, 1.

CHAPTER 10: THIMEROSAL AND THE MYSTERY OF MINAMATA'S DANCING CATS

PAGE

118 *On the morning of January 28, 1928:* Claire Hooker, "Diptheria, Immunisation and the Bundaberg Tragedy," *Health and History* 2000;2(1): 52–78.

120 *That spring, a five-year-old girl:* Timothy George, *Minamata* (Cambridge: Harvard University Asia Center, 2002), 5.

120 *Residents spoke of a "cat dancing disease":* Akio Mishima, *Bitter Sea: The Human Cost of Minamata Disease* (Toyko: Kosei, 1992), 43.

120 *floating fish corpses:* Ibid., 40.

120 *Birds dropped from the sky:* Ibid., 152.

121 *The level of mercury consumed by residents:* Masazumi Harada, *Minamata Disease* (Kunamoto, Japan: Kumamoto Nichinichi Shinbun Culture and Information Center, 2004), 50.

121 *In 1971, nearly five hundred Iraqis:* F. Bakir, "Methylmercury Poisoning in Iraq," *Science* 1973;181(4096): 231.

121 *Labels marked "DANGER" and "POISON":* Jane Hightower, *Diagnosis Mercury* (Washington, D.C.: Island, 2008), 148.

121 *field studies in the Faroe Islands:* Philippe Grandjean et al., "Cognitive Deficit in 7-Year-Old Children with Prenatal Exposure to Methylmercury," *Neurotoxicology and Teratology* 1997;19(6): 417–28.

121 *and in the Seychelles:* Philip Davidson et al., "Effects of Prenatal and Postnatal Methylmercury Exposure from Fish," *Journal of the American Medical Association* 1998;280(8): 701–7.

121 *in the Faroe Islands study:* Grandjean, "Cognitive Deficit in 7-Year-Old Children with Prenatal Exposure to Methylmercury," 417.

121 *All the while, the number of childhood vaccines:* "History of Vaccine Schedule," The Children's Hospital of Philadelphia, n.d., http://www.chop.edu/service/vaccine-education-center/vaccine-schedule/history-of-vaccine-schedule.html. See also: Centers for Disease Control and Prevention, "Notice to Readers: Recommended Childhood Immunization Schedule—

United States, January 1995," *Morbidity and Mortality Weekly Report* January 6, 1995;43(51): 959–60; Centers for Disease Control and Prevention, "Notice to Readers: Recommended Childhood Immunization Schedule—United States, January 15, 1999," *Morbidity and Mortality Weekly Report* 1999;48(01): 8–16.

122 *Vaccinologists also knew that ethylmercury:* Michael Shannon, "Methylmercury and Ethylmercury: Different Sources, Properties and Concerns," *AAP News* 2004;25(1): 23.

122 *In the fall of 1997, a Democratic congressman:* United States Congress House of Representatives, "Mercury Environmental Risk and Comprehensive Utilization Reduction Initiative," 105th Cong., 1st Sess. H.R. 2910 (Washington, D.C., GPO, 1997).

123 *Thimerosal began to be used as a preservative:* U.S. Food and Drug Administration, "Thimerosal in Vaccines," n.d., http://www.fda.gov/BiologicsBloodVaccines/SafetyAvailability/VaccineSafety/UCM096228.

123 *an FDA safety review in 1976:* Leslie K. Ball, Robert Ball, and R. Douglas Pratt, "An Assessment of Thimerosal Use in Childhood Vaccines," *Pediatrics,* 2001; 107: 1148.

123 *The use of the preservative had never:* Freed et al., "The Process of Public Policy Formulation: The Case of Thimerosal in Vaccines," 1153.

123 *In December 1998, it asked:* Food and Drug Administration, "Notice: Mercury Compounds in Drugs and Foods; Request for New Information," *Federal Register* 1998;63(239): 68775–77.

123 *By the following April:* Freed et al., "The Process of Public Policy Formulation: The Case of Thimerosal in Vaccines," 1154.

123 *The confusion was heightened:* Ibid.

124 *There was considerably less consensus:* Ibid.

124 *Included among those was Neal Halsey:* Neal Halsey, interviews and e-mails with author, April–June 2010.

124 *"There was no safety data":* Neal Halsey, interview with author, April 26, 2010.

124 *From the outset, Halsey:* Neal Halsey, interviews and e-mails with author, April–June 2010.

125 *In a series of meetings:* Freed et al., "The Process of Public Policy Formulation: The Case of Thimerosal in Vaccines," 1155.

125 *And then, in the course of a handful of days:* Ibid., 1155–56.

125 *At the last minute, Surgeon General:* Ibid.

126 *The CDC's read as follows:* Centers for Disease Control and Prevention, "Notice to Readers: Thimerosal in Vaccines: A Joint Statement of the American Academy of Pediatrics and the Public Health Service."

127 *The AAP's statement echoed:* Paul Offit, "Thimerosal and Vaccines—A Cautionary Tale," *New England Journal of Medicine* 2007;357(13): 1278–79.

127 *Pediatricians were given no guidance:* Freed et al., "The Process of Public Policy Formulation: The Case of Thimerosal in Vaccines," 1158–59.

127 *at least one death, of a three-month-old:* The American Medical Association, "Impact of the 1999 AAP/USPHS Joint Statement on Thimerosal in Vaccines on Infant Hepatitis B Vaccination Practices," *Journal of the American Medical Association* 2001;285: 1568–70.

128 *One* Los Angeles Times *article:* Jane Allen, "Shots in the Dark," *Los Angeles Times*, October 18, 1999, S1.

128 *A* Denver Post *story quoted at length:* Al Knight, "Pinning Down the Risks of Vaccinations," *Denver Post*, August 5, 1999, B9.

128 *The most outspoken proponent of this view:* Paul Offit, interview with author, February 3, 2009.

128 *In support of his argument Offit pointed:* Ibid.

129 *Halsey was among those who thought Offit:* Neal Halsey, interview with author, April 26, 2010.

129 *That fall, Halsey published an editorial:* Neal A. Halsey, "Limiting Infant Exposure to Thimerosal in Vaccines and Other Sources of Mercury," *Journal of the American Medical Association* 1999;282: 1763–66.

129 *Offit responded with a letter:* P. A. Offit, "Letters: Preventing Harm from Thimerosal in Vaccines," *Journal of the American Medical Association* 2000;283: 2104.

129 *Stanley Plotkin, a colleague of Offit's:* S. A. Plotkin, "Letters: Preventing Harm from Thimerosal in Vaccines," *Journal of the American Medical Association* 2000;283: 2104–5.

130 *In 1997, he staged a backyard demonstration:* Mary Ann Akers, "Dan Burton, Protecting the House from Terrorists (Alone)," washingtonpost.com, June 19, 2009, http://voices.washingtonpost.com/sleuth/2009/06/_rep_dan_burton_r-ind.html.

130 *a demonstration that prompted Calvin Trillin:* Calvin Trillin, "The Trouble with Transcripts," *Time*, May 18, 1998.

130 *a* Washington Post *reporter wrote:* Akers, "Dan Burton, Protecting the House From Terrorists (Alone)."

130 *prompted Republican leader Newt Gingrich:* George Lardner and Juliet Eilperin, "Burton Apologizes to GOP," *The Washington Post*, May 7, 1998, A1.

130 *He was so terrified of catching AIDS:* Frank Pellegrini, "Fool on the Hill," *Time*, May 8, 1998.

130 *When Burton began committee hearings:* Huntly Collins, "Life Giver or Life Taker?," *Philadelphia Inquirer*, October 3, 1999, A1.

131 *"There has never been":* Saad Omer, interview with author, April 2, 2009.

131 *In 1991, Maurice Hilleman:* Maurice R. Hilleman, "Vaccine Task Force Assignment Thimerosal (Merthiolate) Preservative—Problems, Analysis, Suggestions for Resolution," Memo to Gordon Douglas, Merck Pharmaceuticals, Whitehouse Station, New Jersey, 1991.

131 *Merck never told health officials:* Meyer Levin, "'91 Memo Warned of Mercury in Shots," *Los Angeles Times*, February 8, 2005.

132 *"If only we had done this":* Neal Halsey, interview with author, April 26, 2010.

CHAPTER 11: THE MERCURY MOMS

PAGE

133 *She was a thirty-year-old nursing student:* David Kirby, *Evidence of Harm* (New York: St. Martin's, 2005), 9–10.

133 *Finally, her husband:* Ibid., 10.

133 *Suddenly, he was constantly sick:* Elizabeth McBreen, "Spectrum's Person of the Year 2009," *Spectrum*, February/March 2010, http://www.spectrumpublications.com/index.php/magazine/spectrums_person_of_the_year_2009/. See also: Kirby, *Evidence of Harm*, 11–15.

134 *That November, Lyn met with a group:* Kirby, *Evidence of Harm*, 16.

134 *She was going to determine exactly:* McBreen, "Spectrum's Person of the Year 2009."

134 *Tom landed at the powerhouse Wall Street:* Kirby, *Evidence of Harm*, 17–18.

135 *At just barely three pounds:* Ibid., 18.

135 *what's referred to as Very Low Birth Weight:* K. N. Siva Subramanian et al., "Extremely Low Birth Weight Infant," *eMedicine for WebMD*, June 18, 2009; "Very Low Birth Weight," Lucile Packard Children's Hospital, n.d.

135 *Bill remained under constant watch:* Kirby, *Evidence of Harm*, 18.

135 *It would take another two years:* Ibid., 22.

136 *The more she thought about it:* Ibid.

136 *Rimland's notion that autism was primarily a neurological condition:* Steve Edelson, "The Autism Research Institute and Defeat Autism Now!: Who We Are, and What We Do," *Autism Research Review International* 2007;21(3).

137 *"turning their backs on the medical establishment":* Patricia Morris Buckley, "Dr. Bernard Rimland Is Autism's Worst Enemy," *San Diego Jewish Journal*, October 2002.

137 *In 1994, Eric and Karen London:* "History of National Alliance for Autism Research," Autism Speaks, n.d., http://www.autismspeaks.org/naar_history.php.

137 *Hollywood producer Jonathan Shestack and his wife:* Sara Solovitch, "The Citizen Scientists," *Wired*, September 2001.

137 *When she started CAN:* Ibid.

138 *In March 1992, fifty-year-old Lorraine Pace:* Bethany Kandel, "Neighborhoods Try to Zero In on 'Cancer Clusters,' " *USA Today*, December 15, 1992, 2A.

139 *After analyzing its data:* Peter Marks, "U.S. to Finance Project to Study Breast Cancer on Long Island," *The New York Times*, November 25, 1993.

139 *at a cost to taxpayers of more than $30 million:* Dan Fagin, "Tattered Hopes—A $30 Million Federal Study of Breast Cancer and Pollution on LI Has Dis-

appointed Activists and Scientists," *Newsday*, July 28, 2002, A3. See also: Dan Fagin, "So Many Things Went Wrong: Costly Search for Links Between Pollution and Breast Cancer Was Hobbled from the Start, Critics Say," *Newsday*, July 29, 2002, A6; Dan Fagin, "Still Searching: A Computer Mapping System Was Supposed to Help Unearth Information About Breast Cancer and the Environment," *Newsday*, July 30, 2002, A6.

139 *an exhaustive federal study found:* D. M. Winn, "Science and Society: The Long Island Breast Cancer Study Project," *Nature Reviews Cancer* 2005;5(12): 986–94.

140 *"I started reading about mercury":* Lyn Redwood, interview with author, April 21, 2010.

141 *One of the group's first coups:* Kirby, *Evidence of Harm*, 87.

141 *In his opening remarks:* United States Congress House of Representatives' House Government Reform Committee, "Autism—Present Challenges, Future Needs—Why the Increased Rates?," April 6, 2000, Federal News Service.

141 *a Louisiana housewife named Shelley Reynolds:* Ibid.

142 *It was through that network:* Kirby, *Evidence of Harm*, 34–35.

142 *In 1999, shortly after he was laid off:* Ibid., 75.

142 *they had to speak "their language":* Ibid., 64.

142 *"Autism: A Novel Form of Mercury Poisoning":* S. Bernard et al., "Autism: A Novel form of Mercury Poisoning," *Medical Hypotheses* 2001;56(4): 462–71.

143 *"can oblige authors to distort their true views":* "Online Submission and Editorial System," *Medical Hypotheses*, n.d., http://ees.elsevier.com/ymehy/.

143 *"predict whether ideas and facts":* "Aims and Scopes," *Medical Hypotheses*, n.d., http://www.medical-hypotheses.com/aims.

143 *a table comparing ninety-four traits:* Bernard et al., "Autism: A Novel Form of Mercury Poisoning," 463.

144 *the one published study that had looked:* Philip S. Gentile et al., "Trace Elements in the Hair of Autistic and Control Children," *Journal of Autism and Developmental Disorders* 1983;(12(2): 205–06.

144 *"parental reports of autistic children":* Bernard et al., "Autism: A Novel form of Mercury Poisoning," 463.

CHAPTER 12: THE SIMPSONWOOD CONFERENCE AND THE SPEED OF LIGHT: A BRIEF HISTORY OF SCIENCE

PAGE

146 *it wasn't until 1955:* Douglas Pisano, James G. Kenimer and John J. Jessop, "Biologics," in *FDA Regulatory Affairs*, ed., Douglas Pisano and David Mantus (Boca Baton, Florida: CRC Press, 2003), 131.

146 *it wasn't until 1962:* "Milestones in U.S. Food and Drug Law History—Significant Dates in U.S. Food and Drug Law History," U.S. Department of Health and Human Services Food and Drug Administration, n.d., http://www.fda.gov/AboutFDA/WhatWeDo/History/Milestones/ucm128305.htm.

147 *In 1974, when the parents:* William Curran, "Public Warnings of the Risk in Oral Polio Vaccine," *Public Health and the Law* 1975;65(5): 501–2.

147 *sixteen years had passed:* Davies, "2 Polio Victims Win Vaccine Suit but Cutter Is Held Not Negligent."

147 *The American Academy of Pediatrics was among the first:* Martin Smith, "National Childhood Vaccine Injury Compensation Act," *Pediatrics* 1988;82: 264–69.

147 *vaccine-related tort claims:* Mary Beth Neraas, "The National Childhood Vaccine Injury Act of 1986: A Solution to the Vaccine Liability Crisis?," *Washington Law Review* 1988;63(149): 149–69.

147 *The panic that followed:* Smith, "National Childhood Vaccine Injury Compensation Act."

148 *The central feature of the NCVIA:* J. A. Singleton et al., "An Overview of the Vaccine Adverse Event Reporting System (VAERS) as a Surveillance System," *Vaccine* 1999;17(22): 2908–17.

148 *A Special Master is someone:* "Special Master," Cornell University Law School Legal Information Institute, n.d., http://topics.law.cornell.edu/wex/special_master.

148 *The law also established the Vaccine Adverse Event Reporting System:* "Vaccine Adverse Event Reporting System (VAERS) Questions and Answers," U.S. Food and Drug Administration, n.d., http://www.fda.gov/BiologicsBloodVaccines/SafetyAvailability/ReportaProblem/VaccineAdverseEvents/QuestionsabouttheVaccineAdverseEventReportingSystemVAERS/default.htm.

149 *If there was one thing citizens, scientists, and safety monitors agreed on:* Immunization Safety Review Committee, Board on Health Promotion and Disease Prevention, Institute of Medicine, *Immunization Safety Review: Thimerosal-Containing Vaccines and Neurodevelopmental Disorders* (Washington, D.C.: National Academies Press, 2004), http://www.nap.edu/catalog/10208.html.

149 *the Institute of Medicine (IOM), an independent organization:* "About the IOM," Institute of Medicine of the National Academics, n.d., http://www.iom.edu/About–IOM.aspx.

149 *"the significance of the issue in a broader societal context":* Immunization Safety Review Committee, Board on Health Promotion and Disease Prevention, Institute of Medicine, *Immunization Safety Review: Thimerosal-Containing Vaccines and Neurodevelopmental Disorders,* 2.

149 *the committee's efforts led to invitations:* Kirby, *Evidence of Harm,* 185.

150 *a United Methodist Church conference center:* "About Us," The Lodge at Simpsonwood, n.d., http:www.simpsonwood,org/pages/details/1553.

150 *Included among the dozens of participants:* "Scientific Review of Vaccine Safety Datalink Information," Simpsonwood Retreat Center, Norcross, Georgia, July 7–8, 2000, unpublished transcript.

150 *"We who work with vaccines":* Ibid., 1–2.

150 *Verstraeten opened his talk with a quip:* Ibid., 31–32.

151 *Since there was no definitive data:* Ibid., 73–75.

151 *he'd been unable to question:* Ibid., 49–50.

151 *There was also the indiscriminate list:* Ibid., 31.

151 *Verstraeten's results were all over the map:* Thomas Verstraeten et al., "Safety of Thimerosal-Containing Vaccines: A Two-Phased Study of Computerized Health Maintenance Organization Databases," *Pediatrics* 2003;12(5): 1039–48.

152 *Still, Verstraeten said, at the very least:* "Scientific Review of Vaccine Safety Datalink Information," unpublished transcript, 162.

152 *Even if we put the vaccine in single vials:* Ibid., 190.

152 *"It is amazing":* Ibid., 229.

152 *Lyn Redwood was among those in attendance:* Kirby, *Evidence of Harm*, 131.

153 *An extreme example of underreporting:* JA Singleton et al., "An Overview of the Vaccine Adverse Event Reporting System (VAERS) as a Surveillance System," *Vaccine* 1999;17(22): 2913.

154 *Its Civil Registration System:* Anders Hviid et al., "Association Between Thimerosal-Containing Vaccine and Autism," *Journal of the American Medical Association* 2003;290(13): 1764.

154 *Until March 1992, the three-part, whole cell pertussis vaccine:* Ibid., 1763.

155 *abstract "first principles":* Aristotle, *Posterior Analytics*, Book I, Part 6.

155 *Ibn-al Haytham, whose evidence-based experiments:* David C. Lindberg, *Theories of Vision from al-Kindi to Kepler* (Chicago: University of Chicago Press, 1976), 58–59, 62–65.

155 *through the theory of falsifiability:* Karl R. Popper, *The Logic of Scientific Discovery* (New York: Routledge, 1992), 40–42. First published 1959 by Hutchinson Education.

155 *Karl Popper's anecdote:* Ibid., 27.

156 *In 1697, a Dutch sea captain:* Peter Young, *Swan* (London: Reaktion, 2008), 27.

156 *Einstein became obsessed:* Walter Isaacson, *Einstein: His Life and Universe* (New York: Simon & Schuster, 2007), 93.

156 *a series of equations formulated:* Ibid., 110–11.

157 *In 1905, Einstein:* Ibid., 92–93.

157 *In order to reconcile this contradiction:* Ibid., 137–39.

158 *False conclusions are drawn all the time:* Carl Sagan, *The Demon-Haunted World* (New York: Random House, 1996), 21.

CHAPTER 13: THE MEDIA AND ITS MESSAGES

PAGE

160 *the BBC-TV news magazine show* Panorama: Sarah Barclay, "MMR: Every Parent's Choice," *Panorama*, BBC-TV, February 3, 2002, transcript, http://news.bbc.co.uk/hi/english/static/audio_video/programmes/panorama/transcripts/transcript03_02_02.txt.

160 *Wakefield had worked on that paper as well:* Andrew Wakefield and Scott Montgomery, "Measles, Mumps, Rubella Vaccine: Through a Glass, Darkly," *Adverse Drug Reactions* 2000;19(4): 1–19, http://www.whale.to/v/mmr16.html.

160 *This new study:* V. Uhlmann et al., "Potential Viral Pathogenic Mechanism for New Variant Inflammatory Bowel Disease," *Molecular Pathology* 2002;55: 84–90.

161 *Instead, the paper left out information so basic: Cedillo v. Sec'y of Health and Human Services*, 48–60.

161 *It appeared as if the research team:* Ibid., 56–60.

161 *The for-profit lab at which they obtained their results:* Ibid., 51.

161 *raised questions as to whether widespread fraud:* Ibid., 50–51; Transcript of record, *Cedillo v. Sec'y of Health and Human Services*, 2026–29.

162 *"He had the MMR and he's autistic":* Barclay, "MMR: Every Parent's Choice."

162 *in the weeks after the* Panorama *special aired:* Ian Hargreaves, Justin Lewis, and Tammy Spears, "Towards a Better Map: Science, the Public and the Media," Economic and Social Research Council (U.K.), 2003, 22.

162 *Later that year, when researchers from Cardiff University:* Ibid.

163 *"Since most health experts":* Ibid., 25.

163 *The net effect of this manufactured equivalence:* Ibid., 41. See also: Justin Lewis and Tammy Spears, "Misleading Media Reporting," *Nature* 2003;3: 913–18.

163 *After analyzing the children:* K. M. Madsen et al., "A Population-Based Study of Measles, Mumps, and Rubella Vaccination and Autism," *New England Journal of Medicine* 2002;347: 1477.

163 *"It's a sad day in America":* SafeMinds, "Statement of Mercury Policy Project and Safe Minds on the Homeland Security Bill's Thimerosal Shielding Rider, Safe Minds' Rebuttal to Senator Frist and Gramm's Comments on Senate Floor 11-18-02," November 19, 2002, http://www.ewire.com/display.cfm/Wire_ID/1415.

164 *"Vaccine Health Officials":* SafeMinds, "Vaccine Health Officials Manipulate Autism Records to Quell Rising Fears over Mercury in Vaccines," September 2, 2003, http://www.accessmylibrary.com/coms2/summary_0286-24247227_ITM.

164 *SafeMinds released an "assessment":* SafeMinds, "Denmark Study on Autism and MMR Vaccine Shows Need for Biological Research / Assessment of the Denmark MMR-Autism Study," November 6, 2002, http://www.vaccineinfo.net/immunization/injury/autism/DanishMMRAutismStudy.shtml.

165 *The same week that the* NEJM *piece ran:* Arthur Allen, "The Not-So-Crackpot Autism Theory," *The New York Times Magazine*, November 11, 2002, 66.

165 *In an effort to clarify his views:* Neal Halsey, "Letter to the Editor of the New York Times (sent November 11—NYT declined to publish in favor of the above correction)," November 11, 2002, http://www.vaccinesafety.edu/Vaccines-no-autism.htm.

166 *Once again, the conclusion:* K. Wilson et al., "Association of Autistic Spectrum Disorder and the Measles, Mumps, and Rubella Vaccine," *Archives of Pediatrics and Adolescent Medicine* 2003;157: 628.

167 *In August 2003, a comparison of children:* Paul Stehr-Green et al., "Autism and Thimerosal-Containing Vaccines: Lack of Consistent Evidence for an Association," *American Journal of Preventive Medicine* 2003;25(2): 101–6.

167 *In September and October, two other studies:* Madsen et al., "A Population-Based Study of Measles, Mumps, and Rubella Vaccination and Autism"; Hviid et al., "Association Between Thimerosal-Containing Vaccine and Autism."

167 *a detailed analysis by Tom Verstraeten:* Thomas Verstraeten et al., "Safety of Thimerosal-Containing Vaccines: A Two-Phased Study of Computerized Health Maintenance Organization Databases," *Pediatrics* 2003;112(5): 1039–48.

167 *Each individual study might have been:* Donald G. McNeil, Jr., "Study Casts Doubt on Theory of Vaccines' Link to Autism," *The New York Times*, September 4, 2003, 21.

167 *a December 29* Wall Street Journal *editorial:* "The Politics of Autism," *The Wall Street Journal*, December 29, 2003.

168 *the* Journal *felt compelled to respond:* "Autism and Vaccines," *The Wall Street Journal*, February 16, 2004.

CHAPTER 14: MARK GEIER, WITNESS FOR HIRE

PAGE

170 *After hearing rumors that the CDC:* Kirby, *Evidence of Harm*, 346–50.

170 *One if the committee recognized a link:* Ibid., 356.

170 *This eighth and final report:* Immunization Safety Review Committee, *Immunization Safety Review: Vaccines and Autism* (Washington, D.C.: National Academies Press, 2004), http://www.nap.edu/catalog/10997.html.

171 *The "flawed, incomplete" report:* Joe Giganti, "SafeMinds Outraged That IOM Report Fails American Public," SafeMinds, May 18, 2004.

171 *"This report went beyond any other report":* Maggie Fox, "Study Says Vaccine Not Cause of Autism," Reuters, May 19, 2004.

172 *"I don't think it was well described":* Marie McCormick, interview with author, February 27, 2009.

172 *"I've seen what [autism] looks like":* Ibid.

173 *began to refer to McCormick as "Church Lady":* Kirby, *Evidence of Harm,* 188.

173 *"I am out for blood here":* Ibid., 188.

173 *"Part of the criticism of the vaccine advisory panels":* McCormick, interview with author, February 27, 2009.

174 *In support of its conclusions:* Immunization Safety Review Committee, *Immunization Safety Review: Vaccines and Autism,* 6, 86–110.

174 *In contrast, the evidence:* Ibid., 6–7, 65.

174 *A* New York Times *reporter who visited the Geiers':* Gardiner Harris and Anahad O'Connor, "On Autism's Cause, It's Parents v. Research," *The New York Times,* June 25, 2005, 1.

174 *Mark Geier had been a mainstay:* David Geier and Mark Geier, "The True Story of Pertussis Vaccination: A Sordid Legacy?," *Journal of the History of Medicine* 2002;57: 249–84.

174 *"children with severe injuries":* Ibid., 272–73.

175 *In a 1987 case, Geier testified: Talley v. Wyeth,* No. 87-349-C (E.D. Okla., February 24, 1988).

175 *as the result of his mistake: Graham v. Wyeth Laboratories,* 666 F.Supp. 1483 (D. Kansas 1987).

175 *"clearly lacks the expertise to evaluate": Daly v. Sec'y of Health and Human Services,* 1991 WL 154573 (Ct. Fed. Cl., July 26, 1991).

175 *The following year, one judge questioned: Jones v. Lederle Laboratories, American Cyanamid Co.,* 785 F.Supp. 1123, 1126 (E.D. N.Y., 1992).

175 *"Were Dr. Geier an attorney": Aldridge v. Sec'y of Health and Human Services,* footnote, 9.

175 *"seriously intellectually dishonest": Marascalco v. Sec'y of Health and Human Services,* 5.

175 *"not reliable, or grounded": Haim v. Sec'y of Health and Human Services,* footnote, 15.

175 *a Special Master named Laura Millman wrote: Weiss v. Sec'y of Health and Human Services,* No. 03-190V, 2003 (Ct. Fed. Cl., October 9, 2003), *review denied,* 59 Fed. Cl. 624 (Ct. Fed. Cl., January 23, 2004).

176 *Their procedures were "nontransparent":* Immunization Safety Review Committee, *Immunization Safety Review: Vaccines and Autism,* 7.

176 *In one study, the Geiers claimed:* D. A. Geier and M. R. Geier, "An Assessment of the Impact of Thimerosal on Childhood Neurodevelopmental Disorders," *Pediatric Rehabilitation* 2003;6(2): 97–102.

176 *the Geiers had relied on the U.S. Department of Education's:* D. A. Geier and M. R. Geier, "A Comparative Evaluation of the Effects of MMR Immunization and Mercury Doses from Thimerosal-Containing Childhood Vaccines on the Population Prevalence of Autism," *Medical Science Monitor* 2004;10(3): 133–39.

176 *The description for the former took all of seventy-six words:* American Psychiatric Association, *Diagnostic and Statistical Manual of Mental Disorders—III* (Arlington, Virginia: American Psychiatric Association, 1980).

176 *The latter required 698 words to explain:* American Psychiatric Association, *Diagnostic and Statistical Manual of Mental Disorders—III-R* (Arlington, Virginia: American Psychiatric Association, 1987).

176 *In another paper, the Geiers referred to figures:* Geier and Geier, "An Assessment of the Impact of Thimerosal on Childhood Neurodevelopmental Disorders."

177 *They also claimed to have determined:* M. R. Geier and D. A. Geier, "Thimerosal in Childhood Vaccines, Neurodevelopmental Disorders, and Heart Disease in the United States," *Journal of American Physicians and Surgeons* 2003;8(1): 6–11.

CHAPTER 15: THE CASE OF MICHELLE CEDILLO

PAGE

178 *the law established a specially formulated table:* Derry Ridgway, "No-Fault Vaccine Insurance: Lessons from the National Vaccine Injury Compensation Program," *Journal of Health Politics, Policy, and Law* 1999;24(1): 62.

179 *a unanimous Supreme Court ruling: Shalala v. Whitecotton*, 514 U.S. 268 (1995).

179 *The trade-offs for this relaxed evidentiary standard:* Mary Beth Neraas, "The National Childhood Vaccine Injury Act of 1986: A Solution to the Vaccine Liability Crisis?," *Washington Law Review* 1988;63(149): 149–69.

179 *Today, the court allows:* "National Vaccine Injury Compensation Program," U.S. Department of Health and Human Services, n.d., http://www.hrsa.gov/vaccinecompensation/.

179 *By 1999, when the average number of cases:* "NVICP Statistics Report, September 2, 2010," U.S. Department of Health and Human Services, September 2, 2010, http://www.hrsa.gov/vaccinecompensation/statistics_report.htm.

179 *In 1999, a single autism case was filed:* Ibid.

180 *"Attorneys began to start focusing on it":* Lyn Redwood, interview with author, April 21, 2010.

180 *On July 3, 2002, Gary Golkiewicz:* Gary Golkiewicz, "Autism General Order #1," 2002 WL 31696785 (Fed. Cl. Spec. Mstr., July 3, 2002), 1.

180 *By the end of 2004, a total of 4,321 autism cases:* "NVICP Statistics Report, September 2, 2010."

180 *had agreed to address the "general causation" theories:* Golkiewicz, "Autism General Order #1."

181 *"Michelle was a happy, robust baby":* Transcript of record, *Cedillo v. Sec'y of Health and Human Services*, 223.

181 *"We took her everywhere":* Ibid., 226.

181 *On December 20, 1995:* Ibid., 226–27.

181 *she came down with a fever:* Ibid., 227.

181 *"talking less since ill in Jan.": Cedillo v. Sec'y of Health and Human Services*, 6.

181 *"It would appear that there was some neurological harm":* Ibid., 166.

181 *a developmental psychologist gave the Cedillos:* Transcript of record, *Cedillo v. Sec'y of Health and Human Services*, 240–42.

181 *"We were completely overwhelmed":* Ibid., 241.

182 *she came upon the story of Cindy Goldenberg:* Ibid., 242–43.

182 *"incorporate life skills that lead to personal empowerment":* Cindy Goldenberg, "Cindy Goldenberg Clairvoyant Spiritual Psychic Author Speaker," n.d., http://cindygoldenberg.com/.

182 *Goldenberg began a multiyear quest:* "Mother Fights for Recognition of Vaccine and Autism Link," Mothers United for Moral Support, n.d., http://www.nathan.com/momlink.htm. (Originally appeared in *Vaccine Reaction* newsletter, published by the National Vaccine Information Center.)

182 *before she found Sudhir Gupta:* Ibid.

182 *intravenous immunoglobulin (IVIG):* Noah Scheinfeld, "Intravenous Immunoglobulin," *eMedicine from WebMD*, June 22, 2010, http://emedicine.medscape.com/article/210367-overview.

182 *she was not a candidate for IVIG:* Transcript of record, *Cedillo v. Sec'y of Health and Human Services*, 242–43.

183 *when Theresa first heard about Andrew Wakefield's theories:* Ibid., 39–42.

183 *the Cedillos had become convinced:* Ibid., 37–42.

183 *That December, they filed a claim: Cedillo v. Sec'y of Health and Human Services*, 17.

183 *Since brain injuries (or encephalopathies):* "Vaccine Injury Table," U.S. Department of Health and Human Services, n.d., http://www.hrsa.gov/vaccinecompensation/table.htm.

183 *She alternated between bouts of extreme diarrhea:* Transcript of record, *Cedillo v. Sec'y of Health and Human Services*, 36–38, 48, 232–34.

183 *potentially an attempt to attack the pain:* Ibid., 40.

183 *she'd stay up for twelve or eighteen hours:* Ibid., 245.

183 *her body produced so much saliva:* Ibid., 247.

184 *stretches as long as three days:* Ibid., 256.

184 *have a feeding tube permanently installed:* Ibid., 262

184 *Michelle needed two people to attend to her:* Ibid., 273.

184 *Preparing for each of the many trips:* Ibid., 325–26.

184 *along with "baby," "mama," and "daddy":* Ibid., 225.

184 *"We tried":* Ibid., 235.

184 *"It makes people uncomfortable":* Jane Johnson, interview with author, May 4, 2009.

185 *"We both agreed that we didn't ever want to":* Transcript of record, *Cedillo v. Sec'y of Health and Human Services*, 241.

185 *"I heard several presentations":* Ibid., 252.

185 *By that point, the Cedillos:* Transcript of record, *Cedillo v. Sec'y of Health and Human Services*, 40, 42–43, 252–54.

186 *He suggested that Michelle:* Ibid., 390–92.

186 *a biopsy of Michelle's gut tissue:* Ibid., 253–54; *Cedillo v. Sec'y of Health and Human Services*, 8.

186 *in consultation with Wakefield:* Transcript of record, *Cedillo v. Sec'y of Health and Human Services*, 2899–900.

186 *for the purpose of testing tissue samples: Cedillo v. Sec'y of Health and Human Services*, 42.

186 *"[We wanted] to determine":* Transcript of record, *Cedillo v. Sec'y of Health and Human Services*, 397.

186 *"I think we could help him":* Kirby, *Evidence of Harm*, 57.

186 *Wakefield described for his audibly amused audience:* Andrew Wakefield, "Presentation to MIND Institute," University of California, Davis, March 20, 1999, transcript. See also: Nick Triggle, "MMR Scare Doctor 'Acted Unethically,' Panel Finds," *BBC News*, January 28, 2010, http://news.bbc.co.uk/2/hi/health/8483865.stm; "Fitness to Practise Panel Hearing," 54–55.

187 *Sam Debold was born in 1997:* Vicky Debold, interview with author, July 28, 2009.

188 *"on an irrational, unconscious level":* Bryna Siegel, *The World of the Autistic Child* (New York: Oxford University Press, 1996), 323.

188 *"It's the most depressing book":* Vicky Debold, interview with author, July 28, 2009.

188 *"You're an educated person":* Ibid.

188 *scheduled to speak at an upcoming:* Pennsylvania Parents for Vaccine Awareness conference, Erie, Pennsylvania, May 19, 2000.

188 *"Andy said, 'Look, let's calm down here' ":* Vicky Debold, interview with author, July 28, 2009.

189 *"Dr Wakefield's research was no longer in line":* Matthew Beard, "Consultant Who Linked MMR Jab to Autism Quits After 'Political Pressure," *The Independent* (London), December 3, 2001, 4. See also: Sarah Ramsey, "Controversial MMR-Autism Investigator Resigns from Research Post," *The Lancet* 2001;358(9297): 1972.

189 *"I have been asked to go":* Beard, "Consultant Who Linked MMR Jab to Autism Quits After 'Political Pressure.' "

189 *"We get these parents ringing up every day":* Laurance, "Not Immune to How Research Can Hurt."

190 *On January 14, 2002, two months after: Cedillo v. Sec'y of Health and Human Services*, 19–20.

191 *"[P]etitioners' representatives have stated":* Golkiewicz, "Autism General Order #1," 3.

191 *both sides were given sixteen months:* "Master Scheduling Order," Omnibus Autism Proceeding, included as Exhibit E of Golkiewicz, "Autism General Order #1."

CHAPTER 16: COGNITIVE BIASES AND AVAILABILITY CASCADES

PAGE

192 *Lehrer describes a patient named Elliot:* Jonah Lehrer, *How We Decide* (New York: Houghton Mifflin Harcourt, 2009), 13–18.

193 *just two of literally dozens of cognitive biases:* For a list of commonly accepted cognitive biases and a brief definition of each, see: "List of cognitive biases," *Wikipedia*, n.d., http://en.wikipedia.org/wiki/List_of_cognitive_biases.

196 *Vicky Debold says she was motivated:* Vicky Debold, interview with author, July 28, 2009.

196 *Theresa Cedillo was enticed by the prospect:* Transcript of record, *Cedillo v. Sec'y of Health and Human Services*, 39–42, 252–53.

196 *"I looked out at an audience":* Jane Johnson, interview with author, May 4, 2009.

196 *"This is the federal government giving every kid":* Jane Johnson, interview with author, April 22, 2010.

196 *"You wouldn't be saying and doing":* Vicky Debold, interview with author, July 28, 2009.

196 *a concept that was first articulated in a 1999 paper:* Timur Kuran and Cass R. Sunstein, "Availability Cascades and Risk Regulation," *Stanford Law Review* 1999;51(4): 683–768.

196 *"self-reinforcing process of collective belief formation":* Kuran and Sunstein, "Availability Cascades and Risk Regulation," 683.

197 *In the 2008 book* Nudge: Cass Sunstein and Richard Thaler, *Nudge* (New Haven: Yale University Press, 2008).

197 *"As all women":* Cass Sunstein and Richard Thaler, "Easy Does It: How to Make Lazy People Do the Right Thing," *The New Republic*, April 9, 2008; available as "The Amsterdam Urinals" on *Nudge*, http://nudges.wordpress.com/the-amsterdam-urinals/. See also: Sunstein and Thaler, *Nudge*, 3–4.

197 *how nonscientists viewed scientific claims:* Ralph M. Barnes, Audrey L. Alberstadt, and Lesleh E. Keilholtz, "How to Think About Scientific Claims: A Study of How Non-Scientists Evaluate Science Claims," *Skeptic*, January 1, 2009;14(4): 48–55.

197 *"poor source credibility":* B. Fischhoff and R. Beyth-Marom, "Hypothesis Evaluation from a Bayesian Perspective," *Psychological Review* 1983;90: 239–60.

197 *"even more strongly committed":* Barnes, Alberstadt, and Keilholtz, "How to Think About Scientific Claims."

198 *In 1987, nearly three-quarters of Americans:* "Internet Sapping Broadcast News Audience," Pew Research Center for the People and the Press, June 11, 2000, http://people-press.org/report/?pageid=203.

198 *Now that figure has fallen below one-third:* Ibid.

198 *Kuran and Sunstein refer to these people:* Kuran and Sunstein, "Availability Cascades and Risk Regulation," 683.

198 *A 2007 study titled "Inferring the Popularity of an Opinion from Its Familiarity":* Kimberlee Weaver et al., "Inferring the Popularity of an Opinion from Its Familiarity: A Repetitive Voice Can Sound like a Chorus," *Journal of Personality and Social Psychology* 2007;92(5): 821–33.

198 *"to infer that a familiar opinion":* Ibid., 821.

199 *Lyn Redwood became so convinced:* Kirby, *Evidence of Harm*, 139–40.

199 *It wasn't long, Redwood says, before national reporters:* Lyn Redwood, interview with author, April 21, 2010.

199 *"This is fraud":* Kelly Patricia O'Meara, "Vaccines May Fuel Autism Epidemic," *Insight*, June 24, 2003, http://www.wnd.com/?pageid=19204.

199 *"junk scientists and charlatans":* Myron Levin, "Taking It to Vaccine Court—Parents Say Mercury in Shots Caused Their Children's Autism, and They Want Drug Firms to Pay," *Los Angeles Times*, August 7, 2004, 1.

200 *"have to answer later for their failures":* SafeMinds, "SafeMinds Outraged That IOM Report Fails American Public," PR Newswire, May 18, 2004.

200 *resulted in Harvard University increasing security:* Harris and O'Connor, "On Autism's Cause, It's Parents vs. Researchers."

200 *"I'd like to know how you people sleep":* Ibid.

200 *"Forgiveness is between":* Ibid.

CHAPTER 17: HOW TO TURN A LACK OF EVIDENCE INTO *EVIDENCE OF HARM*

PAGE

203 *the forty-two-year-old's work history was most notable:* "About the Author," EvidenceOfHarm.com, n.d., http://evidenceofharm.com/bio.htm.

203 *a friend in Los Angeles suggested:* David Kirby, interview with author, April 14, 2009.

203 *"More Options, and Decisions"*: David Kirby, "More Options, and Decisions, for Men with Prostate Cancer, *The New York Times*, October 3, 2000, F7.

204 *"New Resistant Gonorrhea"*: David Kirby, "New Resistant Gonorrhea Migrating to Mainland U.S.," *The New York Times*, May 7, 2002, F5.

204 *he could turn the idea into a feature:* David Kirby, interview with author, April 14, 2009.

204 *"She was cool"*: Ibid.

204 *"I was like, Oh, mercury is in vaccines"*: Ibid.

204 *In the 2000 election:* "Heavy Hitters: Eli Lilly & Co," OpenSecrets.org—Center for Responsive Politics, n.d., http://www.opensecrets.org/orgs/summary.php?ID=D000000166&Type=P.

204 *Ohio's Dennis Kucinich:* Arianna Huffington, "Finding the Answer to Washington's Hottest Whodunit," *Arianna Online*, December 4, 2002, http://ariannaonline.huffingtonpost.com/columns.php?id=48.

205 *John McCain likened it:* Mark Shields, "The Business Lobby's Campaign Against McCain," *CNN Inside Politics*, CNN, November 27, 2002.

205 *"It was a big whodunit"*: David Kirby, interview with author, April 14, 2009.

205 *"It was too controversial"*: Ibid.

205 *"He said, 'You can't just write' "*: Ibid.

205 *the Republicans had all but promised:* Steven Higgs, "The Mystery of the Eli Lilly Rider," *Counter Punch*, January 22, 2010.

205 *some were carrying poster-board signs:* Kirby, *Evidence of Harm*, 252.

206 *According to Redwood, when Kirby contacted her:* Lyn Redwood, interview with author, April 21, 2010.

206 *One editor explained her reasoning:* McBreen, "Spectrum's Person of the Year 2009."

206 *"We decided to combine forces"*: Lyn Redwood, interview with author, April 21, 2010.

206 *"every single e-mail"*: David Kirby, interview with author, April 14, 2009.

206 *"He came and hung out"*: Lyn Redwood, interview with author, April 21, 2010.

207 *the CDC's 1999 statement:* Centers for Disease Control and Prevention, "Notice to Readers: Thimerosal in Vaccines: A Joint Statement of the American Academy of Pediatrics and the Public Health Service."

207 *He granted that the report "favor[ed] rejection"*: Kirby, *Evidence of Harm*, 351.

207 *Meanwhile, the CDC has been unable:* Ibid., xii.

208 *"admittedly subjective point of view"*: Ibid., xiv.

208 *"an attractive woman"*: Ibid., 10.

208 *"tough businesswoman"*: Ibid., 21.

208 *"fierce streak of determination"*: Ibid., 25.

208 *"remarkable aptitude"*: Ibid., 44.

208 *"the kind of guy you would want"*: Ibid., 228.

208 *"barked at" and "banished"*: Ibid., 23.

208 *"poked and prodded"*: Ibid., 13.

209 *"grew purple"*: Ibid., 38.

209 *"insurgent candidate"*: Ibid., 254.

209 *"Curiously, the first case of autism"*: Ibid., xv.

209 *Eli Lilly* "reportedly *earn[ed] a profit"*: Ibid., 2.

209 *"the American health establishment"*: Ibid., xiii.

209 *resulted in an American Academy of Pediatrics statement:* "Study Fails to Show a Connection Between Thimerosal and Autism," American Academy of Pediatrics, May 16, 2003.

210 *"were made to sound like dimwits"*: Kirby, *Evidence of Harm*, 260.

210 *"felt the same disdain"*: Ibid., 259.

210 *"veiled swipe at their colleagues"*: Ibid., 260.

210 *"looks down their noses"*: Ibid., 230.

210 *"dismiss [the AAPS]"*: Ibid., 230.

210 *It has compared electronic medical records:* Association of American Physicians and Surgeons, "Statement of the Association of American Physicians and Surgeons on Computer-Based Patient Records," n.d., http://www.aapsonline.org/confiden/ncvhs.htm.

210 *cigarette taxes actually led:* Michael Marlow, "Anatomy of Public Health Research: Tobacco Control as a Case Study," *Journal of the Association of American Physicians and Surgeons* 2009;14(3): 79–80.

210 *"deliberately using the techniques of neurolinguistic programming"*: Association of American Physicians and Surgeons, "Oratory—or Hypnotic Induction?," October 25, 2008.

211 *"considered by many experts"*: Kirby, *Evidence of Harm*, 230.

211 *"moving the paradigm away"*: Jim Donnelly, "MIND Immune Dysfunction in Autism: Researchers Suggest Autism Can Be Detected in Newborns?," EOHarm Yahoo! group, May 5, 2005, http://groups.yahoo.com/group/EOHarm/message/1496.

211 *"once causation is established"*: Jim Moody, "An Introduction and a Question," EOHarm Yahoo! group, April 26, 2005, http://groups.yahoo.com/group/EOHarm/message/344.

212 *"Please feel free to share"*: David Kirby, "EVIDENCE OF HARM: NYT Science Times ad," EOHarm Yahoo! group, April 28, 2005, http://groups.yahoo.com/group/EOHarm/message/628.

212 *"Two years ago this was the province"*: John Gilmore, "Sound the Trumpet—Update," EOHarm Yahoo! group, April 28, 2005, http://groups.yahoo.com/group/EOHarm/message/719.

212 *"Somehow," Kirby says, Deirdre Imus:* David Kirby, interview with author, April 14, 2009.

212 *reached more than three million Americans:* "Imus Audience Slips in New York but He Still Packs a Punch," *Business Week*, April 26, 2005, http://nybw.businessweek.com/the_thread/brandnewday/archives/2005/04/imus_audience_slips_in_new_york_but_he_still_packs_a_punch.html.

212 *Imus gave Kirby immediate credibility:* Don Imus, "David Kirby Interviewed by Don Imus," *Imus in the Morning*, MSNBC, March 10, 2005, unofficial transcript.

212 *the only health-related story:* David Kirby, "Sex and Medicine—Party Favors: Pill Popping as Insurance," *The New York Times*, June 21, 2004, F1.

213 *He asked about the book's name:* Imus, "David Kirby Interviewed by Don Imus."

214 *"happened to check Amazon":* David Kirby, interview with author, April 14, 2009.

215 *"rang[ing] from severely flawed":* Tim Russert, *Meet the Press*, NBC, August 7, 2005, transcript, http://www.msnbc.msn.com/id/8714275/.

216 *In June 2010, the FDA accused Haley:* U.S. Food and Drug Administration Division of Inspections, Compliance, Enforcement, and Criminal Investigations, "Warning Letter CIN-10-107927-14 to Boyd Haley," June 17, 2010, http://www.fda.gov/ICECI/EnforcementActions/WarningLetters/ucm216216.htm.

216 *The paper by Deth: King v. Sec'y of Health and Human Services*, No. 03-584V at 73 (Ct. Fed. Cl., March 12, 2010). See also: Transcript of record, *King v. Sec'y of Health and Human Services*, 3967–68.

216 *one of Deth's claims:* M. Waly et al., "Activation of Methionine Synthase by Insulin-Like Growth Factor-1: A Target for Neurodevelopmental Toxins and Thimerosal," *Molecular Psychiatry* 2004;9: 358 (PML 257).

216 *James explicitly wrote: King v. Sec'y of Health and Human Services*, 75.

217 *"The press was great":* David Kirby, interview with author, April 14, 2009.

217 *A full-page write-up:* Polly Morrice, " 'Evidence of Harm': What Caused the Autism Epidemic?," *The New York Times Book Review*, April 17, 2005, 20.

217 *On Thanksgiving 2009, the site ran an illustration:* "Pass the Maalox: An AoA Thanksgiving Nightmare," *Age of Autism*, November 29, 2009. (The blog entry, which had previously been available at http://www.ageofautism.com/2009/11/pass-the-maalox-an-aoa-thanksgiving-nightmare.html, was taken down in December 2009 and replaced by a blank white page.)

218 *"Autism, Vaccines, and the CDC":* Robert F. Kennedy, Jr., and David Kirby, "Autism, Vaccines and the CDC: The Wrong Side of History," *The Huffington Post*, January 27, 2009, http://www.huffingtonpost.com/robert-f-kennedy-and-david-kirby/autism-vaccines-and-the-c_b_161395.html.

218 *"The Autism Vaccine Debate"*: David Kirby, "The Autism Vaccine Debate—Anything but Over," *The Huffington Post*, November 30, 2007, http://www.huffingtonpost.com/david-kirby/the-autismvaccine-debate-_b_74853.html.

218 *"Up to 1-in-50 Troops"*: David Kirby, "Up to 1-in-50 Troops Seriously Injured . . . By Vaccines?," *The Huffington Post*, August 14, 2008, http://www.huffingtonpost.com/david-kirby/up-to-1-in-50-troops-seri_b_119048.html.

218 *In the epilogue to his book:* Kirby, *Evidence of Harm*, 411.

218 *Then in 2005, he told Don Imus:* Imus, "David Kirby Interviewed by Don Imus."

218 *His latest is that myelin:* David Kirby, "Metals, Myelin, Mitochondria, and Mouse Virus—Possible Paths to ASD," presentation, AutismOne conference, Westin O'Hare, Chicago, May 29, 2010. See also: David Kirby, "Metals, Myelin, and Mitochondria—Several Paths to Autism?," presentation, AutismOne conference, Westin O'Hare, Chicago, May 23, 2009.

218 *"It is used in vaccines as an adjuvant"*: Kirby, "Metals, Myelin, and Mitochondria—Several Paths to Autism?"

218 *In 2000, when Hannah was nineteen months old:* David Kirby, "Government Concedes Vaccine-Autism Case in Federal Court—Now What?," *The Huffington Post*, February 25, 2008, http://www.huffingtonpost.com/david-kirby/government-concedes-vacci_b_88323.html.

219 *a study that had appeared in the* Journal of Child Neurology: Jon S. Poling et al., "Developmental Regression and Mitochondrial Dysfunction in a Child with Autism," *Journal of Child Neurology* 2006;21(2): 170–72.

219 *the study was submitted for publication:* Roger A. Brumback, "The Appalling Poling Saga," *Journal of Child Neurology* 2008;23(9): 1090–91.

219 *who called Jon Poling's behavior:* Ibid.

219 *"It's a no-fault system"*: Jon Poling, interview with author, May 23, 2009.

220 *David Kirby stood in front of hundreds of people:* Kirby, "Metals, Myelin, and Mitochondria—Several Paths to Autism?"

CHAPTER 18: A CONSPIRACY OF DUNCES

PAGE

221 Rolling Stone *and the online magazine Salon.com simultaneously published:* Robert F. Kennedy, Jr., "Deadly Immunity," *Rolling Stone*, June 20, 2005, http://www.webcitation.org/5glaWmdym; Robert F. Kennedy, Jr., "Deadly Immunity," Salon.com, June 16, 2005, http://dir.salon.com/story/news/feature/2005/06/16/thimerosal/.

221 *"I was drawn into the controversy"*: Ibid.

223 *Before we all leave:* "Scientific Review of Vaccine Safety Datalink Information," unpublished transcript, 255.

223 *Dr. Bob Chen, head of vaccine safety:* Kennedy, "Deadly Immunity."

224 *And I really want to risk offending:* "Scientific Review of Vaccine Safety Datalink Information," unpublished transcript, 247.

224 *There is now the point:* Ibid., 248.

224 *My message would be:* Ibid., 249.

224 *Dr. John Clements, vaccines advisor:* Kennedy, "Deadly Immunity."

225 *Finally, the thing that concerns me:* "Scientific Review of Vaccine Safety Datalink Information," unpublished transcript, 191.

225 *The medical/legal findings:* Ibid., 229.

225 *"we are in a bad position":* Kennedy, "Deadly Immunity."

226 *didn't warrant inclusion:* "Correction, June 17, 2005," Salon.com, http://www.salon.com/letters/corrections/2005/index.html#thimerosal; "Correction, June 21, 2005," Salon.com, http://www.salon.com/letters/corrections/2005/index.html#rosen; "Correction, June 22, 2005," Salon.com, http://www.salon.com/letters/corrections/2005/index.html#IOMpanel; "Correction, June 24, 2005," Salon.com, http://www.salon.com/letters/corrections/2005/index.html#clements_katz; "Correction, July 1, 2005," Salon.com, http://www.salon.com/letters/corrections/2005/index.html#clarification. See also: "Note," "Clarification," and "Correction," RollingStone.com, n.d., http://www.webcitation.org/5glaWmdym, previously available at http://www.rollingstone.com/politics/story/7395411/deadly_immunity/.

226 *"It is important to note":* "Kennedy Report Sparks Controversy," Rollingstone.com, July 14, 2005, previously available at http://www.rollingstone.com/news/story/7483530/kennedy_report_sparks_controv.

226 *Kennedy told MSNBC's Joe Scarborough:* Joe Scarborough, "A Coverup for a Cause of Autism?," *Morning Joe*, MSNBC, July 21, 2005, transcript, http://www.msnbc.msn.com/id/8243264/.

CHAPTER 19: AUTISM SPEAKS

PAGE

228 *both groups had been relatively open-minded:* National Alliance for Autism Research, "National Association for Autism Research: Committed to Accelerating Biomedical Autism Research to Unlock the Mysteries of Autism Spectrum Disorders," *The Exceptional Parent*, April 2002, http://findarticles.com/p/articles/mi_go2827/is_4_32/ai_n7046769/; "About Cure Autism Now," Cure Autism Now, n.d., http://web.archive.org/web/20020210150114/www.cureautismnow.org/aboutcan/aboutcan.cfm.

229 *which Liz Birt founded:* Kirby, *Evidence of Harm*, 85.

229 *which began when Lisa Ackerman:* Sam Miller, "Autism Is Treatable, She Insists; O.C. Mother Leads Uprising Against Accepted Views," *Orange County Register*, July 2, 2008, A1.

229 *which was started by parents named Teri and Ed Arranga:* "About Us," Autism-One, n.d., http://www.autismone.org/content/about-us.

229 *and the National Autism Association:* "About Us: National Autism Association Inc.," Facebook, n.d., http://apps.facebook.com/causes/beneficiaries/349/info.

229 *as Eric London puts it:* Eric London, interview with author, April 7, 2010.

229 *when London wrote a critique:* Eric London, "The ABCs of MMRs and DTPs," *Narrative*, Summer/Fall 1998.

230 *"I said, Look, there's room":* Eric London, interview with author, April 7, 2010.

230 *when* The New England Journal of Medicine *published:* Madsen et al., "A Population-Based Study of MMR Vaccination and Autism."

230 *NAAR was accused of shilling:* F. Edwards Yazbak, "Epidemiological Autism Studies: Why Parents of Children with Autism Are So Upset!," November 19, 2002, http://www.autismautoimmunityproject.org/upset.html.

230 *Much was made of the fact:* National Association for Autism Research, "Pharmaceutical Industry and NAAR—The Truth About NAAR's Relationship with the Pharmaceutical Industry and Vaccine Research Focusing on Autism," Autism Speaks, March 11, 2003, http://www.autismspeaks.org/inthenews/naar_archive/pharmaceutical_industry_naar.php. See also: National Association for Autism Research, "The Limited Support NAAR Has Received from Pharmaceutical Companies That Make Childhood Vaccines," Autism Speaks, March 3, 2003, http://www.autismspeaks.org/docs/pharmachartvacc.pdf.

230 *Eventually, the attacks became so fierce:* Ibid.

230 *its Autism Tissue Program:* "Autism Tissue Program: About the Program—Overview," Autism Speaks, n.d., http://www.autismtissueprogram.org/site/c.nlKUL7MQIsG/b.5183785/k.E3D8/Overview.htm.

231 *"He developed typically":* Katie Wright, interview with author, July 27, 2010.

231 *"He was having ten bowel movements":* Ibid.

231 *When the pain got especially bad:* David Kirby, "David Kirby Interviews Katie Wright!," Foundation for Autism Information and Research, April 19, 2007, http://www.youtube.com/watch?v=IUNO25l1zFs, dHY5K_MP7w, +TVoJIVqu2Q, and I_IPuYf98uF.

231 *"They didn't connect them":* Katie Wright, interview with author, July 27, 2010.

232 *"Too many parents go to bed each night":* Suzanne Wright, "Willing the World to Listen," *Newsweek*, February 27, 2005.

232 *From the outset, the Wrights made clear:* Jane Gross and Stephanie Strom, "Debate over Cause of Autism Strains a Family and Its Charity," *The New York Times*, June 18, 2007, A1.

232 *That November, NAAR merged with the new charity:* Autism Speaks, "Autism Speaks and NAAR Announce Plans to Combine Operations," November 30, 2005.

232 *Within a year, a similar "consolidation":* Autism Speaks, "CAN and Autism Speaks Announce Plans to Combine Operations," November 29, 2006.

232 *Statements detailing the moves:* Autism Speaks, "Autism Speaks and NAAR Complete Merger," February 13, 2006; Autism Speaks, "Autism Speaks and CAN Complete Merger," February 5, 2007.

233 *Sallie Bernard . . . made sure to signal:* Autism Speaks, "Autism Speaks and CAN Complete Merger."

233 *Eric London struck an equally independent note:* Ibid.

233 *at a high-profile fund-raiser:* Gross and Strom, "Debate over Cause of Autism Strains a Family and Its Charity."

233 *"The child I knew":* Katie Wright, interview with author, July 27, 2010.

234 *Katie's mother met Jane Johnson:* Ibid.

234 *In January 2006, Krigsman performed:* Kirby, "David Kirby Interviews Katie Wright!"

234 *"He was the first doctor who really understood":* Katie Wright, interview with author, July 27, 2010.

234 *Krigsman's proposed treatment:* Kirby, "David Kirby Interviews Katie Wright!"

235 *one government-funded trial:* Trine Tsouderos and Patricia Callahan, "Autism's Risky Experiments; Some Doctors Claim They Can Successfully Treat Children, But the Alternative Therapies Lack Scientific Proof," *Chicago Tribune*, November 22, 2009, 1.

235 *Krigsman's problems dated to 2001:* Transcript of record, *Cedillo v. Sec'y of Health and Human Services*, 498–500, 558–60.

235 *Krigsman's difficulties with medical authorities:* Ibid., 501–03; *Cedillo v. Sec'y of Health and Human Services*, 138; Texas State Board of Medical Examiners, "Agreed Licensure Order for the Application of Arthur Charles Krigsman," September 16, 2005, http://www.casewatch.org/board/med/krigsman/complaint.shtml.

235 *When finally he was permitted:* Ibid.

236 *In February 2004,* The Times *(London):* Brian Deer, "MMR: The Truth Behind the Crisis," *The Sunday Times* (London), February 22, 2004, 12.

236 *The most shocking revelation:* Brian Deer, "Revealed: The First Wakefield MMR Patent Claim Describes 'Safer Measles Vaccine,'" *Brian Deer*, n.d., http://www.briandeer.com/wakefield/vaccine-patent.htm.

236 *as one member of the British Parliament put it:* Brian Deer, "Evan Harris Calls for Inquiry in Commons Debate About MMR Children's Treatment," *Brian Deer*, n.d., http://www.briandeer.com/mmr/lancet-commons.htm.

236 *ten of Wakefield's twelve co-authors:* Simon Murch et al., "Retraction of an Interpretation," *The Lancet* 2004;363(9411): 747–49.

236 *"If we had known the conflict of interest":* Oliver Wright, Nigel Hawkes, and Sam Lister, "Lancet Criticises MMR Scientist Who Raised Alarm, *The Times* (London), February 21, 2004.

236 *the General Medical Council:* Brian Deer, "Key Ally of MMR Doctor Rejects Autism Link," *The Sunday Times* (London), March 7, 2004, 1.

236 *That December, the council formally announced:* Brian Deer, "Doctors in MMR Scare Face Public Inquiry," *The Sunday Times* (London), December 12, 2004, 5.

237 *In 2005, the GMC released:* Brian Deer, "MMR Scare Doctor Faces List of Charges," *The Sunday Times* (London), September 11, 2005, 13.

237 *embody the qualities that David Aaronovitch:* David Aaronovitch, "A Conspiracy-Theory Theory," *Wall Street Journal*, December 19, 2009. See also: David Aaronovitch, *Voodoo Histories* (New York: Riverhead, 2010).

237 *"I have [already] lost my job":* Carole Caplin, "Why Is No One Allowed to Question MMR?," *The Daily Mail* (London), February 29, 2004.

238 *In 2004, with the help:* Mary Ann Roser, "Charting a Different Course on Autism," *Austin American-Statesman*, May 4, 2008, A1.

238 *"We tried the diet":* Katie Wright, interview with author, July 27, 2010.

CHAPTER 20: KATIE WRIGHT'S ACCIDENTAL MANIFESTO

PAGE

239 *a new thread was started on Yahoo!'s EOHarm:* B. J. Blacker, "REPOSTING—PLEASE!!! direct me to source thanks," EOHarm Yahoo! group, March 26, 2007, http://dir.groups.yahoo.com/group/EOHarm/message/47586.

239 *If they are utilizing bio-medical treatments:* Andrea, "Re: REPOSTING—PLEASE!!! direct me to source thanks," EOHarm Yahoo! group, March 27, 2007, http://dir.groups.yahoo.com/group/EOHarm/message/47635.

240 *"Remember when Katie backed out":* Andrea, "Re: REPOSTING—PLEASE!!! direct me to source thanks," EOHarm Yahoo! group, March 27, 2007, http://dir.groups.yahoo.com/group/EOHarm/message/47638.

240 *"What you all should be asking is":* Holly Bortfeld, "Re: REPOSTING—PLEASE!!! direct me to source thanks," EOHarm Yahoo! group, March 27, 2007, http://dir.groups.yahoo.com/group/EOHarm/message/47636.

240 *"I can completely understand":* Katie Wright, "katie wright," EOHarm Yahoo! group, March 27, 2007, http://dir.groups.yahoo.com/group/EOHarm/message/47643.

241 *"Why must we mince words":* Henry Coleman, "Re: katie wright," EOHarm

Yahoo! group, March 27, 2007, http://dir.groups.yahoo.com/group/EOHarm/message/47657.

241 *By treating their grandson:* askotnicki28, "Re: katie wright," EOHarm Yahoo! group, March 27, 2007, http://dir.groups.yahoo.com/group/EOHarm/message/47674.

241 *Who's side of the fence are you on?:* Lori, "Re: katie wright," EOHarm Yahoo! group, March 27, 2007, http://dir.groups.yahoo.com/group/EOHarm/message/47703.

242 *"Do you believe your son was damaged by vaccines?":* Lori, "Re: katie wright," EOHarm Yahoo! group, March 27, 2007, http://dir.groups.yahoo.com/group/EOHarm/message/47667.

242 *"I did not mean to be evasive":* Katie Wright, "environmental factors," EOHarm Yahoo! group, March 28, 2007, http://dir.groups.yahoo.com/group/EOHarm/message/47726.

242 *The following day, David Kirby:* David Kirby, "Autism Speaks: Will Anyone Listen?," *The Huffington Post*, March 28, 2007, http://www.huffingtonpost.com/david-kirby/autism-speaks-will-anyone_b_44414.html.

243 *He failed to mention that the last doses:* Centers for Disease Control and Prevention, "Timeline: Thimerosal in Vaccines (1999–2008), n.d., http://cdc.gov/vaccinesafety/concerns/thimerosal/thimerosal_timeline.html.

243 *His claim that two recently published genetic studies:* Peter Szatmari, "Mapping Autism Risk Loci Using Genetic Linkage and Chromosomal Rearrangements," *Nature Genetics* 2007;39: 319–28; "Scientists Confirm Genetic Distinction Between Heritable and Sporadic Cases of Autism," *Science Daily*, March 21, 2007.

244 *On April 5, Katie Wright and Alison Singer:* Oprah Winfrey, "The Faces of Autism," *The Oprah Winfrey Show*, Harpo Productions, April 5, 2007.

244 *seven million American viewers every day:* Ann Oldenburg, "$7M Car Giveaway Stuns TV Audience," *USA Today*, September 13, 2004.

244 *broadcast in 140 countries:* Michael Conlon, "Oprah Throws Party for U.S. Olympic Medalists," Reuters, September 3, 2008.

244 *"They were pretty clear":* Alison Singer, interview with author, May 15, 2009.

244 *"be a big girl and step it up":* Holly Bortfeld, "Re: REPOSTING—PLEASE!!! direct me to source thanks," EOHarm Yahoo! group, March 27, 2007, http://dir.groups.yahoo.com/group/EOHarm/message/47644.

244 *"I hear you wanted to say something about vaccines":* Alison Singer, interview with author, May 15, 2009.

244 *When they came back on the air:* Winfrey, "The Faces of Autism."

245 *There* had *been studies showing vaccines were safe:* Madsen et al., "A Population-Based Study of Measles, Mumps, and Rubella Vaccination and Autism." See also: Verstraeten et al., "Safety of Thimerosal-Containing Vaccines: A

Two-Phased Study of Computerized Health Maintenance Organization Databases."

245 *which was six more than what she'd claimed:* Katie Wright, "environmental factors."

245 *Even if you counted the MMR vaccine and the DPT vaccine:* Centers for Disease Control and Prevention, "Recommended Immunization Schedule for Persons Aged 0 Through 18 Years," *Morbidity and Mortality Weekly Report,* January 2, 2009, 57(51&52), Q-1, Q-4.

245 *Shortly after the interview began:* Kirby, "David Kirby Interviews Katie Wright!"

246 *the statements she made during the post-screening:* Ginger Taylor, "Katie Wright Speaks for Me!," *Adventures in Autism,* June 2, 2007, http://adventuresinautism.blogspot.com/2007/06/katie-wright-speaks-for-me.html.

246 *"I'd just grown very frustrated":* Katie Wright, interview with author, July 27, 2010.

247 *"We didn't realize that Katie":* Eric London, interview with author, April 7, 2010.

247 *Katie Wright is not a spokesperson for Autism Speaks:* Bob and Suzanne Wright, "Statement from Bob and Suzanne Wright, Co-founders of Autism Speaks," Autism Speaks, n.d., http://www.autismspeaks.org/wrights_statement.php.

247 *a gossip column on foxnews.com:* Roger Friedman, "Celebrity Autism Group in Civil War," foxnews.com, June 7, 2007, http://www.foxnews.com/story/0,2933,278814,00.html.

247 *to the front page of* The New York Times: Gross and Strom, "Autism Debate Strains a Family and Its Charity."

247 *"My daughter feels very strongly":* Bob Wright, interview with author, September 2, 2010.

248 *an entry on the* Age of Autism *blog:* Katie Wright, "Why the Autism Speaks Scientific Advisory Committee Needs to Resign," *Age of Autism,* April 26, 2010, http://www.ageofautism.com/2010/04/why-the-autism-speaks-scientific-advisory-committee-needs-to-resign.html.

248 *Alison Singer resigned:* Press release, Autism Speaks, "Autism Speaks Withdraws Support for Strategic Plan for Autism Research," January 15, 2009, http://www.autismspeaks.org/press/autism_speaks_withdraws_support_for_strategic_plan.php.

248 *Five months later, Eric London resigned:* Autism Science Foundation, "NAAR Founder, Dr. Eric London, Resigns from Autism Speaks," June 30, 2009, http://www.autismsciencefoundation.org/ericlondon.html.

CHAPTER 21: JENNY McCARTHY'S MOMMY INSTINCT

PAGE

249 *she was crowned the magazine's Playmate of the Year: Playboy*, June 1994.

249 *when MTV hired her to co-host: "Singled Out,"* Internet Movie Database, n.d., http://www.imdb.com/title/tt0112164/.

249 *That year, she appeared on two more* Playboy *covers: Playboy*, July 1996; *Playboy*, December 1996.

249 *In 1997, she had two eponymous shows:* J. Beveridge, "Pretty Penny for Jenny," *The Sunday Mail* (Queensland, Australia), September 21, 1997, 102.

250 *was paid $1.3 million by HarperCollins:* Ibid.

250 Jenny *tanked:* Ellen Gray, "McCarthy Bares Soul in Autobiography," *The Charlotte Observer*, November 5, 1997, 6e.

250 *In 1999, she married an actor:* Janet Mock and Julia Wang, "Celebrity Central: Jenny McCarthy," People.com, n.d., http://www.people.com/people/jenny_mccarthy.

250 *In 2004, she released* Belly Laughs: Jenny McCarthy, *Belly Laughs* (Cambridge: Da Capo, 2004).

250 *Her next book, 2005's* Baby Laughs: Jenny McCarthy, *Baby Laughs* (New York: Dutton, 2005).

250 *In his zero star write-up:* Roger Ebert, "'Dirty Love' Is So Scummy It Can't Even Be Called Bad," *Ventura County Star*, September 23, 2005, 5.

250 *Even* Life Laughs, *the third book:* Jenny McCarthy, *Life Laughs* (New York: Dutton, 2006).

251 *"You're an Indigo":* Jenny McCarthy, "A Mother's Awakening," IndigoMoms.com, June 2006, http://web.archive.org/web/20061019001706/indigomoms.com/jenny0606.html.

251 *a group of spiritually advanced children:* Meg Blackburn Losey, *The Children of Now: Crystalline Children, Indigo Children, Star Kids, Angels on Earth, and the Phenomenon of Transitional Children* (Franklin Lakes, New Jersey: New Page, Press 2006).

251 *Parents of Crystals recognize each:* Kabir Jaffe and Ritama Davidson, *Indigo Adults* (Lincoln, Indiana: iuniverse.com, 2005).

251 *"[that] things in my life started to make sense":* McCarthy, "A Mother's Awakening."

251 *"The reason why I was drawn to Indigo":* Jenny McCarthy, "Question to Jenny," IndigoMoms.com, August 3, 2006, http://web.archive.org/web/20061026234851/www.indigomoms.com/phpbb/viewtopic.php?t=122.

251 *That summer, McCarthy launched:* IndigoMoms.com, n.d., http://web.archive.org/web/20061203045246/www.indigomoms.com/index2.html.

251 *and an e-commerce section:* "Resources," IndigoMoms.com, n.d., http://web.archive.org/web/20061112124412/www.indigomoms.com/resources.html.

252 *In 2005, McCarthy contacted Lisa Ackerman:* Lisa Ackerman, "TACA and Jenny McCarthy," October 5, 2008, http://www.tacanow.org/jenny/jenny-mccarthy-autism.htm.

252 *Shortly thereafter, McCarthy told Ackerman:* Miller, "Autism Is Treatable, She Insists."

252 *in stark contrast to the Crystal Child one:* Jay Leno, *The Tonight Show with Jay Leno*, NBC, April 21, 2006.

252 *Now, McCarthy said, her mistreatment:* Jenny McCarthy, *Louder than Words* (New York: Dutton, 2007).

252 *"I say, Okay, let's look at your choices":* Jenny McCarthy, interview with author, March 6, 2009.

252 *True to her word, on September 18, 2007:* Oprah Winfrey, "Mothers Battle Autism," *The Oprah Winfrey Show*, Harpo Productions, September 18, 2007.

252 *That afternoon, with Lisa Ackerman looking on:* Miller, "Autism Is Treatable, She Insists."

253 *her latest journey had begun with a flash:* Winfrey, "Mothers Battle Autism."

253 *"taking away all the beautiful characteristics":* IndigoMoms.com, n.d., http://web.archive.org/web/20061026234851/www.indigomoms.com/phpbb/viewtopic.php?t=122.

253 *First thing I did—Google:* Winfrey, "Mothers Battle Autism."

256 *One fan asked McCarthy:* "Q&A with Jenny McCarthy," Oprah.com, September 18, 2007, http://www.oprah.com/relationships/Questions-About-Autism.

256 *McCarthy had repeated her story:* Larry King, "Interview with Jenny McCarthy," *Larry King Live*, CNN, September 26, 2007; "A Mother's Journey: Star Fights for Son," *Good Morning America*, ABC, September 24, 2007.

256 People . . . *ran an excerpt from* Louder than Words: Jenny McCarthy, "My Autistic Son: A Story of Hope," *People*, September 7, 2007, http://www.people.com/people/article/0,,20057803,00.html.

256 *"Had to":* Miller, "Autism Is Treatable, She Insists."

257 *gives McCarthy credit for single-handedly:* Ibid.

257 *Larry King devoted his full hour-long broadcast:* Larry King, "Jenny McCarthy's Autism Fight," *Larry King Live*, CNN, April 2, 2008.

258 *Two months later, McCarthy and Jim Carrey led:* Talk About Curing Autism, "Green Our Vaccines Rally with Jenny McCarthy and Jim Carrey: Rally Introduction and General Information," May 28, 2008, http://www.talkaboutcuringautism.org/jenny/dc-rally/green-our-vaccines-rally.htm.

258 *In the TACA press release:* Talk About Curing Autism, "Green Our Vaccines

Rally with Jenny McCarthy and Jim Carrey and Keynote Address by Robert F. Kennedy Jr.," PR Newswire, June 4, 2008.

258 *McCarthy's rally-related appearances:* "A Mother's Mission: McCarthy and Carrey Search for Autism Answers," *Good Morning America*, ABC, June 4, 2008; Greta Van Susteren, "Jenny McCarthy Speaks About Autism," *On the Record with Greta Van Susteren*, Fox News, June 6, 2008.

258 *She'd taken over Generation Rescue:* See generationrescue.org.

258 *By the end of the year, she'd published* Mother Warriors: Jenny McCarthy, *Mother Warriors: A Nation of Parents Healing Autism Against All Odds* (New York: Dutton, 2008).

258 *signed a deal with the licensing agency:* Becky Ebenkamp, "Jenny McCarthy's Too Good Line Coming to Stores," *Brandweek*, September 8, 2008, http://www.brandweek.com/bw/content_display/news-and-features/licensing/e3i425366f6874393a7ec111e1578f285b8.

258 *She'd also launched Teach2Talk Academy:* Karl Taro Greenfield, "The Autism Debate: Who's Afraid of Jenny McCarthy?," *Time*, February 25, 2010.

259 *In April 2010, Teach2Talk Academy was closed:* "Inside Jenny McCarthy's Personal Turmoil: Star Closes School for Autistic Children Amidst Split from Jim Carrey," *Shine from Yahoo!*, April 9, 2010, http://shine.yahoo.com/channel/parenting/inside-jenny-mccarthys-personal-turmoil-star-closes-school-for-autistic-children-amidst-split-from-jim-carrey-1271964/.

259 *Scheflen appreciated that putting autistic children:* Sarah Clifford Scheflen, "Video Modeling as a Fun, Effective Way to Communication and Learning," presentation, AutismOne conference, Westin O'Hare, Chicago, May 23, 2009.

259 *The only universally acknowledged treatment:* Scott Myers and Chris Plauché Johnson, "Management of Children with Autism Spectrum Disorders," *Pediatrics* 2007;120(5): 1162–82.

259 *"How many people are back from last year?":* Jenny McCarthy, "Keynote Address," presentation, AutismOne conference, Westin O'Hare, Chicago, May 23, 2009.

260 *Glutathione is a naturally occurring antioxidant:* Alphonso Pompella et al., "The Changing Faces of Glutathione, a Cellular Protagonist," *Biochemical Pharmacology* 2003;66(8): 1499–1503.

260 *which covered a philosophy that involves wishing:* John Hicks, "The Law of Attraction," presentation, AutismOne conference, Westin O'Hare, Chicago, May 20, 2009.

261 *"To take advantage of the natural ups and downs":* John Allen Paulos, *Innumeracy* (New York: Hill & Wang, 1998, 2001), 82–83.

262 *has firsthand experience with the lure:* James R. Laidler, "Through the Looking Glass: My Involvement with Autism Quackery," Autism Watch—Your Sci-

entific Guide to Autism, n.d., http://www.autism-watch.org/about/bio2.shtml.

263 *a chemical cleansing process originally designed:* Gary Wulfsberg, *Inorganic Chemistry* (Sausalito, California: University Science Books, 2000), 225.

263 *When Liz Birt's son, Matthew, was chelated:* Kirby, *Evidence of Harm*, 182.

263 *he went "berserk":* Tsouderos and Callahan, "Autism's Risky Experiments."

263 *who moved with his mother to Pennsylvania:* "Autistic Boy Dies During Therapy," *The Bath Chronicle* (U.K.), September 1, 2005, 2.

263 *Abubakar went into massive cardiac arrest and died:* Alex Kumi, "Autistic Boy Dies After US Therapy Visit," *The Guardian* (London), August 26, 2005, 5.

263 *ear, nose, and throat specialist named Roy Kerry:* Tom Leonard, "Parents Sue Doctor over Death of Autistic Boy," *The Daily Telegraph* (London), July 11, 2007, 18.

263 *Lisa Ackerman told parents:* Ackerman, "Starting the Biomedical Treatment Journey."

264 *a designation that is obtained by attending:* Tsouderos and Callahan, "Autism's Risky Experiments."

264 *Kirkman did a voluntary recall:* Larry Newman, "Notice to Our Customers," Kirkman Labs, January 10, 2010.

264 *had recently been proposing as a potential cause:* Kirby, "Metals, Myelin and Mitochondria: Several Paths to Autism?"

CHAPTER 22: MEDICAL NIMBYISM AND FAITH-BASED METAPHYSICS

PAGE

265 *his exposure to "Vaccine Roulette":* Jay Gordon, interview with author, April 3, 2009.

265 *"No one knows your child better than you do":* See http://drjaygordon.com/.

265 *"I'm a member in good standing":* Jay Gordon, interview with author, April 3, 2009.

265 *"We need someone who's willing to yell":* Ibid.

266 *In the Foreword to one of McCarthy's books:* McCarthy, *Mother Warrior*, xiii.

266 *Gordon wrote a story in* The Huffington Post: Jay Gordon, "Vaccines," *The Huffington Post*, March 31, 2009, http://www.huffingtonpost.com/jay-gordon/vaccines_b_181047.html.

266 *prefers to be called by his first name only:* "About Sears," AskDrSears.com, n.d., http://www.askdrsears.com/about.asp.

266 *"I became passionate about educating parents":* E-mail to author, "Subject: bob sears," August 15, 2009.

266 *which includes an "alternative" vaccination schedule:* Robert Sears, *The Vaccine Book* (New York: Little, Brown, 2007), 239.

266 *about a topic in which he does not have specialized training:* Sears has a medical degree from the Georgetown University School of Medicine and completed an internship and residency in pediatrics at Children's Hospital Los Angeles. See "About Sears," AskDrSears.com.

266 The Autism Book: Robert Sears, *The Autism Book* (New York: Little, Brown, 2010).

267 " *'Front Line' Response to* The Vaccine Book": Brian Bowman, " 'Front Line' Response to *The Vaccine Book,*" *Pediatrics*, 2008;123(1): 164–69.

267 *"natural" immunity is more effective:* Sears, *The Vaccine Book*, 105–6.

267 *don't regret their decisions:* Ibid., 22.

267 *under the heading "The Way I See It":* Ibid., 96–97.

268 *As Anne Harrington writes:* Harrington, *The Cure Within.*

268 *a Freudian approach to illness:* Ibid., 80.

268 *"laboratory based rigor":* Ibid., 244.

269 *a guest on Winfrey's show:* Oprah Winfrey, "The Secret," *The Oprah Winfrey Show*, Harpo Productions, February 8, 2007.

269 *featured on Oprah.com:* "Discovering The Secret," Oprah.com, February 8, 2007, http://www.oprah.com/spirit/Discovering-The-Secret.

269 The Secret *is based on something called the "law of attraction":* Rhonda Byrne, *The Secret* (New York: Atria, 2006), 4.

269 *"everything that's coming into your life":* Ibid.

269 *illness is an example of being "in labor with yourself":* Oprah Winfrey, "The Big Wake-up Call for Women, with Dr. Christine Northrup," *The Oprah Winfrey Show*, Harpo Productions, October 16, 2007.

269 *prototypical example of this phenomenon is thyroid problems:* Ibid.

269 *Somers detailed her health regimen:* Oprah Winfrey, "Suzanne Somers—The Bioidentical Hormone Follow-up," *The Oprah Winfrey Show*, Harpo Productions, January 29, 2009; Weston Kosova and Pat Wingert, "Live Your Best Life Ever!," *Newsweek*, May 30, 2009.

270 *whose net worth is estimated at $2.7 billion:* "The World's Billionaires: #234 Oprah Winfrey," Forbes.com, March 11, 2009, http://www.forbes.com/lists/2009/10/billionaires-2009-richest-people_Oprah-Winfrey_O0ZT.html.

270 *In that instance:* Kosova and Wingert, "Live Your Best Life Ever!"

270 *she reiterated that point in a statement:* Ibid.

271 *Consider the case of Julieanna Metcalf:* "Victims of Vaccine-Preventable Disease," *Vaccinate Your Baby*, n.d., http://www.vaccinateyourbaby.org/why/victims.cfm.

271 *The outbreak that ensnared:* Liz Szabo, "Missed Vaccines Weaken 'Herd Immunity' in Children," *USA Today*, January 6, 2010.

271 *a seven-year-old boy who was later revealed:* David Sugerman et al., "Measles

Outbreak in a Highly Vaccinated Population, San Diego, 2008," *Pediatrics* 2010;125(4): 748.

272 *as his mother explained in a* Time *magazine article:* "How My Son Spread the Measles."

272 *Within days, the measles virus had spread:* Cheryl Clark, "Measles Cases Now Total 11; Health Officials Await Results for 12th Child," *The San Diego Union-Tribune*, February 16, 2008, B1; "Ruining It for the Rest of Us," *This American Life*, WBEZ, Chicago, December 19, 2008.

272 *passengers on a plane headed to Honolulu:* Clark, "Measles Cases Now Total 11; Health Officials Await Results for 12th Child."

272 *a ten-month-old child was hospitalized:* Ibid.

272 *According to Gordon, the hospitalization:* Jay Gordon, interview with author, April 3, 2009.

272 *An additional forty-eight children:* Sugarman et al., "Measles Outbreak in a Highly Vaccinated Population, San Diego, 2008," 747.

272 *In total, the outbreak cost:* Ibid.

272 *the largest measles outbreak in California since 1991:* Ibid., 748; Nora Zamichow, "Program Set to Immunize Preschoolers—San Diego Is One of Six US Cities Picked for the Immunization program," *Los Angeles Times*, February 3, 1992, B1.

272 *That nationwide outbreak began in 1989:* Walter Orenstein et al., "Measles Elimination in the United States," *The Journal of Infectious Diseases* 2004; 189: S1.

272 *"healthy snacks" and "supplements":* *Dr. Sears Family Essentials*, n.d., http://www.drsearsfamilyessentials.com/.

272 *I'd like to put together a nice press kit for you:* Rachelle Duval, e-mail to author, "Subject: follow up/from Seth Mnookin inre: Simon & Schuster book about vaccines," June 24, 2009.

273 The Vaccine Book *has already sold more than 100,000 copies:* Liza Gross, "A Broken Trust: Lessons from the Vaccine-Autism Wars," *PLoS Biology* 2009;7(5): 6.

273 *"The recent measles outbreak":* Robert Sears, "Dr. Bob Sears Offers Advice in March 21st New York Times Health Section on Vaccine Choices Parents Make," AskDrSears.com, March 27, 2008, http://www.askdrsears.com/blog/.

CHAPTER 23: BABY BRIE

PAGE

274 *Ralph Romaguera first met Danielle Broussard:* Danielle Romaguera, interview with author, July 7, 2009.

274 *Ralph was there working as a photographer:* Danielle Romaguera, e-mail to

author, "Subject: Re: from Seth Mnookin/book about vaccines," August 4, 2010.

274 *Brie was not quite four weeks old:* Danielle Romaguera, interview with author, July 7, 2009.

275 *Worst-case scenario, she said:* "Victims of Vaccine-Preventable Diseases," *Vaccinate Your Baby.*

275 *But on Sunday Brie sounded worse:* Danielle Romaguera, interview with author, July 7, 2009.

276 *Finally, Dawn Sokol, a pediatric infectious disease specialist:* "Dr. Dawn Sokol Joins Nemours Division of Infectious Disease," *Orlando Medical News,* August 2, 2007, http://www.orlandomedicalnews.com/grand-rounds-august-cms-481.

276 *"This baby has pertussis":* Danielle Romaguera, interview with author, July 7, 2009.

276 *during its initial incubation period:* Florens G. A. Versteeg et al., "Pertussis: A Concise Historical Review Including Diagnosis, Incidence, Clinical Manifestations and the Role of Treatment and Vaccinations in Management," *Reviews in Medical Microbiology* 2005;16(3): 80.

277 *bring with them paroxysms and cyanosis:* Ibid.

277 *"have the strength to have a whoop":* Ibid.

277 *Throughout history, the disease has been given:* Ibid.

277 *Up until the 1940s, whooping cough:* "Pertussis Disease Q&A," Centers for Disease Control and Prevention, n.d., http://www.cdc.gov/vaccines/vpd-vac/pertussis/dis-faqs.htm.

277 *the total number of cases and the total number of deaths:* Steve Black, "Epidemiology of Pertussis," *Pediatric Infectious Disease Journal* April 1997;16(4): S85–89.

277 *In Japan, pertussis vaccine uptake fell:* E. J. Gangarosa et al., "Impact of Anti-Vaccine Movements on Pertussis Control: The Untold Story," *The Lancet* January 1, 1998;351: 357–58.

277 *between ten thousand and twelve thousand new cases:* Smith, "National Childhood Vaccine Injury Compensation Act," 265.

277 *the Swedish Medical Society abandoned:* Gangarosa, "Impact of Anti-Vaccine Movements on Pertussis Control: The Untold Story," 357.

277 *from just over one thousand in 1976:* Dennis A. Brooks and Richard Clover, "Pertussis Infection in the United States," *Journal of the American Board of Family Medicine* 2006;19: 603–11.

278 *"The first thought that came into my head was":* Danielle Romaguera, interview with author, July 7, 2009.

278 *The best way to think of an ECMO:* "Extracorporeal Membrane Oxygen-

ation (ECMO)," Cincinnati Children's Hospital Medical Center, Surgical Options, n.d., http://www.cincinnatichildrens.org/health/heart-encyclopedia/treat/surg/ecmo.htm.

278 *The Romagueras are intensely religious people:* Danielle Romaguera, interview with author, July 7, 2009.

281 *"It upset me":* Ibid.

282 *In the spring of 2009,* The Hollywood Reporter: Nellie Andreeva, "Jenny McCarthy Inks Deal with Winfrey's Harpo," *The Hollywood Reporter*, May 3, 2009.

282 *Danielle learned she was pregnant again:* Romaguera, e-mail to author, "Subject: Re: from Seth Mnookin/book about vaccines."

CHAPTER 24: CASUALTIES OF A WAR BUILT ON LIES

PAGE

284 *Together, they decided that the Omnibus proceeding:* Golkiewicz, "Autism General Order #1," 2–4.

284 *three separate "causation" theories:* Ibid., 6.

284 *not to present any test cases for the MMR theory: Cedillo v. Sec'y of Health and Human Services*, 13.

284 *they were given a full year:* "Master Scheduling Order," Omnibus Autism Proceeding.

284 *final rulings were not scheduled:* Ibid.

284 *It wasn't until 2005 that the discovery process:* "Petitioners' Filing re: Submission of Expert Reports in Support of General Causation," Omnibus Autism Proceeding, Autism Master File, June 14 2005, http://www.uscfc.uscourts.gov/sites/default/files/autism/6 14 05 pet filing re timing of expert reports.pdf.

284 *the PSC received a further extension:* "Ruling Concerning Issue of Time," Omnibus Autism Proceeding, Autism Master File, August 11, 2005, http://www.uscfc.uscourts.gov/sites/default/files/autism/Ruling Concerning Issue of Time.pdf.

285 *Valentine's Day 2006:* "Petitioners' Initial Disclosure of Experts," Omnibus Autism Proceeding, Autism Master File, February 14, 2006, http://www.uscfc.uscourts.gov/sites/default/files/autism/PET INITIAL EXPERT.pdf.

285 *three interdependent parts: Cedillo v. Sec'y of Health and Human Services*, 19–20.

285 *compared to those worn by heavy machine operators:* Gardiner Harris, "Opening Statements in Case on Autism and Vaccinations," *The New York Times*, June 12, 2007, 21.

285 Vaccine *author Arthur Allen described:* Arthur Allen, "Autism in Court, Day

1," *Vaccine the Book*, June 11, 2007, http://vaccinethebook.typepad.com/mt/2007/06/autism_in_court.html.

286 *"an elderly, hard-of-hearing toxicologist"*: Ibid.

286 *Chin-Caplan kept her questions straightforward:* Transcript of record, *Cedillo v. Sec'y of Health and Human Services*, 62–116.

286 *During the cross-examination:* Ibid., 117–222.

286 *Would you like to see the paper:* Ibid., 142–50.

287 *Aposhian seemed not to realize:* Ibid., 150–52.

288 *"The hypothesis was made"*: Ibid., 207.

288 *whose primary employer was a Shanghai-based company:* Ibid., 867–68.

288 *began to giggle inexplicably:* "Omnibus Hearing: Byers," *Autism Diva*, June 16, 2007, http://autismdiva.blogspot.com/2007/06/omnibus-hearing-byers.html. See also: Transcript of record, *Cedillo v. Sec'y of Health and Human Services*, 986.

288 *"You're making faces at me"*: Transcript of record, *Cedillo v. Sec'y of Health and Human Services*, 996.

288 *"She's much more attractive."*: Ibid., 997.

288 *"educated by Dr. Aposhian"*: Ibid., p. 984–85.

288 *He'd already appeared as an expert witness in at least 185 separate cases: Snyder v. Sec'y of Health and Human Services*, 19.

289 *Along with Andrew Wakefield: Hazelhurst v. Sec'y of Health and Human Services*, 112.

289 *"suffers from the stigma"*: *Snyder v. Sec'y of Health and Human Services*, 19.

289 *in a textbook titled* Child Neurology: Marcel Kinsbourne and Frank B. Wood, "Chapter 18—Disorders of Mental Development," in *Child Neurology*, 7th ed. (Philadelphia: Lippincott Williams & Wilkins, 2005).

289 *"You developed a chart"*: Transcript of record, *Cedillo v. Sec'y of Health and Human Services*, 1169–71.

290 *appeared measured and deliberate:* Ibid., 2295–493.

290 *Stephen Bustin was devastatingly blunt:* Ibid., 1933–2070.

290 *the testimony of Eric Fombonne:* Ibid., 1239–1466.

290 *Kinsbourne had relied on his interpretation:* Ibid., 1040–42, 1062.

290 *Fombonne used contemporaneous notes:* Ibid., 1327–37.

290 *he pointed to numerous examples:* Ibid., 1337–56.

290 *"hand flapping"*: Ibid., 1344, 1350.

290 *a failure to make eye contact:* Ibid., 1345–46.

290 *respond to her name:* Ibid., 1350.

290 *play with toys:* Ibid., 1346–47.

290 *use communicative gestures:* Ibid., 1339–40,

290 *use words, or even use nonverbal sounds:* Ibid., 1340–42.

290 *her head circumference was abnormally large:* Ibid., 1336–37.

291 *it was Theresa Cedillo's conversation with Wakefield:* Ibid., 391–94.
291 *Since 2003, Michelle had been under the care of Arthur Krigsman:* Ibid., 43–44, 256, 325, 356, 363, 394–95.
291 "sine qua non *in [Michelle's] quest for entitlement": Cedillo v. Sec'y of Health and Human Services,* 41.
291 *were tainted by their association with Wakefield:* Ibid., 42–44.
291 *a former graduate student of Wakefield's named Nicholas Chadwick:* Transcript of record, *Cedillo v. Sec'y of Health and Human Services,* 2282–95.
291 *said that he'd alerted Wakefield:* Ibid., 2282–90.
292 *"I attended two of those meetings":* Transcript of record, *Snyder v. Sec'y of Health and Human Services,* 843a.
292 *"The final point I want to make":* Transcript of record, *Cedillo v. Sec'y of Health and Human Services,* 2910.
292 *named Michelle Cedillo its "Child of the Year":* "Age of Autism Award: Michelle Cedillo, Child of the Year," *Age of Autism,* December 24, 2007, http://www.ageofautism.com/2007/12/age-of-autism-a.html.
293 *"evidentiary records are now* closed": "Vaccine Act Interim Costs," Omnibus Autism Proceeding, Autism Master File, September 29, 2008, 2.
293 *"interim attorneys' fees": Cedillo v. Sec'y Health and Human Services,* "Vaccine Act Interim Costs," Omnibus Autism Proceeding, Autism Master File, November 18, 2008, 1.
294 *"The evidentiary record": Cedillo v. Sec'y of Health and Human Services,* 19.
294 *In addition to testimony:* Ibid.
294 *the three Masters who ruled on the dual-causation test cases:* Ibid., 14.
294 *"far better qualified":* Ibid., 2.
294 *characterized Vera Byers's testimony: Snyder v. Sec'y of Health and Human Services,* 24.
294 *"reasonably coherent":* Ibid., 31.
294 *"was unwilling to say measles":* Ibid., 18–19.
294 *"I have not required a level of proof": Cedillo v. Sec'y of Health and Human Services,* 172.
295 *"This is a case in which the evidence is so one-sided":* Ibid., 173.
295 *The record of this case:* Ibid., 173–74.
297 *whom Chin-Caplan accused of being "arbitrary and capricious":* Krigsman and Chin-Caplan, "Autism and Vaccines in the US Omnibus Hearings: Legal and Gastrointestinal Perspectives of the Michelle Cedillo Case."

EPILOGUE

PAGE

299 *Its job was to determine whether Wakefield:* Nick Triggle, "MMR Scare Doctor 'Acted Unethically,' Panel Finds," *BBC News*, January 28, 2010, http://news.bbc.co.uk/2/hi/health/8483865.stm.

299 *the longest and most expensive in the council's:* Michael White, "Blair, Wakefield, Climate Change—Beware of Scapegoats," *The Guardian* (London), January 29, 2010.

299 *wearing a dark gray suit:* Fiona Macrae and David Wilkes, "Damning Verdict on MMR Doctor," *The Daily Mail* (London), January 30, 2010, photograph accompanying the article.

299 *shouting "Bastards!":* Ibid.

299 *speckled with signs reading:* Sophie Borland, " 'Dishonest and Irresponsible': Doctor Who Triggered MMR Vaccine Scare Is Struck Off," *The Daily Mail* (London), May 25, 2010.

299 *spontaneous chants:* Macrae and Wilkes, "Damning Verdict on MMR Doctor."

300 *"My only concern":* Nick Allen, "MMR-Autism Link Doctor Andrew Wakefield Defends Conduct at GMC Hearing," *The Daily Telegraph* (London), March 27, 2008.

300 *Wakefield failed to call on even a single parent:* Orac, "Andrew Wakefield: Struck Off!," *Respectful Insolence,* May 24, 2010, http://scienceblogs.com/insolence/2010/05/andrew_wakefield_struck_off.php.

300 *identified as "Child 12":* Rochelle Poulter, "Statement of Rochelle Poulter—Parent of Child 12 in the GMC Hearing," n.d., http://www.rescuepost.com/files/uk-statement-rochelle-poulter.doc.

300 *Matthew's health problems had begun:* Macrae and Wilkes, "Damning Verdict on MMR Doctor."

300 *Soon thereafter, Poulter contacted Wakefield's research team:* "Fitness to Practise Panel Hearing," 32.

300 *Wakefield himself acknowledged:* Ibid., 33.

301 *"not to have MRI or LP":* Ibid., 36.

301 *Despite an employment contract:* Ibid., 34, 36.

301 *Wakefield signed the hospital request forms:* Ibid., 34.

301 *"contrary to the clinical interests of Child 12":* Ibid., 36.

301 *"I insisted that the hearing be informed":* Poulter, "Statement of Rochelle Poulter—Parent of Child 12 in the GMC Hearing."

301 *"At all material times":* "Fitness to Practise Panel Hearing," 4.

301 *"dishonest," "misleading," and "in breach of your duty":* Ibid., 7.

302 *"dishonest," "irresponsible," and "resulted in a misleading description":* Ibid., 44.

302 *"did not meet the criteria for either autism":* Ibid., 41.

302 *developing an oral measles vaccine:* Ibid., 50–51.

302 *"caused blood to be taken":* Ibid., 54.

302 *"paid those children":* Ibid.

302 *"described the incident":* Ibid.

302 *"expressed an intention":* Ibid.

302 *"You showed a callous disregard":* Ibid., 55.

302 *released a terse statement:* Jane Johnson, "Dr. Wakefield Has Resigned from Thoughtful House," Thoughtful House Yahoo! group, February 17, 2010, http://health.groups.yahoo.com/group/thoughtfulhousecaterforchildren/message/16382.

302 *The following day, Arthur Krigsman said:* Ibid., February 19, 2010, http://health.groups.yahoo.com/group/thoughtfulhousecaterforchildren/message/16427.

302 *"in terms of the local medical community":* Jane Johnson, interview with author, April 22, 2010.

303 *I called Wakefield at his home:* Andrew Wakefield, interview with author, May 10, 2010.

303 *he asked if we could conduct the remainder:* Andrew Wakefield, interview with author, August 31, 2009.

303 *"Basically it's about why":* Andrew Wakefield, interview with author, May 10, 2010.

304 *On Monday, May 24, 2010:* Sarah Bossley, "Andrew Wakefield Struck Off Register by General Medical Council," *The Guardian* (London), May 24, 2010.

304 *he received a standing ovation:* Joanna Weiss, "Autism's 'Unblessed' Scientists," *Boston Globe*, June 1, 2010, 13.

304 *where he also headlined a rally:* "Rally Platform Speakers," American Rally for Personal Rights, n.d., http://www.americanpersonalrights.org/index.php?option=com_content&view=article&id=81&Itemid=55.

304 *gave two presentations:* "Conference Schedule," Autism Redefined 2010—AutismOne, n.d., http://conference.autismone.org/documents/autismone_schedules.pdf.

304 *posed for pictures with Bob Sears:* "Dr. Sears, Dr. Wakefield, and TACA at Autism One," *Age of Autism*, May 29, 2010, http://www.ageofautism.com/2010/05/dr-sears-dr-wakefield-and-taca-at-autism-one.html.

304 *"I spent Saturday at an incredible conference":* Jay Gordon, "AutismOne," *Jay Gordon, MD FAAP*, May 30, 1010, http://drjaygordon.com/miscellaneous/autismone.html.

305 *the most attention as of late is Ashland, Oregon:* "The Vaccine War," *Frontline*, PBS, April 27, 2010, http://www.pbs.org/wgbh/pages/frontline/vaccines/.

305 *the fifth-highest average-per-capita income:* "N.J. Has Four of Nation's 20 Highest-Income Countries," Associated Press, May 20, 2009.

305 *an exemption rate more than three times:* Erin Allday, "Not Enough Bay Area Kids Vaccinated, Docs Say," *San Francisco Chronicle*, August 1, 2009.

305 *"in large part because of parents choosing":* "Map: High Risk Schools in Southern California," latimes.com, March 29 2009, http://www.latimes.com/news/local/la-me-immunization29-2009mar29-map,0,426776.htmlstory.

305 *the Ocean Charter School:* Rong-Gong Lin II and Sandra Poindexter, "Schools' Risks Rise as Vaccine Rate Declines," *Los Angeles Times*, March 29, 2009, A1.

305 *Between 2005 and 2010:* Jonathan D. Rockoff, "More Parents Seek Vaccine Exemption—Despite Assurances, Fear of Childhood Shots Drives Rise," *The Wall Street Journal*, July 6, 2010.

305 *Meg Fisher, the head of the AAP's section:* Joanna Weiss, "Seeking Common Ground in the Autism-Vaccine Debate," *Boston Globe*, February 27, 2010, 11.

305 *six unvaccinated children in southeastern Pennsylvania:* Don Sapatkin, "Hib Disease Deaths Put Focus on Vaccine Shortage," *The Philadelphia Inquirer*, April 1, 2009, A1.

306 *a mumps outbreak that began:* Rong-Gong Lin II, "County Sees Rise in Mumps Cases," *Los Angeles Times*, May 16, 2010, A39.

306 *In October, the California Department of Public Health:* The California Department of Public Health, "Pertussis Report," October 12, 2010.

306 *when the pertussis vaccine was just entering widespread use:* Jeffrey P. Baker, "Immunization and the American Way: 4 Childhood Vaccinations," *American Journal of Public Health*, 2000;90(2): 202.

306 *a columnist for* The Boston Globe: Weiss, "Autism's 'Unblessed' Scientists."

307 *Deth's testimony as an expert witness: King v. Sec'y of Health and Human Services*, 69–77.

307 *"There were also a number of other specific points":* Ibid., 77.

307 *was the hottest year ever recorded:* National Aeronautics and Space Administration, "2009: Second Warmest Year on Record; End of Warmest Decade," n.d., http://www.nasa.gov/topics/earth/features/temp-analysis-2009.html.

307 *said global warming is* not *a problem:* Frank Newport, "Americans' Global Warming Concerns Continue to Drop," Gallup.com, March 11, 2010, http://www.gallup.com/poll/126560/americans-global-warming-concerns-continue-drop.aspx.

307 *both houses of the Kentucky legislature:* Leslie Kaufman, "Darwin Foes Add Warming to Targets," *The New York Times*, March 3, 2010, A1.

307 *Louisiana has already passed:* Ibid.

307 *in 2009, the Texas Board of Education:* Ibid.

AFTERWORD

PAGE

309 *the* British Medical Journal *ran the first in a series of articles*: Brian Deer, "How the Case Against the MMR Vaccine Was Fixed," *British Medical Journal*, 2011;342: c5347.

309 *he'd committed outright fraud*: Fiona Godlee, Jane Smith, and Harvey Marcovitch, "Wakefield's article linking MMR vaccine and autism was fraudulent," *British Medical Journal* 2011;342: c.7452.

309 *Mark and David Geier's Lupron protocol "endangers autistic children"*: "In the matter of Mark R. Geier, M.D., Respondent, before the Maryland State Board of Physicians: Order for Suspension of License to Practice Medicine," Maryland State Board of Physicians, Case Numbers: 2007-0083; 2008-0454; 2009-0308, 15.

309 *Wakefield announced he would headline a rally titled "The Masterplan"*: Seth Mnookin, "It's Official: Wakefield Joins the Ranks of Truthers, New World Order Conspiracists," *The Panic Virus Blog* (The Public Library of Science), June 7, 2011.

310 *"Andrew Wakefield is Nelson Mandela and Jesus Christ rolled up into one"*: Susan Dominus, "The Denunciation of Dr. Wakefield," *The New York Times Magazine*, April 24, 2011, MM36. (A version of this article appeared online on April 20, 2011, with the headline, "The Crash and Burn of an Autism Guru.")

310 *Their research on "the role and treatment of elevated male hormones"*: David A. Geier, Rev. Lisa K. Sykes, and Mark R. Geier, "The Role and Treatment of Elevated Male Hormones in Autism Spectrum Disorders," *Autism Science Digest*, April 2011: 71–75.

310 *the Blue Mountain School, a small private school*: Sarah Bruyn Jones, "Whooping Cough Outbreak Closes Blue Mountain School in Floyd Co.," *The Roanoke Times*, April 5, 2011.

310 *many of the parents of the infected students had chosen not to have their children vaccinated*: New River Health District director Molly O'Dell, interview with author, April 11, 2011.

310 *The largest outbreak was in Minnesota*: Centers for Disease Control and Prevention, "Measles—United States, January–May 20, 2011," *Morbidity and Mortality Weekly Report*, May 27, 2011;60(20): 66668.

310 *anti-vaccine activists had targeted a community of Somali immigrants*: Steve Karnowski, "Autism Fears, Measles Spike Among Minn. Somalis," Associated Press, April 2, 2011.

310 *That outbreak began when a deliberately unvaccinated child returned from Africa*: Maura Lerner, "Measles Whodunit: Tracking 2011's Outbreak," *Minneapolis Star Tribune*, April 8, 2011.

310 *On June 22, the CDC issued an official health alert*: "High Number of Reported Measles Cases in the U.S. in 2011—Linked to Outbreaks Abroad," Centers for Disease Control and Prevention, June 22, 2011, http://www.bt.cdc.gov/HAN/han00323.asp.

310 *From 2001 to 2008, the U.S. had had an average of around 50 measles cases a year*: Centers for Disease Control and Prevention, "Measles—United States, January–May 20, 2011."

310 *Over the first twenty-four weeks of 2011, the CDC had verified 156 measles infections*: "High Number of Reported Measles Cases in the U.S. in 2011—Linked to Outbreaks Abroad," Centers for Disease Control and Prevention.

311 *In the first five months of 2011, the country recorded more than 10,000 infections*: "Epidémie de rougeole en France. Actualisation des données au 20 mai 2011," Institut de veille sanitaire, Département des maladies infectieuses (France), n.d., www.invs.sante.fr/surveillance/rougeole/Point_rougeole_200511.pdf.

311 *A pair of independent studies conducted after 2008 measles outbreaks:* Sanny Y. Chen et al., "Health-Care Associated Measles Outbreak in the United States After an Importation: Challenges and Economic Impact," *The Journal of Infectious Diseases*, 2011;203(11) 1507–1509; Sugarman et al., "Measles Outbreak in a Highly Vaccinated Population, San Diego, 2008," 747.

311 *When there was rumor of a single infected visitor in Maryland*: "Maryland Health Officials Investigating Possible Exposures to Measles. Exposures Possible from Tuesday, May 31 through Sunday, June 5, 2011. Areas include Catonsville, Easton, and Baltimore," Maryland Department of Health and Mental Hygiene, June 8, 2011.

311 *According to a May 2011 report in* American Journal of Preventative Medicine: Allison Kempe et al., "Prevalence of Parental Concerns About Childhood Vaccines: The Experience of Primary Care Physicians," *American Journal of Preventive Medicine*, May 2011;40(5): 548–55.

311 *A CDC-sponsored survey*: Allison Kennedy et al., "Confidence About Vaccines In The United States: Understanding Parents' Perceptions," *Health Affairs*, June 2011;30(6): 1141–50.

312 *A significant majority of new parents*: Ibid.

313 *in a* Washington Post *editorial last June*: Seth Mnookin, "An Early Cure for Parents' Vaccine Panic," *The Washington* Post, June 10, 2011, B1.

Bibliography

BOOKS

Aaronovitch, David. *Voodoo Histories: The Role of the Conspiracy Theory in Shaping Modern History.* New York: Riverhead, 2010.

Allen, Arthur. *Vaccine: The Controversial Story of Medicine's Greatest Lifesaver.* New York: W. W. Norton, 2008. First published 2007.

Bettleheim. Bruno. *The Empty Fortress: Infantile Autism and the Birth of the Self.* New York: Free Press, 1967.

———. *The Informed Heart: Autonomy in a Mass Age.* Chicago: University of Chicago Press, 1961.

Bock, Kenneth, and Cameron Stauth. *Healing the New Childhood Epidemics—Autism, ADHD, Asthma, and Allergies.* New York: Ballantine, 2007.

Boylston, Zabdiel. *An Historical Account of the Small-Pox Inoculated in New England, Upon All Sorts of Persons, Whites, Blacks, and of All Ages and Constitutions.* Boston: T. Hancock, 1730.

Bray, R. S. *Armies of Pestilence: The Effects of Pandemics on History.* Cambridge, U.K.: James Clark, 2004.

Cave, Stephanie, and Deborah Mitchell. *What Your Doctor May Not Tell You About Children's Vaccinations.* New York: Grand Central, 2001.

Cohen, Donald J., and Fred R. Volkmar, eds. *Handbook of Autism and Pervasive Developmental Disorders*, 2nd ed. New York: John Wiley & Sons, 1997.

Colgrove, James. *State of Immunity: The Politics of Vaccination in Twentieth-Century America.* Los Angeles: University of California Press, 2006.

Coulter, Harris H., and Barbara Loe Fisher. *A Shot in the Dark: Why the P in the DPT Vaccination May Be Hazardous to Your Child's Health.* New York: Penguin, 1991.

Crookshank, Edward. *History and Pathology of Vaccination: Vol. I: A Critical Inquiry.* Philadelphia: P. Blakiston, Son, & Co., 1889.

de Kruif, Paul. *Microbe Hunters*, 3rd ed. New York: Mariner, 2002. First published 1926 by Harcourt Brace.

Dennett, Daniel C. *Darwin's Dangerous Idea: Evolution and the Meaning of Life.* New York: Simon & Schuster, 1995.

Fenn, Elizabeth. *Pox Americana: The Great Smallpox Epidemic of 1775–82.* New York: Hill & Wang, 2001.

Festinger, Leon. *When Prophecy Fails: A Social and Psychological Study of a Modern Group That Predicted the Destruction of the World.* London: Pinter & Martin, 2008. First published 1956 by University of Minnesota Press.

George, Timothy. *Minamata: Pollution and the Struggle for Democracy in Postwar Japan.* Cambridge: Harvard University Asia Center, 2002.

Grinker, Richard Roy. *Unstrange Minds: Remapping the World of Autism.* Philadelphia: Basic Books, 2007.

Harrington, Anne. *The Cure Within: A History of Mind-Body Medicine.* New York: W. W. Norton, 2008.

Hightower, Jane. *Diagnosis Mercury: Money, Politics, and Poison.* Washington, D.C.: Island, 2008.

Horton, Richard. *Second Opinion: Doctors, Diseases, and Decisions in Modern Medicine.* London: Granta, 2003.

Immunization Safety Review Committee. *Immunization Safety Review: Vaccines and Autism.* Washington, D.C.: National Academy Press, 2004.

Isaacson, Walter. *Einstein: His Life and Universe.* New York: Simon & Schuster, 2007.

James, William. *The Varieities of Religious Experience.* New York: Touchstone, 2004. First published 1902 by The Modern Library.

Jepson, Bryan, with Jane Johnson. *Changing the Course of Autism: A Scientific Approach for Parents and Physicians.* Boulder: First Sentient, 2007.

Kabat, Geoffrey C. *Hyping Health Risks: Environmental Hazards in Daily Life and the Science of Epidemiology.* New York: Columbia University Press, 2008.

Karlen, Arno. *Man and Microbes: Disease and Plagues in History and Modern Times.* New York: Simon & Schuster, 1996.

Keeton. William T., and James L. Gould. *Biological Science,* 4th ed. New York: W. W. Norton, 1986

Kirby, David. *Evidence of Harm: Mercury in Vaccines and the Autism Epidemic—A Medical Controversy.* New York: St. Martin's, 2006. First published 2005.

Kuhn, Thomas. *The Copernican Revolution: Planetary Astronomy in the Development of Western Thought.* Cambridge: Harvard University Press, 1985. First published 1957.

———. *The Structure of Scientific Revolutions*, 3rd ed. Chicago: University of Chicago Press, 1996. First published 1962.

Lehrer, Jonah. *How We Decide.* New York: Houghton Mifflin Harcourt, 2009.

Mather, Cotton. *Some Account of What Is Said of Inoculating or Transplanting the Small Pox.* Boston: Sold by S. Gerrish at his shop in Corn Hill, 1721.

McCarthy, Jenny. *Louder than Words: A Mother's Journey in Healing Autism.* New York: Dutton, 2007.

———. *Mother Warriors: A Nation of Parents Healing Autism Against All Odds.* New York: Dutton, 2008.

McCarthy, Jenny, and Dr. Jerry Kartizenl. *Healing and Preventing Autism: A Complete Guide.* New York: Dutton, 2009.

Mishima, Akio. *Bitter Sea: The Human Cost of Minamata Disease.* Tokyo: Kosei, 1992.

Mooney, Chris. *The Republican War on Science.* New York: Basic Books, 2005.

Neustadt, Richard, and Harvey Fineberg. *The Swine Flu Affair: Decision-Making on a Slippery Slope.* Washington, D.C.: Department of Health, Education, and Welfare, 1978.

Offit, Paul A. *Autism's False Prophets: Bad Science, Risky Medicine, and the Search for a Cure.* New York: Columbia University Press, 2008.

———. *The Cutter Incident: How America's First Polio Vaccine Led to the Growing Vaccine Crisis.* New Haven: Yale University Press, 2007. First published 2005.

———. *Vaccinated: One Man's Quest to Defeat the World's Deadliest Diseases.* Washington, D.C.: Smithsonian Institute Press, 2007.

Olmsted, Dan, and Mark Blaxill. *The Age of Autism: Mercury, Medicine, and a Manmade Epidemic.* New York: Thomas Dunne, 2010.

Oshinsky, David. *Polio: An American Story.* New York: Oxford University Press, 2006.

Panksepp, Jaak. *Affective Neuroscience: The Foundations of Human and Animal Emotions.* New York: Oxford University Press, 1998.

Paulos, John Allen. *Innumeracy: Mathematical Illiteracy and Its Consequences.* New York: Vintage, 1988.

Peters, Stephanie True. *Epidemic! Smallpox in the New World.* New York: Benchmark, 2005.

Plotkin, Stanley, and Walter Orenstein, eds. *Vaccines.* Philadelphia: W. B. Saunders, 1999.

Popper, Karl R. *The Logic of Scientific Discovery.* London: Routledge, 1992. First published 1959 by Hutchinson Education.

Rimland, Bernard. *Infantile Autism.* Englewood Cliffs, New Jersey: Prentice Hall, 1964.

Sagan, Carl. *The Demon-Haunted World: Science as a Candle in the Dark.* New York: Random House, 1996.

Sapolsky, Robert. *Why Zebras Don't Get Ulcers*, 3rd ed. New York: Holt, 2004. First published 1994 by W. H. Freeman.

Scheibner, Viera. *Vaccination: 100 Years of Orthodox Research shows That Vaccines*

Represent a Medical Assault on the Immune System. Santa Fe, New Mexico: New Atlantean, 1996.

Siegel, Bryna. *The World of an Autistic Child: Understanding and Treating Autistic Spectrum Disorders.* New York: Oxford University Press, 1996.

Sears, Robert W. *The Autism Book: What Every Parent Needs to Know About Early Detection, Treatment, Recovery, and Prevention.* New York: Little, Brown, 2010.

———. *The Vaccine Book: Making the Right Decision for Your Child.* New York: Little, Brown, 2007.

Silverstein, Arthur. *Pure Politics, Impure Science: The Swine Flu Affair.* Baltimore: Johns Hopkins University Press, 1982.

Stratton, Kathleen, et al. *Immunization Safety Review: Measles-Mumps-Rubella Vaccine and Autism.* Washington, D.C.: National Academy Press, 2001.

———. *Immunization Safety Review: Thimerosal-Containing Vaccines and Neurodevelopmental Disorders.* Washington, D.C.: National Academy Press, 2001.

Tammet, Daniel. *Embracing the Wide Sky: A Tour Across the Horizons of the Mind.* New York: Free Press, 2009.

Toumey, Christopher. *Conjuring Science: Scientific Symbols and Cultural Meanings in American Life.* New Brunswick, New Jersey: Rutgers University Press, 1996.

Wakefield, Andrew J. *Callous Disregard: Autism and Vaccines: The Truth Behind a Tragedy.* New York: Skyhorse, 2010.

JOURNAL ARTICLES

Articles with Author

Abbasi, Kamran. "Man, Mission, Rumpus." *British Medical Journal* 2001;322(7281): 306.

Abraham, Thomas. "The Price of Poor Pandemic Communication." *British Medical Journal* 2010;340: c2952.

Accordino, Robert, et al. "Morgellons Disease?" *Dermatologic Therapy* 2008;21(1): 8–12.

Afzal, M. A., et al. "Absence of Measles-Virus Genome in Inflammatory Bowel Disease." *Journal of Medical Virology* 1998;55(3): 243–49.

Armstrong, G. L., et al. "Childhood Hepatitis B Virus Infections in the United States Before Hepatitis B Immunization." *Pediatrics* 2001;108(5): 1123–28.

Baker, Jeffrey P. "Immunization and the American Way: 4 Childhood Vaccinations." *American Journal of Public Health* 2000;90(2): 199–207.

———. "Mercury, Vaccines, and Autism: One Controversy, Three Histories." *American Journal of Public Health* 2008;98(2): 244–53.

———. "The Pertussis Vaccine Controversy in Great Britain, 1974–1986." *Vaccine* 2003;21(25–26): 4003–10.

Bakir, F. "Methylmercury Poisoning in Iraq." *Science* 1973;181(4096): 230–41.

Balicer, Ran, et al. "Is Childhood Vaccination Associated with Asthma? A Meta-Analysis of Observational Studies." *Pediatrics* 2007;120(5): e1269–77.

Ball, Philip. "Predicting Human Activity." *Nature* 2010;465: 692.

Banerjee, Abhijit. "A Simple Model of Herd Behavior." *The Quarterly Journal of Economics* 1992;107(3): 797–817.

Barquet, Nicolau, and Pere Domingo. "The Triumph over the Most Terrible of the Ministers of Death." *Annals of Internal Medicine* 1997;127(8): 635–42.

Bell, Vaughan, et al. "Mind Control' Experiences on the Internet: Implications for the Psychiatric Diagnosis of Delusions." *Psychopathology* 2006;39(2): 87–91.

Bernard, Sally, et al. "Autism: A Novel Form of Mercury Poisoning." *Medical Hypotheses* 2001;56(4): 462–71.

Bikhchandani, Sushil, et al. "A Theory of Fads, Fashion, Custom and Cultural Change as Informational Cascades." *The Journal of Political Economy* 1992;100(5): 992–1026.

Bilder, Deborah. "Prenatal, Perinatal, and Neonatal Factors Associated with Autism Spectrum Disorders." *Pediatrics* 2009;123: 1293–1300.

Bisgard, Kristine, et al. "Infant Pertussis: Who Was the Source?" *The Pediatric Infectious Disease Journal* 2004;23: 985–89.

Blake, John B. "The Inoculation Controversy in Boston: 1721–1722." *The New England Quarterly* 1952;25(4): 489–506.

Blume, Stuart. "Anti-Vaccination Movements and Their Interpretations." *Social Science and Medicine* 2006;62(3): 628–42.

Bowman, Brian P. " 'Front Line' Response to *The Vaccine Book.*" *Pediatrics* 2008;123(1): 164–69.

Boyles, Salynn. "HBV Immunization: Groups Weigh In on Newborn Hepatitis B Vaccinations." *Hepatitis Weekly*, November 29, 1999, 2–3.

Braun, M. M., et al. "Infant Immunization with Acellular Pertussis Vaccines in the United States: Assessment of the First Two Years' Data from the Vaccine Adverse Event Reporting System (VAERS)." *Pediatrics* 2000;106(4): e51.

Brooks, Dennis, and Richard Clover. "Pertussis Infection in the United States: Role for Vaccination of Adolescents and Adults." *Journal of the American Board of Family Medicine* 2006;19(6): 603–11.

Brumback, Roger. "The Appalling Poling Saga." *The Journal of Child Neurology* 2008;23(9): 1090–91.

Buie, Timothy, et al. "Evaluation, Diagnosis, and Treatment of Gastrointestinal Disorders in Individuals with ASDs: A Consensus Report." *Pediatrics* 2010;125: S1–S18.

———. "Recommendations for Evaluation and Treatment of Common Gastrointestinal Problems in Children with ASDs." *Pediatrics* 2010;125: S19–S29.

Charman, Tony. "Autism Research Comes of (a Young) Age." *Journal of the American Academy of Childhood and Adolescent Psychiatry* 2010;49(3): 208–9.

Chen, Robert, et al. "The Vaccine Adverse Event Reporting System." *Vaccine* 1994;12(6): 542–50.

Chen, Robert, and Frank DeStefano. "Vaccine Adverse Events: Causal or Coincidental?" *The Lancet* 1998;351(9103): 611–12.

Chen, Sanny Y., et al. "Health-Care Associated Measles Outbreak in the United States After an Importation: Challenges and Economic Impact." *The Journal of Infectious Diseases* 2011;203(11) 1507–1509.

Clements, C. J., and S. Ratzan. "Misled and Confused? Telling the Public About MMR Vaccine Safety." *Journal of Medical Ethics* 2003;29: 22–26.

Cody, Christopher, et al. "Nature and Rates of Adverse Reactions Associated with DTP and DT Immunizations in Infants and Children." *Pediatrics* 1981;68(5): 650–60.

Cohen, Deborah, and Philip Carter. "WHO and the Pandemic Flu 'Conspiracies.' " *British Medical Journal* 2010;340: c2912.

Colville, A., et al. "Withdrawal of a Mumps Vaccine." *European Journal of Pediatrics* 1994;(153): 467–69.

Cooper, Louis Z., et al. "Protecting Public Trust in Immunization." *Pediatrics* 2008;122: 149–53.

Cotter, Suzanne, and Sarah Gee. "National Measles Outbreak Continues to Escalate." *Epi-Insight* 2010;11(1).

Curran, William. "Public Warnings of the Risk in Oral Polio Vaccine." *Public Health and the Law* 1975;65(5): 501–2.

Curtis, Valeria, and Adam Biran. "Dirt, Disgust, and Disease: Is Hygiene in Our Genes?" *Perspectives in Biology and Medicine* 2001;44(1): 17–31.

Dalrymple, Theodore. "An Injection of Fear." *British Medical Journal* 2010;340: c3216.

Dannetun, Eva, et al. "Parents' Reported Reasons for Avoiding MMR Vaccination." *Scandinavian Journal of Primary Health Care* 2005;23(3): 149–53.

Darmofal, David. "Elite Cues and Citizen Disagreement with Expert Opinion." *Political Research Quarterly* 2005;58(3): 381–95.

Davidson, Philip, et al. "Effects of Prenatal and Postnatal Methylmercury Exposure from Fish." *Journal of the American Medical Association* 1998;280(8): 701–7.

Dayan, Gustavo. "The Cost of Containing One Case of Measles: The Economic Impact on the Public Health Infrastructure—Iowa, 2004." *Pediatrics* 2005;116(1): e1–e4.

Deer, Brian. "How the Case Against the MMR Vaccine was Fixed." *British Medical Journal* 2011;342: c5347.

———. "Wakefield's 'Autistic Enterocolitis' Under the Microscope." *British Medical Journal* 2010;340: c1127.

Desjardins, Tracy, and Alan Scoboria. " 'You and Your Best Friend Suzy Put Slime in Ms. Smollett's Desk': Producing False Memories with Self-Relevant Details." *Psychonomic Bulletin and Review* 2007;14(6): 1090–95.

DeStefano, Frank, et al. "Childhood Vaccinations and Risk of Asthma." *Pediatric Infectious Disease Journal* 2002;21(6): 498–504.

Diekema, Douglas S. "Choices Should Have Consequences: Failure to Vaccinate, Harm to Others, and Civil Liability." *First Impressions* 2009;107(3): 90–94.

Durbach, Nadja. " 'They Might as Well Brand Us': Working-Class Resistance to Compulsory Vaccination in Victorian England." *Social History of Medicine* 2000;13(1): 45–62.

Durkin, Maureen, et al. "Advanced Parental Age and the Risk of Autism Spectrum Disorder." *American Journal of Epidemiology* 2008;168(11): 1268–76.

Eickhoff, T. C., and M. Myers. "Workshop Summary: Aluminum in Vaccines." *Vaccine* 2002;20(Suppl): S1–S4.

Evans, Geoffrey. "Update on Vaccine Liability in the United States." *Clinical Infectious Diseases* 2006;42: S130–37.

Faria, Miguel A. "Vaccines (Part II): Hygiene, Sanitation, Immunization, and Pestilential Diseases." *Journal of American Physicians and Surgeons* March/April 2000; Web, August 18, 2010.

Feeney, Mark, et al. "A Case-Control Study of Measles Vaccination and Inflammatory Bowel Disease." *The Lancet* 1997;350(980): 764–66.

Fine, Paul E. M., and Jacqueline A. Clarkson. "Individual Versus Public Priorities in the Determination of Optimal Vaccination Policies." *American Journal of Epidemiology* 1986;124: 1012–20.

Finger, Reginald, and Jerad Shoemaker. "Preventing Pertussis in Infants by Vaccinating Adults." *American Family Physician* 2006;74(3): 382.

Fischoff, Baruch, and Ruth Beyth-Marom. "Hypothesis Evaluation from a Bayesian Perspective." *Psychological Review* 1983;90(3): 239–60.

Fitzpatrick, M. "Evidence of Harm. Mercury in Vaccines and the Autism Epidemic: Medical Controversy." *British Medical Journal* 2005;330(7500): 1154.

Forshaw, Charles. "The History of Inoculation." *British Medical Journal* 1910;2 (2592): 633–34.

Freed, Gary, et al. "Parental Vaccine Safety Concerns in 2009." *Pediatrics* April 2010;125(4): 654–9.

———. "The Process of Public Policy Formulation: The Case of Thimerosal in Vaccines." *Pediatrics* 2002;109(6): 1153–59.

Frye, Richard, et al. "Conflict of Interest Statement Concerning 'Developmental Regression and Mitochondrial Dysfunction in a Child with Autism.' " *Journal of Child Neurology* 2008;23: 1089–90.

Fullerton, K. E., and S. E. Reef. "Commentary: Ongoing Debate over the Safety of the Different Mumps Vaccine Strains Impacts Mumps Disease Control." *International Journal of Epidemiology* 2002;31(5): 983–84.

Gangarosa, E. J., et al. "Impact of Anti-Vaccine Movements on Pertussis Control: The Untold Story." *The Lancet* 1998;351(9099): 356–61.

Geier, David, et al. "RotaTeq Vaccine Adverse Events and Policy Considerations." *Medical Science Monitor* 2008;14(3): PH9–16.

Geier, David, and Mark Geier. "An Assessment of the Impact of Thimerosal on Childhood Neurodevelopmental Disorders." *Pediatric Rehabilitation* 2003;6(2): 97–102.

———. "A Comparative Evaluation of the Effects of MMR Immunization and Mercury Doses from Thimerosal-Containing Childhood Vaccines on the Population Presence of Autism." *Medical Science Monitor* 2004;10(3): PL33–39.

———. "Thimerosal in Childhood Vaccines, Neurodevelopmental Disorders, and Heart Disease in the United States." *Journal of American Physicians and Surgeons* 2003;8(1): 6–11.

———. "The True Story of Pertussis Vaccination: A Sordid Legacy?" *Journal of the History of Medicine* 2002;57: 249–84.

Geier, David A., Rev. Lisa K. Sykes, and Mark R. Geier. "The Role and Treatment of Elevated Male Hormones in Autism Spectrum Disorders." *Autism Science Digest*, April 2011: 71–75.

Gerber, Jeffrey, and Paul Offit. "Vaccines and Autism: A Tale of Shifting Hypotheses." *Clinical Infectious Diseases* 2009;48(4): 456–61.

Ghosh, S., et al. "Detection of Persistent Measles Virus Infection in Crohn's Disease: Current Status of Experiment Work." *Gut* 2001;48: 748–52.

Gilbert, Steven G., and Kimberly S. Grant-Webster. "Neurobehavioral Effects of Developmental Methylmercury Exposure." *Environmental Health Perspectives* 1995;103(Suppl 6): 135–42.

Glanz, Jason, et al. "Parental Refusal of Pertussis Vaccination Is Associated with an Increased Risk of Pertussis Infection in Children." *Pediatrics* 2009;123(6): 1446–51.

Godlee, Fiona, Jane Smith, and Harvey Marcovitch. "Wakefield's Article Linking MMR Vaccine and Autism Was Fraudulent." *British Medical Journal* 2011;342: c.7452.

Gonzalez, Elizabeth. "TV Report on DPT Galvanizes US Pediatricians." *Journal of the American Medical Association* 1982;248(1): 12–22.

Grandjean, Philippe. "Cognitive Deficit in 7-Year-Old Children with Prenatal Exposure to Methylmercury." *Neurotoxicology and Teratology.* 1997;19(6): 417–28.

Gregory, David S. "Pertussis: A Disease Affecting All Ages." *American Family Physician* 2006;74(3): 420–27.

Griffin, Marie, et al. "What Should an Ideal Vaccine Postlicensure Safety System Be?" *American Journal of Public Health* 2009;99(S2): S345–50.

Griffith, A. H. "Permanent Brain Damage and Pertussis Vaccination: Is the End of the Saga in Sight?" *Vaccine* 1989;7: 199–210.

Grob, G. N. "Origins of DSM-I: A Study in Appearance and Reality." *The American Journal of Psychiatry* 1991;148(4): 421–31.

Gross, Liza. "A Broken Trust: Lessons from the Vaccine-Autism Wars." *PLoS Biology* 2009;7(5): 100014–26.

Halsey, Neal A. "Limiting Infant Exposure to Thimerosal in Vaccines and Other Sources of Mercury." *Journal of the American Medical Association* 1999;282: 1763–66.

Harris, Evan. "After Wakefield: The Real Questions That Need Addressing." *British Medical Journal* 2010;340: c2829.

Hellwege, Jean. "Thimerosal Plaintiffs Push on Despite Setbacks." *Trial* 2004;40(5): 96.

Henderson, Donald A., et al. "Paralytic Disease Associated with Oral Polio Vaccines." *Journal of the American Medical Association* 1964;190(1): 41–48.

———. "Public Health and Medical Responses to the 1957–58 Influenza Pandemic." *Biosecurity and Bioterrorism* 2009;7(3): 265.

Herbert, Martha, and Terri Arranga. "Interview with Dr. Martha Herbert—Autism: A Brain Disorder or a Disorder That Affects the Brain?" *Medical Veritas* 2006;3: 1182–94.

Herbert, Martha, and Chloe Silverman. "Autism and Genetics." *GeneWatch* 2003;16(1).

Hewitson, L, et al. "WITHDRAWN: Delayed Acquisition of Neonatal Reflexes in Newborn Primates Receiving a Thimerosal-Containing Hepatitis B Vaccine: Influence of Gestational Age and Birth Weight." *Neurotoxicology* October 2, 2009; Web, March 9, 2010.

Hooker, Claire. "Diptheria, Immunisation and the Bundaberg Tragedy: A Study of Public Health in Australia." *Health and History* 2000;2: 52–78.

Ibrahim, Samar, et al. "Incidence of Gastrointestinal Symptoms in Children with Autism: A Population-Based Study." *Pediatrics* 2009;124: 680–86.

Iizuka, Masahiro, et al. "Absence of Measles Virus in Crohn's Disease." *The Lancet* 1995;345(8943): 199.

Jefferson, T., et al. "Relation of Study Quality, Concordance, Take Home Message, Funding, and Impact in Studies of Influenza Vaccines: Systematic Review." *British Medical Journal* 2009;338: b354.

Kanner, Leo. "Autistic Disturbances of Affective Contact." *The Nervous Child* 1943: 217–50.

Kaufman, Martin. "The American Anti-Vaccinationists and Their Arguments." *The Bulletin of the History of Medicine* 1967;41(5): 463–78.

Kempe, Allison, et al. "Prevalence of Parental Concerns About Childhood Vaccines: The Experience of Primary Care Physicians." *American Journal of Preventive Medicine,* May 2011;40(5): 548–55.

Kennedy, Allison, et al. "Confidence About Vaccines in the United States: Understanding Parents' Perceptions." *Health Affairs,* June 2011;30(6): 1141–50.

King, Melissa, and Peter Bearman. "Diagnostic Change and the Increased Prevalence of Autism." *International Journal of Epidemiology.* 2009;38: 1224–34.

Kogan, Michael. "Spectrum Disorder Among Children in the United States, A National Profile of the Health Care Experiences and Family Impact of Autism." *Pediatrics* 2008;122(6): e1149–58.

Koplan, J. P., et al. "Pertussis Vaccine: An Analysis of Benefits, Risks and Costs." *New England Journal of Medicine* 1979;301(17): 906–11.

Krajcik, Joseph S., and LeeAnn M. Sutherland. "Supporting Students in Developing Literacy in Science." *Science* 2010;328(5977): 456–59.

Krigsman, Arthur, et al. "Clinical Presentation and Histologic Findings at Ileocolonoscopy in Children with Autistic Spectrum Disorder and Chronic Gastrointestinal Symptoms." *Autism Insights* 2010;2: 1–11.

Krugman, Richard. "Immunization 'Dyspractice': The Need for 'No Fault' Insurance." *Pediatrics* 1975;56(2): 159–60.

Kulenkampff, M., et al. "Neurological Complications of Pertussis Inoculation." *Archives of Disease in Childhood* 1974;49: 46–49.

Kuran, Timur, and Cass Sunstein. "Availability Cascades and Risk Regulation." *Stanford Law Review* 1999;51(4): 683–768.

Laeth, Nasir. "Reconnoitering the Antivaccination Web Sites: News from the Front." *Journal of Family Practice* 2000;49(8): 731–33.

Langmuir, Alexander. "The Epidemic Intelligence Service of the Center for Disease Control." *Public Health Reports* 1980;95(5): 470–77.

Larsson, Heidi Jeanet, et al. "Risk Factors for Autism: Perinatal Factors, Parental Psychiatric History, and Socioeconomic Status." *American Journal of Epidemiology* 2005;161: 916–25.

Leask, Julie-Anne, and Simon Chapman. "An Attempt to Swindle Nature: Press Anti-Immunisation Reportage, 1993–2007." *Australian and New Zealand Journal of Public Health* 1998;22: 17–26.

Lee, Brent, et al. "Measles Hospitalizations, United States, 1985–2002." *Journal of Infectious Diseases* 2004;189(Suppl 1): S210–15.

Lewis, Justin, and Tammy Speers. "Misleading Media Reporting? The MMR Story." *Nature Reviews: Immunology* 2003;3: 913–18.

Liu, Ka-Yuet, et al. "Social Influence and the Autism Epidemic." *American Journal of Sociology* 2010;115(5): 1387–34.

MacDonald, T. T., and P. Domizio. "Autistic Enterocolitis; Is It a Histopathological Entity?" *Histopathology* 2007;50(3): 371–79.

Madsen, Kreesten Meldgaard, et al. "A Population-Based Study of Measles,

Mumps, and Rubella Vaccination and Autism." *New England Journal of Medicine* 2002;347(19): 1477–82.

Marris, Emma. "Mysterious 'Morgellons Disease' Prompts US Investigation." *Nature Medicine* 2006;12(9): 982.

Mazumdar, Soumya, et al. "The Spatial Structure of Autism in California: 1993–2001." *Health and Place* 2010;16(3): 539–46.

McDonald, Kara, et al. "Delay in Diphtheria, Pertussis, Tetanus Vaccination Is Associated with a Reduced Risk of Childhood Asthma." *The Journal of Allergy and Clinical Immunology* 2008;121(3): 626–31.

McLure, Iain. "How Can We Be Confident That the Children with 'Autistic Enterolcolitis' Have Autism?" *British Medical Journal* April 28, 2010; Web, August 9, 2010.

Michaels, David, and Celeste Monforton. "Manufacturing Uncertainty: Contested Science and the Protection of the Public's Health and Environment." *Public Health Matters* 2005;95(S1): S39–S48.

Mills, Edward. "Systematic Review of Qualitative Studies Exploring Parental Beliefs and Attitudes Toward Childhood Vaccination Identifies Common Barriers to Vaccination." *Journal of Clinical Epidemiology* 2005;58(11): 1081–88.

Mrozek-Budzyn, D., et al. "Lack of Association Between Measles-Mumps-Rubella Vaccination and Autism in Children: A Case-Control Study." *Pediatric Infectious Disease Journal* 2010;29(5): 397–400.

Mulholland, E. Kim. "Measles in the United States, 2006." *New England Journal of Medicine* 2006;355(5): 440–43.

Murch, Simon, et al. "Retraction of an Interpretation." *The Lancet* 2004;363(9411): 750.

Myers, Gary. "Prenatal Methylmercury Exposure from Ocean Fish Consumption in the Seychelles Child Development Study." *The Lancet* 2003;361: 1686–92.

Nelson, Karin B., and Margaret L. Bauman. "Thimerosal and Autism?" *Pediatrics* 2003;111: 674–79.

Neraas, Mary Beth. "The National Childhood Vaccine Injury Act of 1986: A Solution to the Vaccine Liability Crisis?" *Washington Law Review* 1988;63(149): 149–69.

Nesbet, Matthew C., and Chris Mooney. "Framing Science." *Science* 2007;316(5821): 56.

Newschaffer, Craig J. et al. "The Epidemiology of Autism Spectrum Disorders," *Annual Review of Public Health.* April 2007;28:235–58.

Offit, Paul A. "The Cutter Incident 50 Years Later." *New England Journal of Medicine* 2005;352(14): 1411–12.

———. "Letters: Preventing Harm from Thimerosal in Vaccines." *Journal of the American Medical Association* 2000;283: 2104

Offit, Paul A., and Rita K. Jew. "Addressing Parents' Concerns: Do Vaccines Contain Harmful Preservatives, Adjuvants, Additives, or Residuals?" *Pediatrics* 2003;112(6): 1394–401.

Offit, Paul A., and Charlotte Moser. "The Problem with Dr. Bob's Alternative Vaccine Schedule." *Pediatrics* 2009;123(1): e164–69.

Olin, Patrick, et al. "Declining Pertussis Incidence in Sweden Following the Introduction of Acellular Pertussis Vaccine." *Vaccine* 2003;21(17–18): 2015–21.

Omer, Saad B., et al. "Geographic Clustering of Nonmedical Exemptions to School Immunization Requirements and Associations with Geographic Clustering of Pertussis." *American Journal of Epidemiology* 2008;168(12): 1389–96.

———. "Policies with Pertussis Incidence Requirements: Secular Trends and Association of State Nonmedical Exemptions to School Immunization." *Journal of the American Medical Association* 2006;296(14): 1757–63.

———. "Vaccine Refusal, Mandatory Immunization, and the Risks of Vaccine-Preventable Diseases." *New England Journal of Medicine* 2009;360(19): 1981–88.

Oransky, Ivan. "Paul Offit." *The Lancet* 1999;366(9502).

Orenstein, W. A., et al. "Measles Elimination in the United States." *Journal of Infectious Diseases* 2004;189(Suppl 1): S1–3.

Osborne, Jonathan. "Arguing to Learn in Science: The Role of Collaborative, Critical Discourse." *Science* 2010;328(5977): 463–66.

Parker, Amy, et al. "Implications of a 2005 Measles Outbreak in Indiana for Sustained Elimination of Measles in the United States." *New England Journal of Medicine* 2006;355(5): 447–55.

Parker, Sarah, et al. "Review of Published Original Data Thimerosal-Containing Vaccines and Autistic Spectrum Disorder." *Pediatrics* 2004;114(3): 793–804.

Patriarca, Peter A., and Judy A. Beeler. "Measles Vaccination and Inflammatory Bowel Disease." *The Lancet* 1995;345(8957): 1062–63.

Pearson, David, et al. "Literacy and Science: Each in the Service of the Other." *Science* 2010;328(5977): 459–63.

Pichichero, Michael, et al. "Mercury Concentrations and Metabolism in Infants Receiving Vaccines Containing Thimerosal: A Descriptive Study." *The Lancet* 2002;360: 1737–41.

Plotkin, Stanley A. "Letters: Preventing Harm from Thimerosal in Vaccines." *Journal of the American Medical Association* 2000;283:2104–5.

Plotkin, Stanley A., and M. Cadoz. "The Acellular Pertussis Vaccine Trials: An Interpretation." *Pediatric Infectious Disease Journal* 1997;16(5): 508–17.

Poling, Jon S. "Correspondence on 'Developmental Regression and Mitochondrial Dysfunction in a Child with Autism.' " *Journal of Child Neurology* 2008;23: 1089.

Poling, Jon S., et al. "Developmental Regression and Mitochondrial Dysfunction in a Child with Autism." *Journal of Child Neurology* 2006;21(2): 170–73.

Potter, C. W. "A History of Influenza." *Journal of Applied Microbiology* 2006;91(4): 573–79.

Ridgway, Derry. "No-Fault Vaccine Insurance: Lessons from the National Vaccine Injury Compensation Program." *Journal of Health Politics, Policy, and Law* 1999;24(1): 59–90.

Rouse, Andrew. "Correspondence: Autism, Inflammatory Bowel Disease, and MMR Vaccine." *The Lancet* 1998;351(9112): 1356.

Rubin, Daniel B., and Sophie Kasimow. "The Problem of Vaccination Noncompliance: Public Health Goals and the Limitations of Tort Law." *Michigan Law Review* 2009;107: 114–19.

Savely, Virginia, et al. "The Mystery of Morgellons Disease." *American Journal of Clinical Dermatology* 2006;7(1): 1–5.

Schaffner, William. "Update on Vaccine-Preventable Diseases: Are Adults in Your Community Adequately Protected?" *Journal of Family Practice* 2008;57(4): S1–S12.

Schwartz, Jason L. "Unintended Consequences: The Primacy of Public Trust in Vaccination." *Michigan Law Review* 2009;107: 100–4.

Scoboria, Alan, et al. "The Effects of Prevalence and Script Information on Plausibility, Belief, and Memory of Autobiographical Events." *Applied Cognitive Psychology* 2006;20: 1049–64.

———. "So That's Why I Don't Remember: Normalizing Forgetting of Childhood Events Influences False Autobiographical Beliefs but Not Memories." *Memory* 2007;15(8): 801–13.

———. "Suggesting Childhood Food Illness Results in Reduced Eating Behavior." *Acta Psychological* 2008;128(2): 304–9.

Singleton, J. A., et al. "An Overview of the Vaccine Adverse Event Reporting System (VAERS) as a Surveillance System." *Vaccine* 1999;17(22): 2908–17.

Smith, Martin. "National Childhood Vaccine Injury Compensation Act." *Pediatrics* 1988;82: 264–69.

Smith, Michael, et al. "Media Coverage of the Measles-Mumps-Rubella Vaccine and Autism Controversy and Its Relationship to MMR Immunization Rates in the United States." *Pediatrics* 2008;121(4): e836–43.

Smith, Philip, et al. "Children Who Have Received No Vaccines: Who Are They and Where Do They Live?" *Pediatrics* 2004;114: 187–95.

Smith Rebecca G., et al. "Advancing Paternal Age Is Associated with Deficits in Social and Exploratory Behaviors in the Offspring: A Mouse Model." *PLoS ONE* 2009;4(12): e8456.

Spinney, Laura. "UK Autism Fracas Fuels Calls for Peer Review Reform." *Nature Medicine* 2004;10: 321.

Spitzer, R. L. "The Diagnostic Status of Homosexuality in the DSM-III: A Reformulation of the Issues." *The American Journals of Psychiatry* 1981;138: 210–15.

Stehr-Green, Paul, et al. "Autism and Thimerosal-Containing Vaccines: Lack of Consistent Evidence for an Association." *American Journal of Preventive Medicine* 2003;25(2): 101–6.

Stewart, Gordon T. "Vaccination Against Whooping Cough: Efficacy Versus Risks." *The Lancet* 1977;1(8005): 234–37.

Stott, Carol, et al. "MMR and Autism in Perspective: the Denmark Story." *Journal of American Physicians and Surgeons* 2004;9(3): 89–91.

Subbotsky, Eugene. "Can Magical Intervention Affect Subjective Experiences? Adults' Reactions to Magical Suggestion." *British Journal of Psychology* 2009;100(3): 517–37.

Sugerman, David E., et al. "Measles Outbreak in a Highly Vaccinated Population, San Diego, 2008: Role of the Intentionally Undervaccinated." *Pediatrics* 2010;125(4): 747–55.

Taylor, Brent, et al. "Autism and Measles, Mumps, and Rubella Vaccine: No Epidemiological Evidence for a Causal Association." *The Lancet* 1999;353 (9169): 2026–29.

Thompson, Nick P., et al. "Crohn's Disease, Measles, and Measles Vaccination: A Case-Control Failure." *The Lancet* 1996;347(8996): 263.

Thompson, Kimberly, and Radbound Tebbers. "Retrospective Cost-Effectiveness Analysis for Polio Vaccination in United States." *Risk Analysis* 2006;26(6): 1423–40.

———. "Is Measles Vaccination a Risk Factor for Inflammatory Bowel Disease?" *The Lancet* 1995;345(8957): 1071–74.

Thompson, William W., et al. "Early Thimerosal Exposure and Neuropsychological Outcomes at 7 to 10 Years." *New England Journal of Medicine* 2007;357(13): 1281–92.

Trevelyan, Barry, et al. "The Spatial Dynamics of Poliomyelitis in the United States: From Epidemic Emergence to Vaccine-Induced Retreat, 1910–1971." *Annals of the Association of American Geographers* 2005;95(2): 269–93.

Tversky, Amos, and Daniel Kahneman. "Belief in the Law of Small Numbers." *Psychological Bulletin* 1971;76(2): 105–10.

Uhlmann, V., et al. "Potential Viral Pathogenic Mechanism for New Variant Inflammatory Bowel Disease." *Journal of Clinical Pathology: Molecular Pathology* 2002;55(2): 84–90.

Van Damme, Wim, et al. "Measles Vaccination and Inflammatory Bowel Disease." *The Lancet* 1997;350(9093): 1774–75.

Varricchio, F. "Understanding Vaccine Safety Information from the Vaccine Adverse Event Reporting System." *The Pediatric Infectious Disease Journal* 2004;23(4): 287–94.

Verstraeten, Thomas. "Thimerosal, the Centers for Disease Control and Prevention, and GlaxoSmithKline." *Pediatrics* 2004;113: 932.

Verstraeten, Thomas, et al. "Safety of Thimerosal-Containing Vaccines: A Two-Phased Study of Computerized Health Maintenance Organization Databases." *Pediatrics* 2003;12(5): 1039–48.

Vickers, David, et al. "Whole-Cell and Acellular Pertussis Vaccination Programs and Rates of Pertussis Among Infants and Young Children." *Canadian Medical Association Journal* 2006;175(10): 1213.

Vila-Rodriguez, Fidel. "Delusional Parasitosis Facilitated by Web-Based Dissemination." *American Journal of Psychiatry* 2008;165(12): 1612.

Wakefield, Andrew J. "Correspondence: Author's reply: Autism, Inflammatory Bowel Disease, and MMR Vaccine." *The Lancet* 1998;351(9106): 908.

———. "Crohn's Disease: Pathogenesis and Persistent Measles Virus Infection." *Gastroenterology* 1995;108(3): 911–16.

———. "That Paper." *Autism File* 2009;33: 38–44.

Wakefield, Andrew J., et al. "Evidence of persistent measles infection in Crohn's disease." *Journal of Medical Virology* 1993;39(4): 345–53.

———. "Ileal-Lymphoid-Nodular Hyperplasia, Non-Specific Colitis, and Pervasive Developmental Disorder in Children." *The Lancet* 1998;351: 637–41.

———. "MMR—Responding to Retraction." *The Lancet* 2004;363(9417): 1327–28.

———. "Response to Doctor Ari Brown and the Immunization Action Coalition." *Medical Veritas* 2009;6: 1907–24.

Wakefield, Andrew J, and Scott Montgomery. "Measles, Mumps, Rubella Vaccine: Through a Glass, Darkly." *Adverse Drug Reactions* 2000;19(4): 1–19.

Wallinga, J., et al. "A Measles Epidemic Threshold in a Highly Vaccinated Population." *PLoS Medicine* 2005;2(11): e316.

Weaver, Kimberlee, et al. "Inferring the Popularity of an Opinion from Its Familiarity: A Repetitive Voice Can Sound Like a Chorus." *Journal of Personality and Social Psychology* 2007;92(5): 821–33.

Whitson, Jennifer, and Adam Galinsky. "Lacking Control Increases Illusory Pattern Perception." *Science* 2008;322(5898): 115–17.

Wilson, Kumanan. "Association of Autism Spectrum Disorder and the Measles, Mumps, and Rubella Vaccine." *Pediatric and Adolescent Medicine* 2003;157: 628–34.

Wolfe, Robert M., et al. "Content and Design Attributes of Antivaccination Web Sites." *Journal of the American Medical Association* 2002;287(24): 3245–48.

Wrangham, Theresa, and Vicky Debold. "Are Federal Research Dollars Being Spent Wisely?" *Autism File* 2009;32: 120–22.

Wright, Nicholas. "Does Autistic Enterocolitis Exist?" *British Medical Journal* 2010;340: c1807.

Zhou, Weigong, et al. "Surveillance for Safety After Immunization: Vaccine Adverse Event Reporting System (VAERS)." *Morbidity and Mortality Weekly Report* 2003;52(ss01): 1–24.

Zimmerman, Richard, et al. "Routine Vaccines Across the Life Span." *Journal of Family Practice* 2007;56(2): S18–S20.

———. "Vaccine Criticism on the World Wide Web." *Journal of Medical Internet Research* 2005;7(2): e17.

Articles Without Author

"Achievements in Public Health, 1900–1999." Centers for Disease Control and Prevention. *Morbidity and Mortality Weekly Report* 1998;(48.29): 621–48.

"Brief Report: Imported Measles Case Associated with Nonmedical Vaccine Exemption—Iowa, March 2004." Centers for Disease Control and Prevention. *Morbidity and Mortality Weekly Report* 2004;53(11): 244–46.

"Confirmed Measles Cases in England and Wales—An Update to End May 2008." Health Protection Agency (U.K.). *Health Protection Report* June 2008:2(25), Web, September 21, 2010.

"The Cutter Polio Vaccine Incident." *Yale Law Journal* 1955;65(2): 262.

"Impact of the 1999 AAP/USPHS Joint Statement on Thimerosal in Vaccines on Infant Hepatitis B Vaccination Practices."Centers for Disease Control and Prevention. *Morbidity and Mortality Weekly Report* 2001;50(6): 94–97.

"Measles—United States, January–May 20, 2011." Centers for Disease Control and Prevention. *Morbidity and Mortality Weekly Report,* May 27, 2011;60(20): 666–68.

"Measles: United States, January–July 2008." Centers for Disease Control and Prevention. *Morbidity and Mortality Weekly Report* 2008;57(33): 893–96.

"Notice to Readers: Availability of Hepatitis B Vaccine That Does Not Contain Thimerosal as a Preservative." Centers for Disease Control and Prevention. *Morbidity and Mortality Weekly Report* 1999; 48(35): 780–82.

"Notice to Readers: Thimerosal in Vaccines: A Joint Statement of the American Academy of Pediatrics and the Public Health Service." Centers for Disease Control and Prevention. *Morbidity and Mortality Weekly Report* 1999;48(26): 563–65.

"Pertussis Vaccines: WHO Position Paper." World Health Organization. *Weekly Epidemiological Record* 2005;80(4): 29–40.

"Prevalence of Autism Spectrum Disorders—Autism and Developmental Disabilities Monitoring Network, 14 Sites, United States, 2002." Centers for Disease Control and Prevention. *Morbidity and Mortality Weekly Report* 2007;56(SS01): 12–28.

"Prevalence of Autism Spectrum Disorders—Autism and Developmental Disabilities Monitoring Network, Six Sites, United States, 2000." Centers for

Disease Control and Prevention. *Morbidity and Mortality Weekly Report* 2007;56(SS01): 1–11.

"Prevalence of Autism Spectrum Disorders—Autism and Developmental Disabilities Monitoring Network, United States, 2006." Centers for Disease Control and Prevention. *Morbidity and Mortality Weekly Report* 58, no. SS10 December 18, 2009: 1–20.

"Preventing Tetanus, Diphtheria, and Pertussis Among Adults: Use of Tetanus Toxoid, Reduced Diphtheria Toxoid and Acellular Pertussis Vaccine." Centers for Disease Control and Prevention. *Morbidity and Mortality Weekly Report. Recommendations and Reports* 2006;55(RR17).

"Strengthening the Credibility of Clinical Research." *The Lancet* 2010;375: 1225.

"Surveillance for Safety After Immunization: Vaccine Adverse Event Reporting System (VAERS)—United States, 1991–2001." Centers for Disease Control and Prevention. *Morbidity and Mortality Weekly Report* 2003;52(SS01): 1–24.

"Thimerosal in Vaccines—An Interim Report to Clinicians." *Pediatrics* 1999;104(3): 570–74.

NEWSPAPER AND MAGAZINE ARTICLES

Articles with Author

Allday, Erin. "H1N1: The Report Card." *Reader's Digest*, February 2010.

———. "Not Enough Bay Area Kids Vaccinated, Docs Say." *San Francisco Chronicle*, August 1, 2009.

Allen, Arthur. "The Not-So-Crackpot Autism Theory." *The New York Times Magazine*, November 11, 2002, 66.

Allen, Jane. "Shots in the Dark." *Los Angeles Times*, October 18, 1999.

Allen, Nick. "MMR-Autism Link Doctor Andrew Wakefield Defends Conduct at GMC Hearing." *The Daily Telegraph* (London), March 27, 2008.

Barnes, Ralph M., Audrey L. Alberstadt, and Lesleh E. Keilholtz. "How to Think About Scientific Claims: A Study of How Non-Scientists Evaluate Science Claims." *Skeptic*, January 1, 2009.

Bettelheim, Bruno. "Joey: A Mechanical Boy." *Scientific American*, March 1959.

Biba, Erin. "H1N1 Flu Shot: 3 Major Fears Debunked." *Wired*, November 2009.

———. "How to Win an Argument About Vaccines." *Wired*, November 2009.

Blair, William. "Capital Expects Further Delay in Polio Program." *The New York Times*, May 13, 1955, 1.

———. "U.S. Lays Defects in Polio Program to Mass Output." *The New York Times*, June 10, 1955, 1.

Blaxill, Mark. "Unjustly Accused." *USA Today*, February 16, 2010.

Bowers, Page. "Itching for Answers to a Mystery Condition." *Time*, July 28, 2006.

Bowman, Lee. "Thousands of Unvaccinated Children Enter Schools." Scripps Howard News Service, September 4, 2008.

Brennan, Patricia. "Lea Thompson Family, Career, Kudos: She Has Them All." *The Washington Post*, June 15, 1986, 8.

Breslin, Meg McSherry. "Daughter's Murder Puts Focus on Toll of Autism." *Chicago Tribune*, June 9, 2006, 1.

———. "Painful Questions of Blame; Parents, Doctors and the Disputed Link Between Vaccines and Autism." *Chicago Tribune*, June 25, 2006, 1.

Brown, Wendy. "Doctor Now Focuses on Disputed Skin Disease." *The Santa Fe New Mexican*, December 14, 2005, A2.

Bruyn Jones, Sarah. "Whooping Cough Outbreak Closes Blue Mountain School in Floyd Co." *The Roanoke Times*, April 5, 2011.

Buckley, Patricia Morris. "Dr. Bernard Rimland Is Autism's Worst Enemy." *San Diego Jewish Journal*, October 2002.

Callahan, Patricia, and Trine Tsouderos. "Autism Doctor: Troubling Record Trails Doctor Treating Autism." *Chicago Tribune*, May 22, 2009, 1.

Carlyle, Erin. "Rare Hib Disease Increases in Minnesota: Is the Anti-Vaccine Movement to Blame?" *Minnesota City Pages*, June 3, 2009.

Charter, David. "I'd Welcome Inquiry, Says MMR Doctor." *The Times* (London), February 23, 2004.

Chong, Jia-Rui. "Morgellons Study Begins in Calif.; Sufferers of Crawling Sensations Hope Data Validate the Disease." *Los Angeles Times*, January 19, 2008, A10.

Cone, Marla. "Autism Clusters Found in California's Major Cities." *Scientific American*, January 6, 2010.

Crook, Amanda. "Mum's MMR Jab Guilt for Tot." *The Manchester Evening News* (U.K.), May 31, 2008.

Davies, Lawrence. "2 Polio Victims Win Vaccine Suit but Cutter Is Held Not Negligent." *The New York Times*, January 18, 1958, 1.

Day, Michael. "Families Defend Anti-MMR Doctor Against 'Witch-Hunt.' " *The Sunday Telegraph* (London), February 22, 2004, 4.

Deardorff, Julie. "A Little Shot of Skepticism Won't Hurt a Bit." *Chicago Tribune*, August 31, 2008, Q7.

Deer, Brian. "Doctors in MMR Scare Face Public Inquiry." *The Sunday Times* (London), December 12, 2004, 5.

———. "Fresh Doubts Cast on MMR Study Data." *The Sunday Times* (London), April 25, 2004, 11.

———. "Key Ally of MMR Doctor Rejects Autism Link." *The Sunday Times* (London), March 7, 2004, 1.

———. "MMR Doctor's Irish Ally Rejects Link to Autism." *The Sunday Times* (London), March 7, 2004, 1.

———. "MMR Jab Scare Research Dealt a 'Killer Blow.' " *The Sunday Times* (London), March 6, 2005, 2.

———. "MMR Scare Doctor Faces List of Charges." *The Sunday Times* (London), September 11, 2005, 13.

———. "MMR Scare Doctor Planned Rival Vaccine." *The Sunday Times* (London), November 14, 2004, 8.

———. "MMR: The Truth Behind the Crisis." *The Sunday Times* (London), February 22, 2004, 12.

———. "Scientists Desert MMR Maverick." *The Sunday Times* (London), March 7, 2004, 11.

———. "Truth of the MMR Vaccine Scandal." *The Sunday Times* (London), January 24, 2010.

Derbyshire, David. "MMR Scare Scientist Warns of Impending Measles Epidemic." *The Daily Telegraph* (London), October 31, 2003.

Devita-Raeburn, Elizabeth. "The Morgellons Mystery." *Psychology Today*, March 1, 2007.

Devlin, Kate. "I Have Been Struck Off to Keep Me Quiet, Says MMR Doctor." *The Daily Telegraph* (London), May 25, 2010, 6.

Dominus, Susan. "The Denunciation of Dr. Wakefield." *The New York Times Magazine*, April 24, 2011, MM36. (A version of this article appeared online on April 20, 2011 at http://www.nytimes.com/2011/04/24/magazine/mag-24Autism-t.html with the headline, "The Crash and Burn of an Autism Guru.")

Engel, Leonard. "The Salk Vaccine: What Caused the Mess?" *Harper's*, August 1955, 27–33.

Estridge, Bonnie. "We Feel Betrayed by This MMR Witch-Hunt." *The Evening Standard* (London), February 24, 2004, A25.

Fagin, Dan. "Tattered Hopes—A $30 Million Federal Study of Breast Cancer and Pollution on LI Has Disappointed Activists and Scientists Alike; What Went Wrong?" *Newsday*, July 28, 2002, A3.

———. "Tattered Hopes—Still Searching—A Computer Mapping System Was Supposed to Help Unearth Information About Breast Cancer and the Environment." *Newsday*, July 30, 2002, A6.

———. "Tattered Hopes—Study in Frustration—Ambitious Search for Links Between Pollution and Breast Cancer on LI Was Hobbled from the Start, Critics Say." *Newsday*, July 29, 2002, A6.

Fagone, Jason. "Will This Doctor Hurt Your Baby?" *Philadelphia*, May 27, 2009.

Ferriman, Annabel. "Health: Why Another Needle, Mummy?; Warnings of a Measles Epidemic Are Causing Parents Concern and Confusion." *The Independent* (London), October 11, 1994, 23.

Firestone, David. "G.O.P. Leaders Promise Repeal of Provisions Hidden in Bill." *The New York Times*, January 11, 2003, A9.

Firestone, David, and Richard Oppel. "THE FINE PRINT: Special Concerns;

Critics Say Security Bill Favors Special Interests." *The New York Times*, November 19, 2002, A28.

Fisher, Barbara Loe. "In the Wake of Vaccines: The Founder of the National Vaccine Information Center Raises Profound Questions About the Relationship Between the Increase in Childhood Vaccinations and the Rise in Chronic Illness." *Mothering*, September 2004, 126.

Fitzpatrick, Michael. "Jabs and Junk Science: Parents-Led Anti-Vaccination Groups Are Becoming Hugely Influential. But the Information They Provide Is Often Extremely Dodgy." *The Guardian* (London), September 8, 2001, 9.

Franke-Ruta, Garance. "George Washington's Bioterrorism Strategy: How We Handled It Last Time." *Washington Monthly*, December 2001.

Frankel, Glenn. "Charismatic Doctor at Vortex of Vaccine Dispute; Experts Argue over Findings, but Specialist Sees Possible MMR Link to Autism." *The Washington Post*, July 11, 2004, A1.

Fraser, Lorraine. "The Damning Evidence That the Medical Establishment Has Chosen to Ignore." *The Mail on Sunday* (London), April 9, 2000, 8.

———. "Parents Left Stunned as MMR Doctor Is Forced Out." *The Sunday Telegraph* (London), December 2, 2001, 4.

Fulbright, Leslie. "A Child's Return from Autism." *San Francisco Chronicle*, May 25, 2005.

Fumento, Michael. "Immune to Reason: Robert F. Kennedy Jr.'s Dangerous Vaccine Conspiracy Theories." *The Wall Street Journal*, June 24, 2005.

———. "There Is No Thimerosal-Autism Conspiracy." *The Wall Street Journal*, July 14, 2005.

Furman, Bess. "One Firm's Vaccine Barred; 6 Polio Cases Are Studied." *The New York Times*, April 28, 1955, 1.

———. "6 Vaccine Makers Get U.S. Licenses." *The New York Times*, April 13, 1955, 1.

———. "U.S. Blames Its Own Tests in Cutter Vaccine Incident." *The New York Times*, August 26, 1955, 1.

Gallegos, Alicia. "South Bend Couple Loses Baby to Pertussis." *South Bend Tribune* (Indiana), February 28, 2010.

Gawande, Atul. "The Cancer-Cluster Myth." *The New Yorker*, February 8, 1999, 34–37.

Giles, Jim. "Desperate Measures: The Lure of an Autism Cure." *New Scientist*, June 29, 2010.

Goldacre, Ben. "Comment and Debate: The MMR Sceptic Who Just Doesn't Understand Science." *The Guardian* (London), November 2, 2005, 32.

Greenfield, Karl Taro. "The Autism Debate: Who's Afraid of Jenny McCarthy?" *Time*, February 25, 2010.

Grinker, Richard Roy. "Disorder Out of Chaos." *The New York Times*, February 9, 2010, A25.

———. "Rare No More; With Research Up and Stigma Down, Autism Sheds More of Its Mystery." *The Washington Post*, February 27, 2007, F1.

Groopman, Jerome. "Faith and Healing." *The New York Times*, January 27, 2008.

Gross, Jane and Stephanie Strom. "Debate over Cause of Autism Strains a Family and Its Charity." *The New York Times*, June 18, 2007, A1.

Hadwen, Walter. "The Fraud of Vaccination." *Truth*, January 3, 1923.

———. "Sanitation v. Vaccination: The Origin of Smallpox." *Truth*, January 17, 1923.

Hall, Celia. "Campaign to Persuade Parents That the MMR Jab Is Safe." *The Daily Telegraph* (London), January 23, 2001.

———. "Parents 'Not Told About Side-Effects of Child Vaccines'." *The Independent* (London), October 6, 1992, 4.

Hall, Landon. "Defrocked Doctor Answers Critics About Vaccines." *The Orange County Register*, August 31, 2010.

Hargreaves, Ian. "The Science of Understanding." *The Times Higher Education* (London), July 4, 2003, 14.

Hari, Johann. "This Deadly Resistance to Vaccination." *The Independent* (London), December 10, 2007, 30.

Harris, Gardiner. "Opening Statements in Case on Autism and Vaccinations." *The New York Times*, June 12, 2007, A21.

Harris, Gardiner, and Anahad O'Connor. "On Autism's Cause, It's Parents vs. Research." *The New York Times*, June 25, 2005, A1.

Healy, Melissa. "Disease: Real or State of Mind?; Morgellons Sufferers Describe Wild Symptoms of a Disorder That Many Doctors Doubt Exists." *Los Angeles Times*, November 13, 2006, F1.

Hope, Jenny. "The Flaw That Let Measles Back in Force." *The Daily Mail* (London), August 2, 1994, 38.

———. "Mass Measles Jabs in Epidemic Alert." *The Daily Mail* (London), July 29, 1994, 24.

Hope, Jenny, and James Chapman. "How the MMR Experts Are Tied to the Drug Firms." *The Daily Mail* (London), March 11, 2003, 2.

Hunt, Liz. "Measles Campaign to Avert Epidemic; Vaccine for Seven Million Children." *The Independent* (London), July 29, 1994, 7.

———. "Vaccine Ban Blow to Fight Against Epidemic." *The Independent* (London), October 27, 1994, 6.

Hunt, Liz, and Andrew Brown. "Muslims Urged to Boycott Rubella Vaccine." *The Independent* (London), October 29, 1994, 2.

Hunt, Liz, and Jan Roberts. "Minister Admits Measles Vaccine Made 500 Children Ill." *The Independent* (London), November 3, 1995, 1.

Illman, John. "Painful Choice of Risks: Measles Kills. But Preventative Vaccination Causes Problems, Too." *The Guardian* (London), November 2, 1994, T12.

Jackson, Deborah. "Please Be Sick After the Party." *The Independent* (London), November 15, 1994, 27.

Jardine, Cassandra. "Dangerous Maverick or Medical Martyr?" *The Daily Telegraph* (London), January 29, 2010, 29.

Karnowski, Steve, "Autism Fears, Measles Spike Among Minn. Somalis." Associated Press, April 2, 2011.

Kaufman, Leslie. "Darwin Foes Add Warming to Targets." *The New York Times*, March 3, 2010, A1.

Kennedy, Robert F., Jr. "Deadly Immunity." *Rolling Stone*, June 20, 2005.

———. "Letter to the Editor: Thimerosal, Children's Vaccines and Autism." *The Wall Street Journal*, July 8, 2005.

Klass, Perri. "Fearing a Flu Vaccine, and Wanting More of It." *The New York Times*, November 10, 2009, D5.

Kluger, Jeffrey. "Jenny McCarthy on Autism and Vaccines." *Time*, April 1, 2009; Web, October 12, 2009.

Kolata, Gina. "Environment and Cancer: The Links Are Elusive." *The New York Times*, December 13, 2005, F1.

———. "The Epidemic That Wasn't." *The New York Times*, August 29, 2002.

Kosova, Weston, and Pat Wingert. "Live Your Best Life Ever! Wish Away Cancer! Get a Lunchtime Face-Lift! Eradicate Autism! Turn Back the Clock! Thin Your Thighs! Cure Menopause! Harness Positive Energy! Erase Wrinkles! Banish Obesity! Live Your Best Life Ever!" *Newsweek*, June 8, 2009.

Krieger, Jane. "A Safe Polio Vaccine After Anxious Months." *The New York Times*, June 12, 1955, E6.

LaMendola, Bob. "Castration Drug Given to Kids as Autism Therapy." *Sun Sentinel* (South Florida), August 3, 2010.

Langdon-Down, Grania. "Law: A Shot in the Dark; The Complications from Vaccine Damage Seem to Multiply in the Courtroom." *The Independent* (London), November 27, 1996.

Laurance, Jeremy. "I Was There When He Dropped His Bombshell." *The Independent* (London), January 29, 2010, 2.

———. "Official Warning: Measles 'Endemic' in Britain." *The Independent* (London), June 21, 2008.

———. "Research Team's Work Led to Withdrawal of Children's Vaccines." *The Times* (London), September 16, 1992.

Laurence, William. "Salk Polio Vaccine Proves Successful; Millions Will Be Immunized Soon; City Schools Begin Shots April 25." *The New York Times*, April 13, 1955, 1.

Lerner, Maura. "Measles Whodunit: Tracking 2011's Outbreak." *The Minneapolis Star Tribune*, April 8, 2011.

Levin, Myron. " '91 Memo Warned of Mercury in Shots." *Los Angeles Times*, February 8, 2005.

———. "Taking It to Vaccine Court—Parents Say Mercury in Shots Caused Their Children's Autism, and They Want Drug Firms to Pay." *Los Angeles Times*, August 7, 2004, 1.

Lewis, Jemima. "Taking a Shot in the Dark." *The Sunday Telegraph* (London), January 31, 2010, 26.

Lin II, Rong-Gong. "County Sees Rise in Mumps Cases." *Los Angeles Times*, May 16, 2010, A39.

———. "State's Whooping Cough Surge May Be Tied to Lagging Immunization Rate." *Los Angeles Times*, June 28, 2010, A3.

Lin II, Rong-Gong, and Sandra Poindexter. "Schools' Risks Rise as Vaccine Rate Declines." *Los Angeles Times*, March 29, 2009, A1.

Lister, Sam. "Disgraced MMR-Scare Doctor Andrew Wakefield Quits US Clinic He Founded." *The Times* (London), February 19, 2010.

Loftus, Elizabeth. "Creating False Memories." *Scientific American*, September 1997.

MacRae, Fiona, and David Wilkes. "Damning Verdict on MMR Doctor: Anger as GMC Attacks 'Callous Disregard' for Sick Children." *The Daily Mail* (London), January 30, 2010.

Manning, Anita. "Mistrust Rises with Autism Rate." *USA Today*, July 6, 2005.

McBreen, Elizabeth. "Spectrum's Person of the Year 2009." *Spectrum Magazine*, February/March 2010.

McCarthy, Jenny. "My Autistic Son: A Story of Hope." *People*, September 20, 2007.

McGovern, Cammie. "Autism's Parent Trap: When False Hope Can Be Fatal." *The New York Times*, June 5, 2006, A19.

McKinley, Jesse. "Illness Kills 5 in California; State Declares an Epidemic." *The New York Times*, June 24, 2010, A15.

Midgley, Carol. "Vaccination: The Dilemma Now Facing Every Parent." *The Daily Mail* (London), September 16, 1992, 14.

Mihill, Chris. "Drive to Combat Measles Threat." *The Guardian* (London), September 30, 1994, 2.

———. "Illness Linked to Measles Vaccine." *The Guardian* (London), April 28, 1995.

Miller, Sam. "Autism Is Treatable, She Insists; O.C. Mother Leads Uprising Against Accepted Views." *The Orange County Register*, July 2, 2008, A1.

Mnookin, Seth. "An Early Cure for Parents' Vaccine Panic." *The Washington Post*, June 10, 2011, B1.

Moisse, Katie. "Study Confirms Link Between Older Maternal Age and Autism." *Scientific American*, February 11, 2010.

Mooney, Chris, and Sheril Kirshenbaum. "Unpopular Science." *The Nation*, August 17, 2009.

Morrice, Polly. " 'Evidence of Harm': What Caused the Autism Epidemic?" *The New York Times Book Review*, April 17, 2005, 20.

Murphy, Kate. "Enduring and Painful, Pertussis Leaps Back." *The New York Times*, February 22, 2005, F5.

Nguyen, Pamela. "Vaccine Refusal Is Putting Everyone in Danger." *Los Angeles Times*, June 1, 2010.

Offit, Paul A. "Companies, Courts, and the Cutter Case." *Pharmaceutical Technology*, December 1, 2005.

Park, Alice. "How Safe Are Vaccines?" *Time*, May 21, 2008.

Phillips, Fiona. "MMR Doc's Just Guilty of Caring." *The Daily Mirror* (London), January 30, 2010, 21.

Phillips, Melanie. "After MMR, How Can We Believe This New Child Vaccine Is Safe?" *The Daily Mail* (London), August 9, 2004, 12.

———. "MMR Safe? Baloney. This Is One Scandal That's Getting Worse." *The Daily Mail* (London), October 31, 2005, 14.

———. "MMR: The Truth." *The Daily Mail* (London), March 11, 2003, 42.

———. "We STILL Don't Know if MMR Is Safe." *The Daily Mail* (London), February 7, 2006, 12.

Pollack, Andrew. "Fear of a Swine Flu Epidemic in 1976 Offers Some Lessons, and Concerns Today." *The New York Times*, May 8, 2009, 11.

Rabin, Roni. "Both Parents' Ages Linked to Autism Risk." *The New York Times*, February 9, 2010, D6.

Rimland, Bernard. "Do Children's Shots Invite Autism?" *Los Angeles Times*, April 26, 2000, B9.

Roan, Shari. "2 Shots May Be Better Than One for Measles, Mumps, Rubella and Chicken Pox," *Los Angeles Times*, June 27, 2010.

Rockoff, Jonathan D. "More Parents Seek Vaccine Exemption: Despite Assurances, Fear of Childhood Shots Drives Rise." *The Wall Street Journal*, July 6, 2010.

Roser, Mary Ann. "Charting a Different Course on Autism." *Austin American-Statesman*, May 4, 2008, A1.

Russell, Sabin. "When Polio Vaccine Backfired." *San Francisco Chronicle*, April 25, 2005, A1.

Sapatkin, Don. "Hib Disease Deaths Put Focus on Vaccine Shortage." *The Philadelphia Inquirer*, April 1, 2009, A1.

Schmeck, Harold. "Ford Urges Flu Campaign to Inoculate Entire U.S." *The New York Times*, March 25, 1976, 1.

Schulte, Brigid. "Figments of the Imagination." *The Washington Post*, January 20, 2008, W10.

Shermer, Michael. "Why People Believe Invisible Agents Control the World." *Scientific American*, May 19, 2009.

Shute, Nancy. "Parents' Vaccine Safety Fears Mean Big Trouble for Children's Health." *U.S. News & World Report*, March 1, 2010.

Simons, Daniel J., and Christopher F. Chabris. "The Trouble with Intuition." *The Chronicle of Higher Education*, May 30, 2010.

Smith, Joan. "The Real MMR Conspiracy." *The Independent on Sunday* (London), September 12, 2004, 25.

Smith, Rebecca. "Dishonest, Callous and Irresponsible: Verdict on Doctor in MMR Row; Wakefield Didn't Care About Children's Pain and Distress, Says GMC." *The Daily Telegraph* (London), January 29, 2010, 7.

Solovitch, Sara. "The Citizen Scientists." *Wired*, September 2001.

Stevens, William. "Despite Vaccine, Perilous Measles Won't Go Away." *The New York Times*, March 14, 1989, C1.

Sunstein, Cass, and Richard Thaler. "Easy Does It: How to Make Lazy People Do the Right Thing." *The New Republic*, April 9, 2008.

Szabo, Liz. "Missed Vaccines Weaken 'Herd Immunity' in Children." *USA Today*, January 6, 2010.

Thompson, Tanya. "Doctor at Centre of MMR Controversy 'Paid Children at Son's Party GBP 5 for Blood Samples': Supporters Gather Outside Hearing to Accuse GMC of 'Witch Hunt.' " *The Scotsman* (Edinburgh), July 17, 2007, 6.

Trebbe, Ann L. "Local TV Honors Its Own; Channels 4 and 7 Sweep the Emmys." *Washington Post*, June 27, 1983, B1.

Tsouderos, Trine. "Autism 'Miracle' Called Junk Science." *Chicago Tribune*, May 21, 2009, 1.

———. "OSR#1: Industrial Chemical or Autism Treatment?" *Chicago Tribune*, January 17, 2010, 1.

Tsouderos, Trine, and Patricia Callahan. "Autism's Risky Experiments; Some Doctors Claim They Can Successfully Treat Children, But the Alternative Therapies Lack Scientific Proof." *Chicago Tribune*, November 22, 2009, 1.

Wallace, Amy. "An Epidemic of Fear: How Panicked Parents Skipping Shots Endangers Us All." *Wired*, November 2009, 1.

———. "A Short History of Vaccine Panic." *Wired*, November 2009.

Warren, Jennifer, and Greg Johnson. "Officials Brace for Start of School in Shadow of Measles Outbreak." *Los Angeles Times*, August 24, 1989, B8.

Weiss, Joanna. "Autism's 'Unblessed' Scientists." *The Boston Globe*, June 1, 2010, 13.

———. "Seeking Common Ground in the Autism-Vaccine Debate." *The Boston Globe*, February 27, 2010, 11.

Wielawski, Irene. "Measles Epidemic Called Sign of System's Ills." *Los Angeles Times*, July 25, 1990, A13.

Woolcock, Nicola, and Nigel Hawkes. "Decline in MMR Uptake Blamed for Measles Death." *The Times* (London), April 3, 2006.

Wright, Oliver, Nigel Hawkes, and Sam Lister. "Lancet Criticises MMR Scientist Who Raised Alarm." *The Sunday Times* (London), February 21, 2004.

Articles Without Author

"A.M.A. Balked on Report; New Head Says Advance Data on Salk Tests Was Withheld." *The New York Times*, May 14, 1955, 7.

"Autism and Vaccines." *The Wall Street Journal*, February 16, 2004.

"The Birth of Vaccination Fraud." *Truth*, January 10, 1923.

"54,000 Physicians See Digest on TV." *The New York Times*, April 13, 1955, 13.

"Government Aid in Paralysis Fight; 133 New Cases." *The New York Times*, July 7, 1916, A1.

"Hard Battle Won By Perseverance." *The New York Times*, April 13, 1955, 21.

"Measles: A Spot of Bother." *The Economist*, October 29, 1994.

"Medicine: Cutter in Court." *Time*, January 27, 1958.

"O'Connor Charges U.S. Gives No Light on Cutter Shots." *The New York Times*, June 11, 1955, 1.

"One Flu Vaccine Maker Is Losing Liability Insurance." *The New York Times*, June 16, 1976, 47.

"Oyster Bay Revolts over Poliomyelitis." *The New York Times*, August 29, 1916, A1.

"The Paralysis Epidemic." *The New York Times*, July 6, 1916, A12.

"Paralysis Kills 22 More Babies in New York City." *The New York Times*, July 8, 1916, 1.

"Polio Questions; Program in Suspense." *The New York Times*, May 8, 1955, E1.

"The Politics of Autism." *The Wall Street Journal*, December 29, 2003.

"The Salk Verdict." *Time*, November 28, 1955.

"Scheele Reports Sharp Cut in Polio." *The New York Times*, November 8, 1955, 64.

"39 Die of Paralysis; Highest Day's Toll." *The New York Times*, July 23, 1916, A7.

"A Welcome Retraction." *The New York Times*, February 5, 2010.

ONLINE

With Author

Ackerman, Lisa. "TACA & Jenny McCarthy" Talk About Curing Autism, October 5, 2008; Web, August 10, 2010.

Akers, Mary Ann. "Dan Burton, Protecting the House from Terrorists (Alone)." Washingtonpost.com, June 19, 2009; Web, August 10, 2010.

Allen, Arthur. "Why Is Oprah Winfrey Promoting Vaccine Skeptic Jenny McCarthy?" *Slate*, May 6, 2009; Web, May 10, 2009.

Bell, Vaughan. "Cigarette Smoking Lady Cops to Read Minds." *Mind Hacks*, March 15, 2009; Web, August 28, 2010.

Blaxill, Mark. "Naked Intimidation: The Wakefield Inquisition Is Only the Tip of the Autism Censorship Iceberg." *Age of Autism*, January 29, 2010; Web, August 10, 2010.

Chew, Kristina. "Not a Happy, but a True, Autism Story." *Autism Vox*, June 10, 2007; Web, January 2, 2008.

———. "Response to the IACC (#25)." *Kristina Chew*, August 21, 2009; Web, October 12, 2009.

Cox, Anthony. "The Upside of Infection." *Black Triangle*, February 3, 2010; Web, August 10, 2010.

Crosby, Jake. "Autism, Cancer and AIDS." *Age of Autism*, June 19, 2009; Web, October 12, 2009.

Dachel, Anne. "Old News About Age and Autism." *Age of Autism*, February 26, 2010; Web, March 9, 2010.

Deer, Brian. "Revealed: The First Wakefield MMR Patent Claim Describes 'Safer Measles Vaccine,' " *Brian Deer*, n.d.; Web, August 11, 2010.

Dorey, Meryl. "Australian Vaccination Network Asks for Your Support." *Age of Autism*, February 6, 2010; Web, April 16, 2010.

———. "Pertussis: The Fear Factor." *Natural Parenting*, n.d.; Web, August 18, 2010.

Edelson, Stephen M. "Interview with Professor Jaak Panksepp." Autism Research Institute, March 11, 1999; Web, October 12, 2009.

Engler, Crystal. "United We Stand, Divided We Fall." *Age of Autism*, August 2, 2009; Web, October 12, 2009.

Fisher, Barbara Loe. "Claims Court Opinion on Autism and Vaccines Is Not the Last Word." *No One Has to Die Tomorrow*, February 16, 2009; Web, January 25, 2010.

———. "2010 Needs a Fearless Conversation About Vaccination." *Age of Autism*, January 7, 2010; Web, January 25, 2010.

———. "Vaccines: Doctors, Judges, and Juries Hanging Their Own." National Vaccine Information Center, January 29, 2010; Web, August 10, 2010.

Fitzpatrick, Michael. "Mercury and Autism: A Damaging Delusion." *Spiked*, May 13, 2005; Web, May 18, 2010.

Gao, Helen. "Vaccine Refusals Are on the Rise." SignOn San Diego, The Watchdog Institute, August 23, 2010; Web, August 23, 2010.

Gordon, Jay. "PBS *Frontline* on Autism Resorts to Pseudo-Documentary, Tabloid Journalism." *The Huffington Post*, April 28, 2010; Web, May 17, 2010.

———. "Pertussis, Tylenol Recall and More." *Jay Gordon MD FAAP*, June 25, 2010, Web, June 30, 2010.

Gorski, David. " 'Medical Voices' on Vaccines: Brave, Brave Sir Robin . . ." *Science-Based Medicine*, May 20, 2010; Web, August 10, 2010.

Greenberg, David. "We Still Don't Understand How Fringe Conservatism Went Mainstream." *Slate*, September 23, 2009; Web, December 28, 2009.

Habakus, Louise Kuo. "Dr. Andrew Wakefield and the Distasteful Practice of the Ignorant 'Pile On.' " *Life Health Choices*, February 6, 2010; Web, April 16, 2010.

Halsey, Neal. "Misleading the Public About Autism and Vaccines." *VaccineSafety*, November 11, 2002; Web, March 9, 2010.

Handley, J. B. "Every Child by Two: A Front Group for Wyeth." *Age of Autism*, August 4, 2008; Web, October 12, 2009.

———. "J. B. Handley: Show Me the Monkeys!" *Age of Autism*, March 24, 2010; Web, April 16, 2010.

———. "Tinderbox: U.S. Vaccine Fears up 700% in 7 years." *Age of Autism*, March 17, 2010; Web, April 16, 2010.

Heckenlively, Kent. "How to Fight Autism." *Age of Autism*, March 24, 2010; Web, April 16, 2010.

———. "On Being Compared to Hitler and the Nazi Movement." *Age of Autism*, March 29, 2010; Web, April 16, 2010.

Higgs, Steven. "J. B. Handley: It's Unequivocal; Vaccines Hurt Some Kids." *The Bloomington Alternative*, April 4, 2010; Web, May 17, 2010.

———. "J. B. Handley: Tobacco Science in Its Early Phase." *The Bloomington Alternative*, April 18, 2010; Web, May 17, 2010.

Joseph. "The Administrative Prevalence of Autism Is a Bass Distribution." *Natural Variation—The Autism Blog*, April 15, 2010; Web, May 17, 2010.

Kennedy, Robert F., Jr. "Attack on Mothers." *The Huffington Post*, June 19, 2007; Web, March 9, 2007.

———. "Central Figure in CDC Vaccine Cover-up Absconds with $2M." *The Huffington Post*, March 11, 2010; Web, April 2, 2010.

———. "Deadly Immunity." Salon.com, June 16, 2005; Web, August 26, 2010.

———. "Letter to the editor." Salon.com, June 22, 2005; Web, April 16, 2010.

———. "Tobacco Science and the Thimerosal Scandal." *Robert F. Kennedy Jr.*, June 22, 2005; Web, April 2, 2010.

Kennedy, Robert F., Jr., and David Kirby. "Autism, Vaccines and the CDC: The Wrong Side of History." *The Huffington Post*, January 27, 2009; Web, August 21, 2010.

Kirby, David. "Autism Speaks: Will Anyone Listen?" *The Huffington Post*, March 28, 2007; Web, August 10, 2010.

———. "The Autism Vaccine Debate—Anything but Over." *The Huffington Post*, November 30, 2007; Web, August 23, 2010.

———. "Capitol Hill Briefing on Autism." *Evidence of Harm*, September 24, 2008; Web, August 18, 2010.

———. "Controversial, Bestselling American Author David Kirby to Speak at Houses of Parliament." *Evidence of Harm*, May 23, 2008; Web, August 18, 2010.

———. "Dr. Rahul K. Parikh, I Am Becoming Embarrassed for You." *Age of Autism*, September 23, 2008; Web, August 10, 2010.

———. "Fever, Vaccines, and 'Mitochondrial Autism'." *Age of Autism*, May 28, 2008; Web, May 12, 2010.

———. "Government Concedes Vaccine-Autism Case in Federal Court—Now What?" *The Huffington Post*, February 25, 2008; Web, August 10, 2010.

———. "The *Lancet* Retraction Changes Nothing." *The Huffington Post*, February 2, 2010; Web, August 10, 2010.

———. "Mercury, Autism and the Coming Storm." *The Huffington Post*, June 25, 2005; Web, August 10, 2010.

———. "Notes from the Big 'Anti-Vaccine' Conference." *The Huffington Post*, June 1, 2009; Web, August 10, 2010.

———. "Up to 1-in-50 Troops Seriously Injured . . . by Vaccines?" *The Huffington Post*, August 14, 2008; Web, August 24, 2010.

———. "The Vaccine-Autism Court Document Every American Should Read." *The Huffington Post*, February 26, 2008; Web, August 10, 2010.

———. "Vaccines and Autism: Questions." *The Huffington Post*, May 20, 2005; Web, August 10, 2010.

Laidler, James R. "Through the Looking Glass: My Involvement with Autism Quackery." *Autism Watch*, December 7, 2004; Web, October 12, 2009.

Lehrer, Jonah. "Cable News." *ScienceBlogs*, January 26, 2010; Web, August 10, 2010.

———. "Loss Aversion." *ScienceBlogs*, February 10, 2010; Web, April 16, 2010.

Leitch, Kevin. "DAN!—On a mission from God." *Left Brain, Right Brain*, October 9, 2006; Web, August 10, 2010.

McCarthy, Jenny. "A Girl's Gotta Do What a Girl's Gotta Do." Oprah.com, May 17, 2009; Web, October 12, 2009.

McCarthy, Jenny, and Jim Carrey. "A Statement from Jenny McCarthy and Jim Carrey: Andrew Wakefield, Scientific Censorship, and Fourteen Monkeys." *Age of Autism*, February 5, 2010; Web, April 16, 2010.

Mnookin, Seth. "It's Official: Wakefield Joins the Ranks of Truthers, New World Order Conspiracists." *The Panic Virus Blog* (The Public Library of Science), June 7, 2011; Web, July 11, 2011.

Moody, Jim. "CDC Media Plan Shocker—We Don't Have the Science—Some claims Against Vaccine Cannot Be Disproved." *Age of Autism*, August 5, 2009; Web, October 12, 2009.

Obradovic, Julie. "An Overview of the Fourteen Vaccine Studies." *Age of Autism*, July 16, 2009; Web, October 12, 2009.

Offit, Paul A. "The 'Wakefield Studies': Studies Hypothesizing That MMR Causes Autism." *American Academy of Pediatrics*, n.d.; Web, December 28, 2009.

Orac. "Dr. Bob Sears: Stealth Anti-Vaccinationist?" *Respectful Insolence*, August 7, 2009; Web, October 12, 2009.

———. "The Intellectual Dishonesty of the 'Vaccines Didn't Save Us' Gambit." *Respectful Insolence*, March 29, 2010; Web, April 16, 2010.

———. "Suing DAN! Practitioners for Malpractice: It's About Time." *Respectful Insolence*, March 5, 2010; Web, April 16, 2010.

Parikh, Rahul. "David Kirby Smacks Me Down." Salon.com, September 25, 2008; Web, March 9, 2010.

———. "Doc Hollywood: If Celebrities Insist on Using Their Fame to Bring Awareness to Health Problems, They Should Follow These Guidelines." *Slate*, December 9, 2009; Web, December 28, 2009.

———. "Inside the Vaccine-and-Autism Scare." Salon.com, September 22, 2008; Web, October 12, 2009.

Rudy, Lisa Jo. "David Kirby's Take on Katie Wright, Autism Speaks, and the Vaccine / Autism Debate." About.com: Autism, June 4, 2007; Web, August 10, 2010.

Schulte, Brigid. "Post Magazine: Morgellons Disease. Live Discussion with Brigid Schulte." Washingtonpost.com, January 22, 2008; Web, August 12, 2010.

Sears, Robert W. "A Response to Dr. Offit's Misleading and Inaccurate Review of The Vaccine Book in Pediatrics, January 2009." *AskDr.Sears*, December 29, 2008; Web, October 12, 2009.

Seidel, Kathleen. "Evidence of Venom: An Open Letter to David Kirby." *Neurodiversity*, May 29, 2005; Web, March 9, 2010.

Singer, Alison. "Vaccine Court Denies All Three 'Thimerosal Causes Autism' Test Cases." Autism Science Foundation, March 12, 2010; Web, April 16, 2010.

Siva, Nayanah. "Andrew Wakefield Tried to Connect the MMR Vaccine to Crohn's Before Implicating It in Autism." *Slate*, June 2, 2010; Web, June 3, 2010.

Stagliano, Kim. "Old Promises Never Kept at Autism Speaks." *Age of Autism*, April 26, 2010; Web, May 17, 2010.

Stone, John. "Histopathologist from the Lancet Study Rebuffs Brian Deer's Article in British Medical Journal." *Age of Autism*, April 30, 2010; Web, May 17, 2010.

———. "The Scandalous History of MMR in the UK." *Age of Autism*, January 25, 2010; Web, August 10, 2010.

Sullivan. "Fees for the Omnibus Autism Proceeding hit $7M." *Left Brain, Right Brain*, December 19, 2009; Web, December 28, 2009.

———. "It's Time for David Kirby to Disavow the Autism Epidemic." *Left Brain/ Right Brain*, August 3, 2009; Web, October 12, 2009.

———. "More on Autism 'Clusters.' " *Left Brain/Right Brain*, February 4, 2010; Web, August 10, 2010.

Taylor, Ginger. "Bob and Suzanne Hurt Another Autism Family." *Adventures in Autism*, June 7, 2007; Web, August 10, 2010.

———. "Katie Wright Speaks for Me!" *Adventures in Autism*, June 2, 2007; Web, August 10, 2010.

Tommey, Polly. "Polly Tommey of Autism File Magazine on 'Discredited Defamation of Dr. Andrew Wakefield.' " *Age of Autism*, January 6, 2010; Web, January 25, 2010.

Wakefield, Andrew. "In His Desperation, Deer Gets It Wrong Once Again." *Rescue Post*, n.d.; Web, August 18, 2010.

Walker, Martin. "Counterfeit Law: And They Think They Have Got Away with It." *Age of Autism*, February 21, 2010; Web, April 16, 2010.

———. "Eye Witness Report from the UK GMC Wakefield, Walker-Smith, Murch Hearing." *Age of Autism*, January 31, 2010; Web, August 10, 2010.

———. "The UK GMC Panel: A Sinister and Tawdry Hearing." *Age of Autism*, January 13, 2010; Web, August 10, 2010.

Wallace, Amy. "Covering Vaccines: Science, Policy and Politics in the Minefield." Annenberg School for Communication, University of Southern California: Reporting on Health, August 30, 2010, Web, August 30, 2010.

Winfrey, Oprah. "What I Know for Sure." Oprah.com, June 15, 2002; Web, August 10, 2010.

Wright, Katie. "Autism Speaks Attends the DAN! Conference." *Age of Autism*, April 20, 2010; Web, May 12, 2010.

———. "The Autism Speaks Baby Sibs Research Consortium, aka Everything That Is Wrong with Autism Research." *Age of Autism*, February 24, 2010; Web, March 9, 2010.

———. "Autism Speaks's Dr. Daniel Geschwind Declares Paul Offit 'A Brave and Articulate Champion of Truth!' " *Age of Autism*, September 9, 2010, Web, September 10, 2010.

———. "Why the Autism Speaks Scientific Advisory Committee Needs to Resign." *Age of Autism*, April 26, 2010; Web, May 18, 2010.

———. "Will the Interagency Autism Coordinating Committee Finally Get It (Katie, Bob and Suzanne) Wright?" *Age of Autism*, February 9, 2010; Web, April 16, 2010.

Without Author

"Age of Autism Award: Michelle Cedillo, Child of the Year." *Age of Autism*, December 24, 2007; Web, December 29, 2008.

"ASAT and Autism Science Foundation: An Open Letter to Autism Speaks." Association for Science in Autism Treatment, September 8, 2010, Web, September 10, 2010.

"Autism Recovery Revisited: Tell the Chicago Tribune They Are Wrong." *Age of Autism*, November 21, 2009; Web, December 28, 2009.

"An Exchange with J. B. Handley, Generation Rescue." *Neurodiversity*, n.d.; Web, March 9, 2010.

"The Jenny McCarthy Conundrum: Is False Hope Better than No Hope?" *Jezebel*, February 27, 2010; Web, March 9, 2010.

"Oratory—Or Hypnotic Induction?" Association of American Physicians and Surgeons, October 25, 2008; Web, May 18, 2010.

"A Short Form FAQ About the Wakefield GMC Case." *Age of Autism*, January 28, 2010; Web, August 10, 2010.

"Statement from Bob and Suzanne Wright, Co-founders of Autism Speaks." Autism Speaks, n.d.; Web, March 9, 2010.

"Statement on UK General Medical Council." Thoughtful House, January 28, 2010; Web, April 16, 2010.

"Thimerosal in Vaccines Questions and Answers." Food and Drug Administration, May 1, 2009; Web, October 11, 2009.

"Three Parents' Statements in UK GMC Hearing." *Age of Autism*, May 8, 2009; Web, August 10, 2010.

"A Tidal Wave of Misinformation at WTAE." *Autism News Beat*, January 26, 2010; Web, August 10, 2010.

"Transcription of Dr. Andrew Wakefield, Author of Callous Disregard, on Imus in the Morning." *Age of Autism*, July 3, 2010; Web, August 10, 2010.

"Truth and Consequences—The Anti-Vaccination Movement Exacts a Price." *Left Brain, Right Brain*, September 23, 2009; Web, December 28, 2009.

"Unvaccinated Minnesota Child Dies from Hib Meningitis." *Vaccinate Your Baby*, January 27, 2009; Web, October 11, 2009.

"Update: Measles—United States, January–July 2008." Centers for Disease Control and Prevention, August 22, 2008; Web, October 11, 2009.

LEGAL

The Omnibus Autism Proceeding

Cedillo v. Sec'y of Health and Human Services, No. 98-916V (Ct. Fed. Cl., February 12, 2009). Digital audio files and transcripts available at ftp://autism.uscfc.uscourts.gov/autism/cedillo.html.

Dwyer v. Sec'y of Health and Human Services, No. 03-1202V (Ct. Fed. Cl., March 12, 2010). Digital audio files and transcripts available at ftp://autism.uscfc.uscourts.gov/autism/dwyer.html.

Hazlehurst v. Sec'y of Health and Human Services, No. 03-654V (Ct. Fed. Cl., February 12, 2009). Digital audio files and transcripts available at ftp://autism.uscfc.uscourts.gov/autism/hazlehurst.html.

King v. Sec'y of Health and Human Services, No. 03-584V (Ct. Fed. Cl., filed March 12, 2010). Digital audio files and transcripts available at ftp://autism.uscfc.uscourts.gov/autism/thimerosal.html.

Mead v. Sec'y of Health and Human Services, No. 03-215V (Ct. Fed. Cl., March 12, 2010).

Snyder v. Sec'y of Health and Human Services, No 01-162V (Ct. Fed. Cl., February 12, 2009). Digital audio files and transcripts available at ftp://autism.uscfc.uscourts.gov/autism/snyder.html.

The docket of the U.S. Court of Federal Claims' Omnibus Autism Proceeding includes several hundred documents, ranging from rulings on the designation of counsel to scheduling orders to briefs filed by outside parties. A full list of orders and rulings can be found here: http://www.uscfc.uscourts.gov/node/2718.

Links to PDFs of the dozens of expert reports have been compiled on *Neurodiversity*, and can be found here: http://www.neurodiversity.com/weblog/article/189/.

Other Vaccine Court Cases

Aldridge v. Sec'y of Health and Human Services, 1992 WL 153770 (Ct. Fed. Cl., June 11, 1992).

Barber v. Sec'y of Health and Human Services, No. 99-434V. 2008 (United States Court of Special Claims).

Daly v. Sec'y of Health and Human Services, 1991 WL 154573 (Ct. Fed. Cl., July 26, 1991).

Haim v. Sec'y of Health and Human Services, 1993 WL 346392 (Ct. Fed. Cl., August 27, 1993).

Marascalco v. Sec'y of Health and Human Services, 1993 WL 277095 (Ct. Fed. Cl., July 9, 1993).

Other Court Cases

Arthur v. Offit et al. No. 09-CV-1398 (U.S. Dist. Ct., E.D. VA), March 10, 2010.

Daubert v. Merrell Dow Pharmaceuticals (92–102), 509 U.S. 579 (1993). Blackmun opinion.

Daubert v. Merrell Dow Pharmaceuticals (92–102), 509 U.S. 579 (1993). Rehnquist opinion.

Escola v. Coca Cola Bottling Co. 24 C2d 453 (Cal. Sup. Ct. 1944).

Gottsdanker v. Cutter Laboratories. 182 Cal. App. 2d 602 (1960).

Reyes v. Wyeth Laboratoriesm, 498 F. 2nd 1264 Ct. App., 5th Circ. (1974).

VIDEO AND AUDIO RECORDINGS

Television

"The Age of Miracles: The New Midlife." Oprah Winfrey, *The Oprah Winfrey Show*, Harpo Productions, February 28, 2008.

"The Big Wake-Up Call for Women with Dr. Christiane Northrup." Oprah Winfrey, *The Oprah Winfrey Show*, Harpo Productions, October 16, 2007.

"The Bioidentical Debate with Suzanne Somers." Oprah Winfrey, *The Oprah Winfrey Show*, Harpo Productions, January 29, 2009.

"Can Diet Heal Autism? Jenny McCarthy's Controversial Opinion." Diane Sawyer, *Good Morning America*, ABC, September 29, 2008.

"Controversial Morgellons Disease and Its Sufferers." Martin Savidge, *Today*. NBC, July 28, 2006.

"A Coverup for a Cause of Autism?" Joe Scarborough, *Morning Joe*, MSNBC, July 21, 2005.

"David Kirby and Dr. Harvey Fineberg." Tim Russert, *Meet the Press*, NBC, August 7, 2005.

"Does the MMR Jab Cause Autism?" *Horizon*, BBC, May 29, 2005.

"DPT: Vaccine Roulette." Lea Thompson, WRC (Washington, D.C.), NBC, April 19, 1982.

"Dr. Andrew Wakefield Continues Quest to See if Link Exists Between Childhood Vaccines and Autism." Matt Lauer, *Dateline*, NBC, August 30, 2009.

"Dr. Andrew Wakefield Loses Medical License over Study He Claims Shows a Link Between Autism and MMR Vaccine; Dr. Wakefield Discusses the Issue." Matt Lauer, *Today*, NBC, May 24, 2010.

"The Faces of Autism." Oprah Winfrey, *The Oprah Winfrey Show*, Harpo Productions, April 5, 2007.

"Interview with Jenny McCarthy." Larry King, *Larry King Live*, CNN, September 26, 2007.

"Jenny McCarthy's Autism Fight." Larry King, *Larry Ling Live*, CNN, April 2, 2008.

"Jenny McCarthy Speaks About Autism." Greta Van Susteren, *On the Record with Greta Van Susteren*, Fox News, June 6, 2008.

"Jenny McCarthy's Warrior Spirit." Oprah Winfrey, *The Oprah Winfrey Show*, Harpo Productions, September 24, 2008.

"Medical Mystery: Morgellons Disease." *Nightline*, ABC January 16, 2008.

"MMR: Every Parent's Choice'." Sarah Barclay, *Panorama*, BBC, February 2, 2002.

"Morgellon's Disease Remains Controversial." *House Call with Doctor Sanjay Gupta*, CNN, September 30, 2006.

"Mothers Battle Autism." Oprah Winfrey, *The Oprah Winfrey Show*, Harpo Productions, September 18, 2007.

"A Mother's Journey: Star Fights for Son." Diane Sawyer, *Good Morning America*, ABC, September 24, 2007.

"A Mother's Mission: McCarthy and Carrey Search for Autism Answers." Diane Sawyer, *Good Morning America*, ABC, June 4, 2008.

"Mysterious Skin Disease Causes Itching, Loose Fibers, Morgellons Has Plenty of Skeptics." Cynthia McFadden, *Good Morning America*, ABC, July 28, 2006.

"The Vaccine Wars." *Frontline*, WGBH (Boston), Public Broadcasting System, April 27, 2010.

"What Is Your Body Telling You?" Oprah Winfrey, *The Oprah Winfrey Show*, Harpo Productions, January 1, 2006.

Radio

"Asperger's Officially Placed Inside Autism Spectrum." Jon Hamilton, *Morning Edition*, National Public Radio, February 10, 2010.

"David Kirby Interviewed by Don Imus." Don Imus, *Imus in the Morning*, WFAN-AM 660 and MSNBC, March 10, 2005.

"Measles Resurgence Tied to Parents' Vaccine Fears." Richard Knox, *National Public Radio*, April 5, 2010; Web, May 17, 2010.

"Vaccines and Their Link to Autism." Jon Hamilton, *Morning Edition*, National Public Radio, November 11, 2002.

Web Video

"Chicago Anti-Vax Rally—Andrew Wakefield 5-26-10." Bruce Critelli, Vimeo, http://vimeo.com/12079650.

"David Kirby Interviews Katie Wright!" David Kirby, Foundation for Autism Information and Research, April 19. 2007, http://www.youtube.com/watch?v=IUNO25l1zFs, dHYsK_MP7w, tTVoJIVqu2Q, and l_lPuYf98uF.

"Frontline Interview: J. B. Handley." *Frontline*, WGBH (Boston), Public Broadcasting System, http://www.pbs.org/wgbh/pages/frontline/vaccines/interviews/handley.html.

"Frontline Interview: Jenny McCarthy." *Frontline*, WGBH (Boston), Public Broadcasting System, http://www.pbs.org/wgbh/pages/frontline/vaccines/interviews/mccarthy.html.

"Frontline Interview: Paul Offit, MD." *Frontline*, WGBH (Boston), Public Broadcasting System, http://www.pbs.org/wgbh/pages/frontline/vaccines/interviews/offit.html.

"Frontline Interview: Robert W. Sears, MD." *Frontline*, WGBH (Boston), Public Broadcasting System, http://www.pbs.org/wgbh/pages/frontline/vaccines/interviews/sears.html.

"Jenny McCarthy's Webcast." Oprah.com, September 26, 2008, http://www.oprah.com/relationships/Watch-Jenny-McCarthys-Webcast-on-Oprahcom.

REPORTS

With Author

Bernard, Sallie. "Analysis of the Danish Autism Registry Data Base in Response to the Hviid et al Paper on Thimerosal in JAMA (October, 2003)." SafeMinds, October 2003.

Blaxill, Mark. "Danish Thimerosal-Autism Study in Pediatrics: Misleading and Uninformative on Autism-Mercury Link." SafeMinds, September 2, 2003.

Hargreaves, Ian, Justin Lewis, and Tammy Speers. "Towards a Better Map: Science, the Public and the Media." Economic and Social Research Council (U.K.), 2003.

Without Author

"Acrodynia, a Form of Mercury Poisoning in Childhood, and Its Similarities to Autism and Other Neurological and Learning Disorders of Children." SafeMinds, 2001.

"Autism and Vaccines Around the World: Vaccine Schedules, Autism Rates, and Under 5 Mortality." Generation Rescue, April 2009.

"Celebrities vs. Science." American Council on Science and Health, 2007.

"Fourteen Studies." Generation Rescue, 2009.

"Global Elimination of Measles." World Health Organization, report by the Secretariat, April 16, 2009.

"Public Praises Science; Scientists Fault Public, Media. Scientific Achievements Less Prominent than a Decade Ago." Pew Research Center for the People and the Press, July 9, 2009.

Governmental

"Autism Spectrum Disorders (Pervasive Developmental Disorders)." National Institute of Mental Health, July 22, 2009.

"Childhood Vaccines: Ensuring an Adequate Supply Poses Continuing Challenges." U.S. General Accounting Office, GAO-02-987, Washington, D.C., September 13, 2002.

"Claims Filed and Compensated or Dismissed by Vaccine." National Vaccine Compensation Program, November 3, 2009.

"Compensation for Vaccine-Related Injuries: A Technical Memorandum." Congress of the United States Office of Technology Assessment, November 1980.

"Conflicts of Interest in Vaccine Policy Making: Majority Staff Report." Committee on Government Reform, U.S. House of Representatives, August 21, 2000.

"DRAFT: Brief SWOT [Strengths, Weaknesses, Opportunities, and Threats] Analysis and Vaccine Safety Communication/Media Strategy." Centers for Disease Control and Prevention, 2008.

"Epidémie de rougeole en France. Actualisation des données au 20 mai 2011," Institut de veille sanitaire, Département des maladies infectieuses (France), n.d.

"Fitness to Practise Panel Hearing." General Medical Council (U.K.), January 28, 2010.

"Group Concluded No Reason for Change in MMR Vaccine Policy." Medical Research Council (U.K.), March 24, 1998.

"Measles and Crohn's Disease." Joint Committee on Vaccination and Immunisation (U.K.), 1995.

"National Vaccine Injury Compensation Program: Claims Filed and Compensated or Dismissed by Disease." Health Resources and Services Administration, November 3, 2009.

"National Vaccine Injury Compensation Program Post-1988 Statistics Report." Health Resources and Services Administration, April 1, 2010.

"Number of Influenza-Associated Pediatric Deaths by Week of Death: 2006–07 Season to Present." Centers for Disease Control and Prevention, n.d.

"Parents' Guide to Childhood Immunization." Centers for Disease Control and Prevention, n.d.

"Report of the Strategy Development Group Subgroup on Research into Inflammatory Bowel Disorders and Autism." Medical Research Council (U.K.), 2000.

"Review of Autism Research." Medical Research Council (U.K.), December 2001.

"Some Common Misconceptions About Vaccines and How to Respond to Them." Centers for Disease Control and Prevention, n.d.

"Vaccine Injury Compensation: Program Challenged to Settle Claims Quickly and Easily." Government Accounting Office, Washington D.C., 1999.

"Vaccine Injury Table." Health Resources and Services Administration, n.d.

OTHER

With Author

Fisher, Barbara Loe. "Statement to the IOM Immunization Safety Committee," January 11, 2001.

Ford, Gerald. "Remarks Announcing the National Swine Flu Immunization Program," March 24, 1976, Gerald R. Ford Presidential Library and Museum.

Halsey, Neal, and Susan Hyman. "Measles-Mumps-Rubella Vaccine and Autistic Spectrum Disorder." Childhood Immunizations Conference, Oak Brook, Illinois, June 12–13, 2000.

Hilleman, Maurice R. "Vaccine Task Force Assignment Thimerosal (Merthiolate) Preservative—Problems, Analysis, Suggestions for Resolution." Memo to Gordon Douglas, Merck Pharmaceuticals, Whitehouse Station, New Jersey, 1991.

Wakefield, Andrew. "Handout to Parent(s)/Guardian." Royal Free Hospital (U.K.), September 16, 1996.

Without Author

"High Number of Reported Measles Cases in the U.S. in 2011—Linked to Outbreaks Abroad." Centers for Disease Control and Prevention, June 22, 2011.

"In the matter of Mark R. Geier, M.D., Respondent, before the Maryland State Board of Physicians: Order for Suspension of License to Practice Medicine," Maryland State Board of Physicians, Case Numbers: 2007-0083; 2008-0454; 2009-0308, 15.

"Joint Statement of AAFP, AAP, ACIP, and the USPHS on Thimerosal in Childhood Vaccines." Institute for Vaccine Safety, June 22, 2000.

"The Limited Support NAAR Has Received from Pharmaceutical Companies That Make Childhood Vaccines." National Alliance for Autism Research, March 11, 2003.

"Maryland Health Officials Investigating Possible Exposures to Measles. Exposures Possible from Tuesday, May 31 through Sunday, June 5, 2011. Areas include Catonsville, Easton, and Baltimore," Maryland Department of Health and Mental Hygiene, June 8, 2011.

"New Research Links Autism and Bowel Disease." Press release from the Royal Free Hospital School of Medicine (U.K.), February 26, 1998.

"Poliomyelitis: Fact Sheet #114." World Health Organization, January 2008.

"Responding to 7 Common Parental Concerns About Vaccines and Vaccine Safety." American Academy of Pediatrics Practice Management Online, April 2005.

"Scientific Review of Vaccine Safety Datalink Information." Simpsonwood Retreat Center, Norcross, Georgia, July 7–8, 2000, unpublished transcript.

"Something Is Rotten in Denmark." SafeMinds, May 2004.

"299.00: Autistic Disorder. Proposed Revision." DSM-5 Development. American Psychiatric Association, n.d., http://www.dsm5.org/ProposedRevisions/Pages/proposedrevision.aspx?rid=94.

ADVERTISEMENTS

"Are we over-vaccinating our kids?" Generation Rescue. *The Oregonian* and *Orange County Register*, September 25, 2007.

"Are we poisoning our kids?" Generation Rescue. *USA Today*, February 12, 2008.

"Autism and mercury poisoning: not a coincidence." Generation Rescue. *New York Times*, June 8, 2005.

"Autism is preventable and reversible." Generation Rescue. *USA Today*, May 24, 2005.

"If you caused A 6,000% Increase In Autism." Generation Rescue. *USA Today*, April 6, 2006.

"A Little Boy Shouldn't Have To Take on an Entire Industry Alone." Generation Rescue. *USA Today*, February 23, 2009.

Index

A number followed by an *n* refers to a footnote on that page.